# Lecture Notes in Mathematics

Edited by A. Dold and B. Eckmann

Series: Mathematics Institute, University of Warwick
Adviser: D.B.A. Epstein



898

# Dynamical Systems and Turbulence, Warwick 1980

Proceedings of a Symposium Held at the University of Warwick 1979/80

Edited by D.A. Rand and L.-S. Young

Springer-Verlag
Berlin Heidelberg New York 1981

**Editors**

David Rand
Mathematics Institute, University of Warwick
Coventry, CV4 7AL, England

Lai-Sang Young
Mathematics Department, Michigan State University
Michigan, MI 48824, USA

AMS Subject Classifications (1980): 34A34, 34C35, 35B32, 35Q10, 39A10, 58D30, 58F10, 58F12, 58F14, 58F17, 58F40.

ISBN 3-540-11171-9 Springer-Verlag Berlin Heidelberg New York
ISBN 0-387-11171-9 Springer-Verlag New York Heidelberg Berlin



Printing and binding: Beltz Offsetdruck, Hemsbach/Bergstr.
2141/3140-543210

# INTRODUCTION.

The papers in these Proceedings were contributed by participants in the 1979/80 Symposium on Dynamical Systems and Turbulence. In this meeting we attempted to bring together a wide variety of scientists from different backgrounds, both theoretical and experimental, with a common interest in the problem of the nature of the dynamics of turbulence and other related topics. During the year these people visited Warwick as long-term visitors, as participants in one of the three main conferences, or as speakers in the regular seminar programme.

The majority of the participants in the largest of these conferences, the Research Conference, were pure mathematicians working in the general area of dynamical systems, and the papers in these Proceedings reflect this bias. However, a number of the papers address problems of direct relevance to fluid mechanics, and both expository papers are directed, in one way or another, towards this area.

It is a pleasure to thank the various people who contributed to the success of the Symposium. In particular, the help and advice of E.C. Zeeman and T.B. Benjamin were crucial to its success, as were the organising skills of Elaine Shiels. Also D. Fowler was kind enough to give up a large amount of his time to help with the organisation.

We are very grateful to Roger Butler for his help in proof-reading and to Elaine Shiels for typing the final version of these Proceedings.

September 1981.

David Rand
Lai-Sang Young.

## CONTENTS

Introduction

EXPOSITORY PAPERS :



CONTRIBUTED PAPERS :

-----------------------

# Lectures on Bifurcation from Periodic Orbits.

Lectures given by D.D. Joseph.
Notes by K. Burns.

These lectures were delivered at Warwick in April 1980 as part of the Symposium.

CONTENTS



These lectures are about bifurcations from a periodic orbit of an evolution equation with periodic forcing. The analysis applies to equations in an arbitrary Hilbert space, not just to finite dimensional problems. The results described here are joint work of G. Iooss and D.D. Joseph [3, 4, 5]. In the lectures I will outline the methods, proofs are given in [3, 4, 5].

## 1. Introduction.

We consider the equation

$$d\underline{V}/dt = F(t, \mu, \underline{V}) \tag{1}$$

Here $\underline{V}(t, \mu)$ lies in a real Hilbert space $(H, \langle\rangle)$, $\mu$ is a real bifurcation parameter, and F is T-periodic i.e., $F(T, \mu, \underline{V}) = F(t+T, \mu, \underline{V})$. Assume that there is a T-periodic solution

$$\underline{V} = \underline{U}(t, \mu) = \underline{U}(t+T, \mu) \ . \tag{2}$$

We rewrite (1) in local form about $\underline{U}$. If $\underline{u} = \underline{V} - \underline{U}$, then

$$d\underline{u}/dt = \underline{f}(t, \mu, \underline{u})$$

where

$$\underline{f}(t, \mu, \underline{u}) = F(t, \mu, \underline{U}+\underline{u}) - F(t, \mu, \underline{U}) \tag{3}$$

We shall study (3) with

$$\underline{f}(t, \mu, .) = \underline{f}_u(t, \mu | .) + \underline{N}(t, \mu, .)$$

where $\underline{f}_u(t, \mu | .)$ is linear and $\underline{N}(t, \mu, \underline{v}) = 0(\|\underline{v}\|^2)$ .

We assume that the periodic orbit $\underline{U}$, that is the orbit $\underline{u} = 0$ of (3), is stable if

$\mu < 0$, and loses stability for $\mu > 0$. To express this precisely consider the linearisation of (3)

$$d\underline{v}/dt = \underline{f}_u(t, \mu | \underline{v}) \tag{4}$$

This is to be thought of as a complex linear equation (with real coefficients) on $H^{\mathbb{C}}$, the complexification of H. Associated with (4) is a linear operator on the space $\mathbf{P}_T^{\mathbb{C}}$ of T-periodic vector fields on $H^{\mathbb{C}}$,

$$J_\mu = -d/dt + \underline{f}_u(t, \mu | .) \tag{5}$$

Eigenvalues of $J_\mu$ are called <u>Floquet exponents.</u> The orbit $\underline{u} = 0$ is stable if all Floquet exponents have negative real part, and unstable if any has positive real part. The loss of stability at $\mu = 0$ is assumed to occur in the simplest way :

Bifurcation Assumptions :

There is a Floquet exponent $\sigma(\mu) = \xi(\mu) + i\eta(\mu)$ such that

(i) $\sigma(0) = i\omega_0 = \frac{2\pi r}{T} \qquad 0 \leq r < 1.$

(ii) $\sigma(\mu)$ and $\overline{\sigma}(\mu)$ are isolated algebraically simple eigenvalues of $J_\mu$.

(iii) $d\xi/d\mu(0) > 0$ .

(iv) All eigenvalues of $J_0$ other than $\sigma(0)$ and $\overline{\sigma}(0)$ have negative real part.

The type of bifurcation that occurs depends on the value of r.

(i) Strong Resonance : if $r = m/n$ and $n = 1, 2, 3,$ or $n = 4$ and a certain inequality holds then nT-periodic solutions bifurcate.

(ii) Y.H. Wan [6] has shown that there is an invariant torus when $n = 4$ and the inequality does not hold.

(iii) Weak resonance : if $r = m/n$, $n \geq 5$, and certain exceptional conditions hold then nT-periodic solutions bifurcate.

(iv) If $r \neq m/n$, $n = 1, 2, 3, 4$ there is a Hopf bifurcation to an invariant torus.

The next section describes how to approximate the original problem (3) with an autonomous equation in $\mathbb{R}^2$. Sections 3 and 4 outline how to solve the approximate equation for Hopf and subharmonic bifurcations. The final section touches on the question of "lock-ons".

It should be mentioned that the asymptotic representations can be constructed directly, without normal forms, by methods of applied analysis (see appendices to Chapter X in [4]).

## 2. Derivation of the Autonomous Equation.

We assume that $r \neq 0, \frac{1}{2}$ (see [3,4,5] for a study of these cases). This means that the periodic orbit $\underline{u} = 0$ loses stability in two real dimensions instead of just one. The first step is to decompose (3) into a part in this plane and a complementary part.

There is an inner product on $\mathbb{P}_T^{\mathbb{C}}$,

$$[\underline{\xi}_1, \underline{\xi}_2] = \frac{1}{T}\int_0^T \langle \underline{\xi}_1(t), \underline{\xi}_2(t)\rangle dt \quad .$$

Let $J_\mu^*$ be the adjoint of $J_\mu$ with respect to $[,]$. It can be verified that

$$J_\mu^* = d/dt + \underline{f}_{u}^{*}(t,\mu|.) \ , \tag{6}$$

where $\underline{f}^*(t,\mu|.)$ is the adjoint of $\underline{f}(t,\mu|.)$ with respect to $\langle , \rangle$. Now $\sigma(\mu)$, $\bar{\sigma}(\mu)$ are eigenvalues of $J_\mu, J_\mu^*$ respectively; let $\underline{\xi}_\mu$, $\underline{\xi}_\mu^*$ be corresponding eigenfunctions. Using (6) and the assumption that $r \neq 0, \frac{1}{2}$, one can show that

$$\langle \underline{\xi}_\mu(t), \underline{\xi}_\mu^*(t)\rangle \equiv \langle \underline{\xi}_\mu(0), \underline{\xi}_\mu^*(0)\rangle$$

$$\langle \bar{\underline{\xi}}_\mu(t), \underline{\xi}_\mu^*(t)\rangle \equiv 0 \ .$$

Normalise $\underline{\xi}_\mu, \underline{\xi}_\mu^*$ so $\langle \underline{\xi}_\mu, \underline{\xi}_\mu^*\rangle = 1$. Now we can write

$$\underline{u} = z\underline{\xi}_\mu + \bar{z}\bar{\underline{\xi}}_\mu + \underline{W}$$

where $z = \langle \underline{u}, \underline{\xi}_\mu^*\rangle$ and $\underline{W}$ is real. Equation (3) becomes

$$dz/dt = \sigma(\mu)z + b \tag{7a}$$

$$d\underline{W}/dt = \underline{f}_u(t,\mu|\underline{W}) + \underline{B} \tag{7b}$$

where

$$b(t,\mu,z,\bar{z},\underline{W}) = \langle \underline{N}(t,\mu,\underline{u}),\ \underline{\xi}_\mu^*(t)\rangle$$

$$\underline{B}(t,\mu,z,\bar{z},\underline{W}) = \underline{N}(t,\mu,\underline{u}) - \langle \underline{N}(t,\mu,\underline{u}),\ \underline{\xi}_\mu^*\rangle\, \underline{\xi}_\mu - \langle \underline{N}(t,\mu,\underline{u}),\ \bar{\underline{\xi}}_\mu^*\rangle\, \bar{\underline{\xi}}_\mu \ .$$

We have $b = b_0 + b_1$, $\underline{B} = \underline{B}_0 + \underline{B}_1$, where $b_0 = b(t,\mu,z,\bar{z},0)$, $b_1 = 0(|z|\,\|\underline{W}\| + \|\underline{W}\|^2)$ $\underline{B}_0 = \underline{B}(t,\mu,z,\bar{z},0)$ $\underline{B}_1 = 0(|z|\,\|\underline{W}\| + \|\underline{W}\|^2)$.

Roughly speaking (7b) will be eliminated and (7a) made autonomous up to $0(|z|^{N+1})$. To do this we change variables

$$y = z + \gamma(t,\mu,z,\bar{z}) = z + \sum_{p+q\geq 2}^{N} z^p\bar{z}^q\,\gamma_{pq}(t,\mu)$$

$$\underline{Y} = \underline{W} + \underline{\Gamma}(t,\mu,z,\bar{z}) = \underline{W} + \sum_{p+q\geq 2}^{N} z^p\bar{z}^q\,\underline{\Gamma}_{pq}(t,\mu) \tag{8}$$

where N is arbitrary, $\gamma_{pq}$ and $\underline{\Gamma}_{pq}$ are T-periodic, and $\underline{\Gamma}_{pq} \perp \underline{\xi}_\mu^*, \bar{\underline{\xi}}_\mu^*$ . We chose $\gamma_{pq}$, $\underline{\Gamma}_{pq}$ later, after (7) has been rewritten in terms of $y, \underline{Y}$. Now

$$\frac{dy}{dt} = \sigma z + b + \frac{\partial\gamma}{\partial t} + \frac{\partial\gamma}{\partial z}(\sigma z + b) + \frac{\partial\gamma}{\partial\bar{z}}(\bar{\sigma}\bar{z} + \bar{b})$$

$$= \sigma y + \{\frac{\partial \gamma}{\partial t} + \sigma z \frac{\partial \gamma}{\partial z} + \bar{\sigma}\bar{z}\frac{\partial \gamma}{\partial \bar{z}} - \sigma\gamma + \tilde{b}\} + b_1(1+\frac{\partial \gamma}{\partial z}) + \bar{b}_1 \frac{\partial \gamma}{\partial \bar{z}}$$

where $\tilde{b}(t,\mu,z,\bar{z}) = b_0(1+\frac{\partial \gamma}{\partial z}) + \bar{b}_0 \frac{\partial \gamma}{\partial \bar{z}}$;

$$\frac{d\underline{Y}}{dt} = \underline{f}_u(t,\mu|\underline{Y}) + \{\frac{\partial \underline{\Gamma}}{\partial t} + \sigma z \frac{\partial \underline{\Gamma}}{\partial z} + \bar{\sigma}\bar{z}\frac{\partial \underline{\Gamma}}{\partial \bar{z}} - \underline{f}_u(t,\mu|\underline{\Gamma}) + \tilde{\underline{B}}\} + \underline{B}_1(1+\frac{\partial \underline{\Gamma}}{\partial z}) + \bar{\underline{B}}_1 \frac{\partial \underline{\Gamma}}{\partial \bar{z}}$$

where $\tilde{\underline{B}}(t,\mu,z,\bar{z}) = \underline{B}_0(1+\frac{\partial \underline{\Gamma}}{\partial z}) + \bar{\underline{B}}_0 \frac{\partial \underline{\Gamma}}{\partial \bar{z}}$.

Expand $\begin{matrix}\tilde{b}\\ \tilde{\underline{B}}\end{matrix} = \sum_{p+q\geq 2}^{N} \begin{matrix}\tilde{b}_{pq}\\ \tilde{\underline{B}}_{pq}\end{matrix}(t,\mu)z^p\bar{z}^q + 0(|z|^{N+1})$

where $\tilde{b}_{pq}$, $\tilde{\underline{B}}_{pq}$ are T periodic and $\tilde{\underline{B}}_{pq} \perp \xi^*, \bar{\xi}^*$. Then

$$\frac{dy}{dt} = \sigma y + \sum_{p+q\geq 2}^{N} \{\frac{\partial \gamma_{pq}}{\partial t} + [\sigma(p-1) + \bar{\sigma}q]\gamma_{pq} + \tilde{b}_{pq}\}z^p\bar{z}^q + 0(|z|\,||w|| + ||w||^2 + |z|^{N+1}),$$

$$\frac{d\underline{Y}}{dt} = \underline{f}_u(t,\mu|\underline{Y}) + \sum_{p+q\geq 2}\{\frac{\partial \underline{\Gamma}_{pq}}{\partial t} - \underline{f}_u(t,\mu|\underline{\Gamma}_{pq}) + [\sigma p + \bar{\sigma}q]\,\underline{\Gamma}_{pq} + \tilde{\underline{B}}_{pq}\}z^p\bar{z}^q + 0(|z|\,||\underline{w}|| + ||\underline{w}||^2 + |z|^{N+1})$$

Finally use (8) on the right hand side to get

$$\frac{dy}{dt} = \sigma y + \sum_{p+q\geq 2}\{\frac{\partial \gamma_{pq}}{\partial t} + [\sigma(p-1) + \bar{\sigma}q]\gamma_{pq} + \bar{\bar{b}}_{pq}\}y^p\bar{y}^q + 0(|y|\,||\underline{Y}|| + ||\underline{Y}||^2 + |y|^{N+1}) \quad (9a)$$

$$\frac{d\underline{Y}}{dt} = \underline{f}_u(t,\mu|\underline{Y}) + \sum_{p+q\geq 2}^{N}\{-J_\mu(\underline{\Gamma}_{pq}) + [\sigma p + \bar{\sigma}q]\underline{\Gamma}_{pq} + \bar{\bar{\underline{B}}}_{pq}\} + 0(|y|\,||\underline{Y}|| + ||\underline{Y}||^2 + |y|^{N+1}) \quad (9b)$$

where $\bar{\bar{b}}_{pq}$ and $\bar{\bar{\underline{B}}}_{pq}$ are functions of $\gamma_{ij}$, $\underline{\Gamma}_{ij}$ with <u>$i+j < p+q$</u> with T-periodic coefficients and such that all terms in 9b are orthogonal to $\xi^*_\mu$, $\bar{\xi}^*_\mu$.

Now $\gamma_{pq}$ $\underline{\Gamma}_{pq}$ are chosen successively for $p+q = 2,3,\ldots,N$ so as to simplify (9). This is the key step. We choose $\underline{\Gamma}_{pq}$ to make $-J_\mu(\underline{\Gamma}_{pq}) + [\sigma p + \bar{\sigma}q] + \bar{\bar{\underline{B}}}_{pq} \equiv 0$ for small $\mu$. This is always possible since $\underline{\Gamma}_{pq}, \bar{\bar{\underline{B}}}_{pq} \in \{\underline{\xi} \in \mathbb{P}_T : \underline{\xi} \perp \underline{\xi}^*_\mu, \bar{\underline{\xi}}^*_\mu\}$ and the bifurcation assumptions mean that for small $\mu$ none of the eigenvalues of $J_\mu$ on this space has real part as small as $Re(\sigma p + \bar{\sigma}q)$. This reduces (9b) to

$$\frac{d\underline{Y}}{dt} = \underline{f}_u(t,\mu|\underline{Y}) + 0(||\underline{Y}||\,|y| + ||\underline{Y}||^2 + |y|^{N+1}). \quad (10a)$$

In order to choose $\gamma_{pq}$, write

$$\bar{b}_{pq}(t,\mu) = \sum_{\ell\in\mathbb{R}} b_{pq\ell}(\mu)\, e^{\frac{2\pi i\ell t}{T}}$$

$$\gamma_{pq}(t,\mu) = \sum_{\ell\in\mathbb{R}} \gamma_{pq\ell}(\mu)\, e^{\frac{2\pi i\ell t}{T}}$$

Then $\frac{\partial \gamma_{pq}}{\partial t} + [\sigma(p-1) + \sigma q]\gamma_{pq} + \bar{b}_{pq} = \sum_{\ell\in\mathbb{Z}} \alpha_{pq\ell}(\mu)\, e^{\frac{2\pi i\ell t}{T}}$

where $\alpha_{pq\ell}(\mu) = \{\frac{2\pi i\ell}{T} + [\sigma(p-1) + \bar{\sigma}q]\}\gamma_{pq\ell}(\mu) + b_{pq\ell}(\mu)$

$$\alpha_{pq\ell}(0) = \frac{2\pi i}{T}\{\ell + r[p-1-q]\}\gamma_{pq\ell}(0) + b_{pq\ell}(0) \quad .$$

We see that we can always choose $\gamma_{pq\ell}$ to make $\alpha_{pq\ell}(\mu) \equiv 0$ for small $\mu$ unless $\ell + r[p-1-q] = 0$. We call $\{(p,q,\ell,r) : \ell + r[p-\ell-q] = 0\}$ the <u>Exceptional Set</u>. It is the union of two disjoint subsets.

I The mean set : $(p,q,\ell,r) = (q+1,q,0,r) \qquad 2 \le 2q+1 \le N$

II The resonant set : $(p,q,\ell,r) = (q+1+nk, q, -km, \frac{m}{n}) \qquad 0 \le m < n,\ k \ge 1 \quad 2 \le 2q+1+nk \le N$ .

The mean set is present for any r, but the resonant set arises only when r is rational.

<u>When (p,q,ℓ,r) is in the exceptional set choose $\gamma_{pq\ell}(\mu) \equiv 0$; otherwise choose $\gamma_{pq\ell}(\mu)$ to make $\alpha_{pq\ell}(\mu) \equiv 0$.</u> This reduces (9a) to

$$\frac{dy}{dt} = \sigma(\mu)y + \sum_{q\ge 1}^{2q+1\le N} y^{q+1}\bar{y}^{q} b_{q+1,q,0}(\mu) + \sum_{k>0}\sum_{q\ge 0}^{2q-1+nk\le N}\{y^{q+1+nk}\bar{y}^{q} b_{q+1+nk,q,-mk}(\mu)\, e^{-2\pi imkt/T}$$
$$+ y^{q}\bar{y}^{q-1+nk} b_{q,q-1+nk,mk}(\mu)\, e^{2\pi imkt/T}\}$$
$$+ 0(|y|\,\|\underline{Y}\| + \|\underline{Y}\|^2 + |y|^{N+1}) \tag{10b}$$

<u>The asympotic representation is obtained by neglecting the order terms in (10a, b). The truncation number N in (10b) is arbitrary.</u> The justification of this approximation will not be attempted here; see [3, 4, 5]. We proceed to study the approximate problem.

It is clear that (10a) gives $\underline{Y}(t,\mu) \equiv 0$. To study (10b) set

$$y = xe^{i\omega_0 t} \tag{11}$$

Substitution in (10b) gives an autonomous equation of the form

$$\frac{dx}{dt} = \mu\hat{\sigma}(\mu)x + \sum_{q\ge 1}^{2q+1<N} + x|x|^{2q} a_q(\mu)$$
$$+ \sum_{k>0}\sum_{q\ge 0} |x|^{2q}\{x^{1+nk} a_{qk} + \bar{x}^{nk-1} a_{q,-k}\} \tag{12}$$

where $\mu\hat{\sigma}(\mu) = \sigma(\mu) - \sigma(0)$ and $a_{q,k}(\mu) \equiv 0$ if r is irrational.

We shall look for the equilibrium solutions of (12). We expect to find fixed points and closed curves. These will be cross sections of subharmonic trajectories and invariant tori for the original problem. The type of solution will depend on which terms on the right hand side of (12) have lowest order in x after $\mu\hat{\sigma}(\mu)x$. If $n = 3$ the term from the resonant set $a_{0,-1}\bar{x}^{n-1}$ is the only term of order 2, and we shall find fixed points for (12). If $n = 4$, $a_{0,-1}\bar{x}^{n-1}$ from the resonant set and $a_1 x|x|^2$ from the mean set both have order 3, and either fixed points or an invariant circle can occur. If $n \geq 5$ then terms from the mean set have lower order, and we expect a closed orbit of (12). Normally it is traversed at a speed $O(\varepsilon^2)$, but if enough exceptional conditions hold this speed can be so low that the terms from the resonant set break up the closed orbit into fixed points. This is weak resonance.

All of the above remarks assume that various terms are $\neq 0$. The exceptional cases where this is not true are ignored here. Also it will be assumed for simplicity that $\hat{\sigma}, a_q, a_{q,k}, a_{q,-k}$ ... are independent of $\mu$. This does not change the essence of the arguments.

## 3. Hopf Bifurcation.

This section outlines how to compute the trajectories on the torus when $n \geq 5$. We introduce an amplitude which is the mean radius of the invariant circle,

$$\varepsilon = \frac{1}{2\pi}\int_0^{2\pi} x(s)e^{-is}ds \quad .$$

We assume the orbit can be written in the form

$$\begin{aligned} x(t,\mu) &= \varepsilon\, e^{is}\chi(s,\varepsilon) \\ \mu &= \varepsilon\tilde{\mu}(\varepsilon) \\ s &= \varepsilon\Omega(\varepsilon)t \end{aligned} \qquad (13)$$

where $\chi$ is $2\pi$ periodic in $\theta$. Note that $2\pi/\varepsilon\Omega(\varepsilon)$ is the period of the closed orbit of (12).

Substitution in (12) gives

$$(i\Omega - \tilde{\mu}\hat{\sigma})\chi + \Omega\frac{d\chi}{ds} = \Sigma\, \chi|\chi|^{2q}a_q\varepsilon^{2q-1} + \sum_{k>0}\sum_{q\geq 0}|\chi|^{2q}\{a_{q,k}e^{ink\theta}\chi^{1+nk}\varepsilon^{2q+nk-}$$
$$+ a_{q,-k}e^{-ink\theta}\chi^{nk-1}\varepsilon^{2q+nk-3}\} \; .$$

Expand in powers of $\varepsilon$ :

$$\begin{aligned} \chi(s,\varepsilon) &= \sum_{j=0}^{\infty} \chi_j(s)\varepsilon^j \; , \\ \tilde{\mu}(\varepsilon) &= \sum_{j=0}^{\infty} \mu_j\varepsilon^j \; , \\ \Omega(\varepsilon) &= \sum_{j=0}^{\infty} \Omega_j\varepsilon^j \; . \end{aligned}$$

The functions $\chi_j(.)$ are $2\pi$-periodic; and

$$1 = \frac{1}{2\pi}\int_0^{2\pi}\chi(s,\varepsilon)ds ,$$

so

$$\frac{1}{2\pi}\int_0^{2\pi}\chi_j(s,\varepsilon)ds = \begin{cases} 1 & j = 0 \\ 0 & j \geq 1 \end{cases} \tag{14}$$

We now solve by evaluating coefficients of successive powers of $\varepsilon$. From the terms of order 0,

$$(i\Omega_0 - \tilde{\mu}_0\hat{\sigma})\chi_0 + \Omega_0\frac{d\chi_0}{ds} = 0 .$$

Taking the mean over $(0,2\pi)$ gives

$$i\Omega_0 = \tilde{\mu}_0\hat{\sigma} .$$

Now it follows from the bifurcation assumptions that $\hat{\sigma}$ has positive real part. Hence, since $\Omega_0$ and $\tilde{\mu}_0$ are both real,

$$\Omega_0 = \tilde{\mu}_0 = 0 .$$

The terms of order 1 in $\varepsilon$ now give

$$(i\Omega_1 - \tilde{\mu}_1\hat{\sigma})\chi_0 + \Omega_1\frac{d\chi_0}{ds} = |\chi_0|^2\chi_0 a_1 . \tag{15}$$

Taking the mean over $(0,2\pi)$ gives

$$i\Omega_1 - \tilde{\mu}_1\hat{\sigma} = a_1\frac{1}{2\pi}\int_0^{2\pi}|\chi_0|^2\chi_0 ds . \tag{16}$$

It can be shown from (14), (15) and (16) that $\chi_0(s) \equiv 1$. From (13) we now obtain

$$i\Omega_1 - \tilde{\mu}_1\hat{\sigma} = a_1 .$$

Taking real and imaginary parts gives

$$\tilde{\mu}_1\hat{\xi} + \alpha_1 = 0$$

$$\Omega_1 - \tilde{\mu}_1\hat{\eta} = \beta_1 .$$

To continue we have to assume that $\Omega_1 \neq 0$. It will be seen in the next section that $\Omega_1 = 0$ is the first of the special conditions leading to weak resonance.

The terms in $\varepsilon^2$ give

$$\Omega_1\frac{d\chi_1}{ds} - a_1(\chi_1 + \bar{\chi}_1) = g_1(s) - (i\Omega_2 - \tilde{\mu}_2\hat{\sigma})$$

where

$$g_1(s) = a_{0,-1}e^{-5is} \quad n = 5$$
$$= 0 \quad n \geq 5 .$$

We see from (14) that we must have

$$\int_0^{2\pi}\{g_1(s) - (i\Omega_2 - \tilde{\mu}_2\hat{\sigma})\}ds = 0 .$$

This is true if and only if

$$\Omega_2 = \tilde{\mu}_2 = 0 .$$

It is easily shown using Fourier series that the equation $\Omega_1 \frac{dy}{ds} - a_1(y+\bar{y}) = \hat{g}(s)$ where $\hat{g}$ is $2\pi$-periodic and $\int_0^{2\pi}\hat{g}(s)ds = 0$ has a unique $2\pi$-periodic solution. We see that

$$\chi_1(s) = Ae^{5is} + Be^{-5is} \qquad n = 5$$
$$= 0 \qquad n \geq 5 .$$

The analysis continues along these lines. It is found that $\tilde{\mu}(.)$ and $\Omega(.)$ are both odd functions, and that $\chi(.,\varepsilon)$ is $\frac{2\pi}{n}$ periodic (constant if r is irrational). This is to be expected since (12) is invariant under rotation through $\frac{2\pi}{n}$. By tracing back through the derivation in section 2, we see that our approximate solution is quasi-periodic with the two frequencies $\frac{2\pi}{T}$ and $\omega_0 + \varepsilon^2\omega(\varepsilon^2) = \omega_0 + \Omega_1 + \varepsilon^2\Omega_3 + \dots$ .

## 4. Subharmonic Bifurcation.

Suppose $x = \delta e^{i\varphi(\delta)}$ is a steady solution of (12). Note that $\delta e^{i\varphi(\delta)} e^{2\pi ik/n}$, $0 \leq k \leq n-1$, are all steady solutions of (12). They are the n-piercing points of a single nT-periodic trajectory. We have

$$0 = \mu\hat{\sigma} + \delta^2 a_1 + \delta^4 a_2 + \dots + \delta^{n-2} e^{-in\varphi} a_{0,-1} + \dots .$$

Assume

$$\varphi(\delta) = \varphi_0 + \varphi_1\delta + \varphi_2\delta^2 + \dots$$
$$\mu = \mu^{(1)}\delta + \mu^{(2)}\delta^2 + \dots .$$

We evaluate the coefficients of increasing powers of $\delta$ .

For $n = 3$ : the terms in $\delta$ give

$$\mu^{(1)}\hat{\sigma} + a_{0,-1}e^{-3i\varphi_0} = 0 .$$

Hence

$$\mu^{(1)} = |a_{0,-1}/\hat{\sigma}| ,$$
$$\varphi_0 = \tfrac{1}{3}\arg(a_{0,-1}/\hat{\sigma}) + \frac{2k-1}{3} \qquad k = 0,1,2$$

(taking $\mu^{(1)} = -|a_{0,-1}/\hat{\sigma}|$ will give the same solution).

The higher order terms can now be calculated. We obtain a single 3T-periodic trajectory. The bifurcation is two sided since $\mu(\delta) = 0(\delta)$.

If $n \geq 4$ : the terms in $\delta$ give

$$\mu^{(1)} = 0 .$$

For n = 4 the terms in $\delta^2$ give

$$\mu^{(2)}\hat{\sigma} + a_1 + e^{-4i\varphi_0} a_{0,-1} = 0 \quad ,$$

so
$$|\mu^{(2)}\hat{\sigma} + a_1|^2 = |a_{0,-1}|^2 \quad .$$

This gives a quadratic equation for $\mu^{(2)}$. If the discriminant is positive we have two different values of $\mu^{(2)}$ which lead to two different 4T-periodic trajectories.

If $n \geq 5$ : the terms in $\delta^2$ give

$$\mu^{(2)}\hat{\sigma} + a_1 = 0 \quad .$$

This is the first special condition for weak resonance; the requirement that $\mu^{(2)}$ be real restricts $\hat{\sigma}$ and $a_1$. It can be verified that this restriction is equivalent to the requirement that $\Omega_1 = 0$ which was used in section 3.

For n = 5 : the terms in $\delta^3$ give

$$\mu^{(3)}\hat{\sigma} + a_{0,-1} e^{-5i\varphi_0} = 0 \quad .$$

This determines $\mu^{(3)}$ and $\varphi_0$. Higher order terms can then be calculated. Since $\mu(\delta) = O(\delta^2)$ the bifurcation is one sided. Since $\mu^{(3)} \neq 0$ $\mu(\delta)$ is not even we obtain two 5T-periodic trajectories.

If $n \geq 6$ : the terms in $\delta^3$ give

$$\mu^{(3)} = 0 \quad .$$

For n = 6 : the terms in $\delta^4$ give

$$\mu^{(4)}\hat{\sigma} + a_2 + a_{0,-1} e^{-6i\varphi_0} = 0 \quad .$$

This gives a quadratic equation for $\mu^{(4)}$; if the discriminant is positive two 6T-periodic trajectories bifurcate.

If $n \geq 7$ : the terms in $\delta^4$ give

$$\mu^{(4)}\hat{\sigma}_0 + a_2 = 0 \quad .$$

This is the second special condition for weak resonance.

The results continue along these lines. As n increases subharmonic trajectories are possible only if more and more special conditions hold. When they do hold (and for even n, if an extra inequality holds) there is one sided bifurcation of two nT-periodic trajectories. In the supercritical case, one solution is unstable and one is stable. In the subcritical case the torus itself is repelling and the periodic solution which is stable to disturbances or the torus is unstable in the larger space.

## 5. Rotation Number and Lock-Ins.

We conclude with a few remarks about the phenomenon of frequency locking when there is an invariant torus. This occurs when all the trajectories on the torus are captured by a single (subharmonic) trajectory.

We introduce the Poincaré (first return map). This is the map from the invariant circle to itself, this map takes a point on the circle to where the trajectory passing through it meets the circle again after going round the torus once (i.e. after time T). Consider its rotation number, $\rho$ (defined for example in [4]; the reader may think of $\rho$ as a frequency ratio). If $\rho$ is irrational there is a change of coordinates which makes the Poincaré map a rotation, and the flow on the torus is quasiperiodic. The Poincaré map has no periodic points. If $\rho = p/q$ is rational, the Poincaré map must have periodic points of order q, to which correspond subharmonic trajectories. Generally there will be two such trajectories one attracting, the other repelling.

It is important to distinguish between the rotation number $\hat{\rho}(\varepsilon)$ for the asymptotic representation computed in section 3, and the rotation number $\rho(\varepsilon)$ for the real flow. It is known that $\rho(\varepsilon)$ is continuous but it is generally not differentiable. What happens is that if $\rho(\varepsilon_0) = p/q$ then $\rho(\varepsilon) \equiv p/q$ on an interval about $\varepsilon_0$. The rotation number locks on to each rational value. This happens because if $\theta_0$ is a periodic point of order q of the Poincaré map, $f_{\varepsilon_0}$, then generically $\partial/\partial\theta(f_\varepsilon^q)|_{\varepsilon=\varepsilon_0,\,\theta=\theta_0} \neq 0$. This enables us to solve for a fixed point of $f_\varepsilon^q$ when $\varepsilon$ is near $\varepsilon_0$, so $\rho(\varepsilon)$ cannot change near $\varepsilon_0$.

In particular the set of values of $\varepsilon$ for which $\rho(\varepsilon)$ is rational has positive measure. It is an important result of M. Herman [2] that the set on which $\rho(\varepsilon)$ is irrational also has positive measure.

The results from section 3 show that the approximate rotation number is of the form

$$\hat{\rho}(\varepsilon) = \omega_0 + \varepsilon^2\omega(\varepsilon^2).$$

It can be concluded from this that the true rotation number lies between two polynomials

$$\rho(\varepsilon) = \hat{\rho}(\varepsilon) \pm K\varepsilon^N ,$$

where N is arbitrary. It follows that the lengths of the flat line segments on which lock-ins occur must tend to zero faster than any power of N as $\varepsilon \to 0$.

## 6. Experiments.

The type of dynamics which I have discussed in these lectures is characteristic of the observed dynamics in some mechanical systems involving fluid motions. The fact that an analysis of the kind given here does seem to fit well the observations of motion in small boxes of liquid heated from below, and in flow systems like the Taylor problem may surprise some readers. The surprise is that an analysis in two dimensions, and low dimensions greater than 2 give results in agreement with observations of continuium systems with "infinitely" many dimensions. In fact, this kind of agreement is associated with the fact that the spectrum of eigennodes in the small scale systems for which agreements is sought is widely separated and the dimension of active eigenvalues is actually small.

I do not want to give a too cryptic explanation of the relevance to real fluid mechanics of the kind of analysis sketched in these lectures. In fact this kind of analysis is recommended for actual computation of bifurcated objects in fluid mechanics near the point of bifurcation [4]. A not too cryptic explanation of relevance can be found in my two review papers (D.D. Joseph, Hydrodynamic Instability and Turbulence, Ed. H. Swinney and J. Gollub, Topics in Physics, Springer, 1980) or in 'Bifurcation in Fluid Mechanics" in the translation of the XIIIth International Congress of Theoretical and Applied Mechanics, (IUTAM), Toronto 1980.

## References.

1. M. Herman, Sur la conjugaison differentiable des diffeomorphismes du cercle a des rotations, Publ. I.H.E.S. 49, 5-234 (1979).

2. M. Herman, Mesure de Lebesgue et nombre de rotation, Lecture Notes in Maths., No. 597 Springer Verlag, 271-293 (1977).

3. G. Iooss, Bifurcation of Maps and Applications, Mathematics Studies No. 36 North Holland (1979).

4. G. Iooss & D.D. Joseph, Elementary Stability and bifurcation Theory, Undergraduate Textbook in Mathematics, Springer (1980).

5. G. Iooss & D.D. Joseph, Bifurcation and Stability of nT-periodic solutions at a point of resonance, Arch. Rational Mech. Anal. 66, 135-172 (1977).

6. Y.H. Wan, Bifurcation into invariant tori at points of resonance, Arch. Rat. Mech. Anal. 68, 343-357 (1978).

D.D. Joseph, Department of Aerospace Engineering & Mechanics, 107 Aeronautical Engineering, University of Minnesota, 110, Union St. S.E., Minneapolis, Minnesota 55455, U.S.A.

K. Burns, Mathematics Institute, University of Warwick, Coventry, CV4 7AL, England.

General introduction to steady state bifurcation.

Lectures given by David Schaeffer
Notes by John Hayden.

## LECTURE 1.

The purpose of the first two lectures is to establish the notation and ideas needed for the study of steady state bifurcation. We start the discussion with the classical problem of the buckling beam considered by Euler in 1744.

Figure 1.

Consider the problem of compressing an elastic beam with end loading. Multiple states occur if the load $\lambda$ is sufficiently large. In Euler's model the beam is assumed incompressible but capable of bending. The potential energy is proportional to the integral of the square of the curvature. We describe the buckled configuration by $\psi(s)$, the angle that the buckled beam makes with the x-axis as a function of arc length.

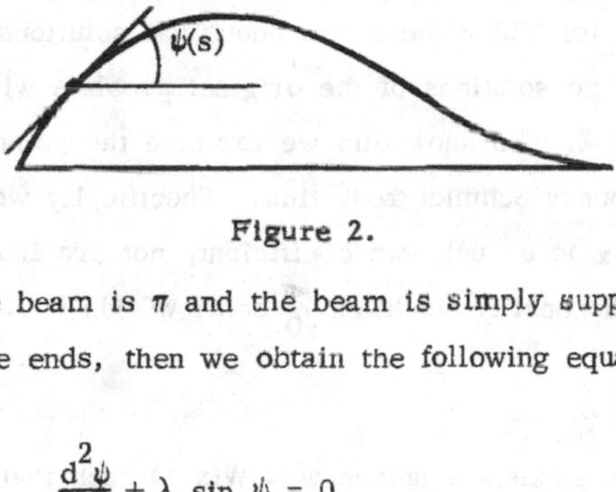


Figure 2.

If the length of the beam is $\pi$ and the beam is simply supported i.e. the curvature vanishes at the ends, then we obtain the following equation for equilibrium :

$$\frac{d^2\psi}{ds^2} + \lambda \sin \psi = 0$$

$$\psi'(0) = \psi'(\pi) = 0$$

See Reiss's article in Antman and Keller [1].

The abstract formulation of the problem is in the Sobolev space

$$\mathfrak{X} = \{u \in H^2(0,\pi) : u'(0) = u'(\pi) = 0\} .$$

$H^2(0,\pi)$ consists of those functions in $L^2(0,\pi)$ whose second order distributional derivatives also belong to $L^2(0,\pi)$. We have a non-linear map

$$\Phi : \mathfrak{X} \times \mathbb{R} \to L^2(0,\pi)$$

$$\Phi(u,\lambda) = u'' + \lambda \sin u \quad .$$

We are interested in the solutions of $\Phi(u,\lambda) = 0$. Observe that $\Phi(0,\lambda) = 0$ is the undeflected solution. Are there any other solutions? We consider only the restricted question, are there any other solutions near the trivial solution? This is answered by the implicit function theorem. We look at $d_u\Phi$, the linearisation of $\Phi$. If $d_u\Phi$ is invertible we may solve for u as a function of $\lambda$ uniquely. But $u \equiv 0$ independent of $\lambda$ is one solution so it is the solution.

In the specific example of the buckling beam the linearisation involves the approximation $\sin u \approx u$. Therefore, we have the following linearised equations.

$$u'' + \lambda u = 0$$

$$u'(0) = u'(\pi) = 0$$

This is a classical eigenvalue problem which has solutions if any only if $\lambda = n^2$, $n \in \mathbb{Z}$. The smallest $\lambda$ for which there are nontrivial solutions is $\lambda = 1$. It turns out that for $\lambda < 1$ there are no solutions of the original problem with buckling while such states do appear for $\lambda > 1$. To show this we examine the point $\lambda = 1$ more carefully, using the so-called Lyapunov-Schmidt reduction. Specifically we look for a solution of the form $u = x \cos + w$ $x$ is an unknown coefficient, not arc length and w is given unambigously by the orthogonality condition $\int_0^{\pi} \cos s \, w(s) ds = 0$.

Lemma.

Given $x, \lambda$ near $(0,1)$ there exists a unique $w = W(x,\lambda)$ such that $\Phi(x \cos + w, \lambda) \in \mathbb{R}\{\cos\}$, the complement of the range.

The proof is by the Lyapunov-Schmidt reduction. Intuitively, $\ker d\Phi$ is one dimensional and so we only lose control of that one dimension and may solve in the complementary space.

Define $F: \mathbb{R} \times \mathbb{R} \to \mathbb{R}$ by

$$F(x,\lambda) = \langle \cos, \Phi(x \cos + W(x,\lambda), \lambda) \rangle_{L^2}$$

Lemma.

$$F(x,\lambda) = 0 \Leftrightarrow \Phi(x \cos + W(x,\lambda), \lambda) = 0 \ .$$

Moreover every solution of $\Phi(u,\lambda)$ may be obtained in this way.

We now study the reduced problem F at x = 0 and $\lambda = 1$. It turns out that $F = F_x = F_{xx} = F_\lambda = 0$ but $F_{xxx} < 0$ and $F_{\lambda x} > 0$ at $(x,\lambda) = (0,1)$. Indeed $F_x = 0$ since $\cos \in \ker d\Phi$, $F_{xx} = 0$ since the equation is odd, and $F_\lambda = 0$ since $F(0,\lambda) \equiv 0$.

We can write

$$0 = F(x,\lambda) = -ax^2 + 2b(\lambda-1)x + 0(x^5, (\lambda-1)^2x) .$$

The solution set is given by

$$x = 0 \text{ or } \lambda = 1 + \frac{a}{b}x^2$$

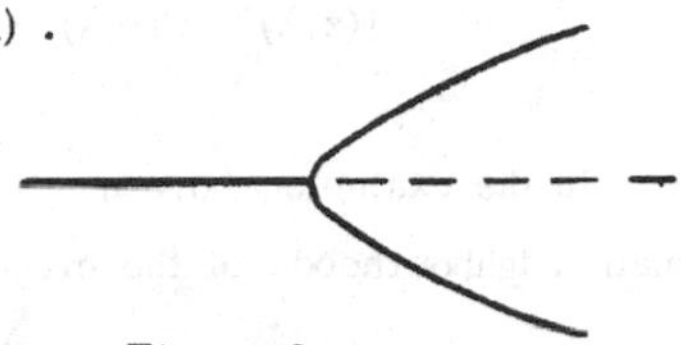

Figure 3.

Here we have neglected the higher order terms because our theory will show that they do not change the qualitative picture.

A criticism of Euler's theory is that under perturbation the solution set breaks into two disconnected pieces :

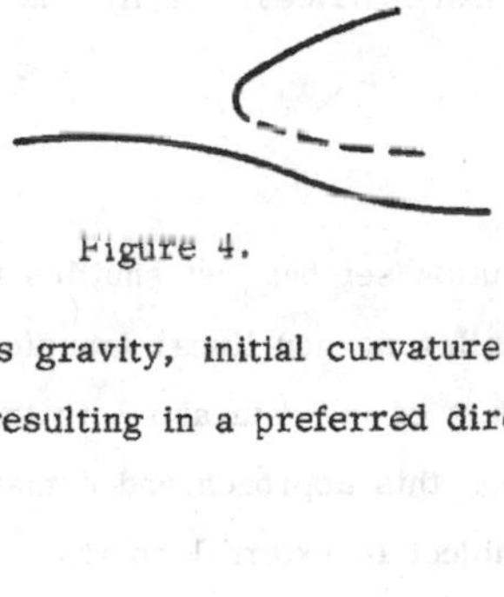

Figure 4.

Typically imperfections such as gravity, initial curvature or a knick in the side of the beam destroy the symmetry, resulting in a preferred direction of bifurcation.

Fundamental question.

Is it possible to describe in finite terms all the qualitative different perturbations of the equation?

The Lyapunov-Schmidt reduction can be used to discuss the question in terms of

$$F(x,\lambda) + \text{perturbation } (x,\lambda) = 0 .$$

To study this question we need a precise notion of qualitatively different. We shall say that perturbations of the equations are qualitatively similar if they differ by an allowable change of coordinates, defined below. See [6,7] for greater detail.

Definition.

A bifurcation problem is a $C^\infty$ mapping $G : \mathbb{R}^n \times \mathbb{R} \to \mathbb{R}^n$. (The problem can arise from a Lyapunov reduction with n-dimensional kernel and we study the solution set

$$\{(x,\lambda) \mid G(x,\lambda) = 0\}, \text{ the bifurcation diagram.})$$

In the example above $n = 1$. All statements concern the behaviour of maps in small neighbourhoods of the origin - in other words G is the germ of a mapping.

Definition.

Two bifurcation problems G, H are contact equivalent if

$$H(x,\lambda) = \tau_{x,\lambda} G(\rho(x,\lambda), \Lambda(\lambda)) .$$

$\tau_{x,\lambda}$ is an invertible family of $n \times n$ matrices. $(\rho, \Lambda) : \mathbb{R}^n \times \mathbb{R} \to \mathbb{R}^n \times \mathbb{R}$ is a diffeomorphism germ.

Remarks.

(i) $\tau$ does not change the solution set but just shuffles the equations. No more generality is obtained by allowing non-linear transformations.

(ii) The dependence on $\lambda$ is part of the data and $\lambda$ is treated specially. We discuss later the difference between this approach and catastrophe theory. We do not mix x in $\lambda$, since $\lambda$ is subject to external control.

The basic problems of the theory.

1. Given a bifurcation problem, when can you characterise all perturbations (up to contact equivalence) by a finite number of parameters? (This amounts to finding a universal unfolding.)

2. When 1. applies, enumerate all qualitatively distinct bifurcation diagrams.

Example.

$G(x,\lambda) = x^3 - \lambda x$ as above (except we have translated the bifurcation point to $\lambda = 0$).

Lemma.

Let $H(x,\lambda)$ be a bifurcation problem such that at $x = \lambda = 0$, $H = H_x = H_{xx} = H_\lambda = 0$, $H_{xxx}H_{\lambda x} < 0$. Then H is contact equivalent to $x^3 - \lambda x$.

The proof is elementary but vaguely tricky and gives no special insight. See [6]. The lemma shows that the canonical form in the example is fairly general.

Solution for the first problem for the example.

Theorem 1.

$F(x,\lambda,\alpha,\beta) = x^3 - \lambda x + \alpha + \beta x^2$ is a universal unfolding of $G(x,\lambda) = x^3 - \lambda x$.

The interpretation of the theorem is $G + \varepsilon P \cong F(.,.,\alpha,\beta)$. i.e. given any small perturbation $G + \varepsilon P$ of G, there exists $\alpha$ and $\beta$ such that $F(.,.,\alpha,\beta)$ is contact equivalent to the perturbed bifurcation problem.

Remark.

To unfold $x^3 - \lambda x$ we need two parameters, contrast this with catastrophe theory where we only have one parameter in the unfolding.

Solution of the second problem.

By Theorem 1 it suffices to search the $\alpha,\beta$ plane.

Theorem 2.

The bifurcation diagram of $F(.,.,\alpha,\beta)$ depends on $\alpha,\beta$ as indicated in the following diagram.

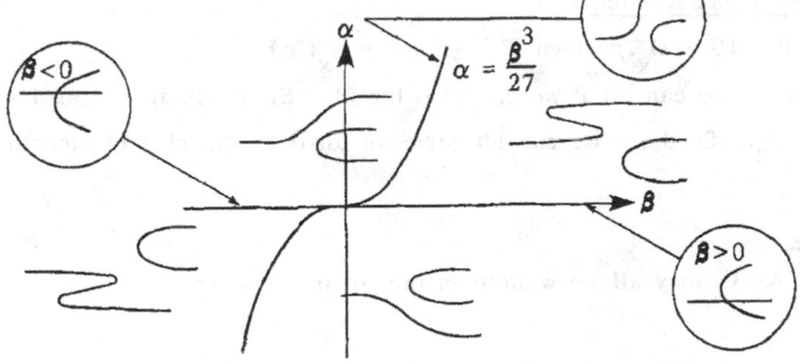


Figure 5. At the origin

Remark.

In general the first problem has an elegant and simple solution in terms of singularity theory. The second problem is much harder and at the present time can only be done on a case by case basis.

We first discuss the proof of these theorems for the particular problem $x^3 - \lambda x$ and them consider the general case.

The approach we adopt for the solution of the first problem is to find which perturbations affect G and which perturbations do not affect G. We define the unfolding by adding perturbations which do affect G. The Malgrange preparation theorem is the big theoretical gun. We first use the Lie algebra technique, given G we ask for which perturbations P is $G + \varepsilon P \cong G$ to first order in $\varepsilon$.

Suppose $G(x,\lambda) + \varepsilon P(x,\lambda) = (I+\varepsilon\tau_{x\lambda})G(x+\varepsilon\rho(x,\lambda),\ \lambda + \varepsilon\,\Lambda(\lambda)) + O(\varepsilon^2)$ then

$P(x,\lambda) = \tau_{x\lambda}(G(x,\lambda) + \frac{\partial G}{\partial x}\rho(x,\lambda) + \frac{\partial G}{\partial \lambda}\Lambda(\lambda) + O(\varepsilon^2)$. For which P do there exist $\tau_{x\lambda}$, $\rho(x,\lambda)$, and $\Lambda(\lambda)$ such that $P = \tau . G + G_x . \rho + G_\lambda . \Lambda$?

In the specific example $G = x^3 - \lambda x$ the equation is

$$P = \tau(x,\lambda)(x^3 - \lambda x) + (3x^2 - \lambda).\rho(x,\lambda) - x\,\Lambda(\lambda)$$

The third term, which depends only on $\lambda$, is much weaker than the others and so we ignore it in the initial analysis.

Fundamental observation :

If $P = \tau G + G_x . \rho$ then $fP = (f\tau).G + G_x(f\rho)$ .

In words, if we can hit P we can also hit fP. Similarly if we can hit $P_1$ and $P_2$ we can hit $P_1 + P_2$. So the P we can hit form an ideal in the ring of smooth function germs.

Lemma.

$x^3, \lambda x, \lambda^2$ may all be written in the form

$$\tau . G + G_x \rho \quad .$$

Proof.

(a) $-2x^3 = 1.(x^3-\lambda x) - (3x^2-\lambda)x,$

(b) $-\lambda x = (x^3-\lambda x) - x^3$, and

Similarly (c) $-\lambda^2 = (3x^2-\lambda).\lambda - 3(\lambda x).x$ .

We have a form of Newton's diagram

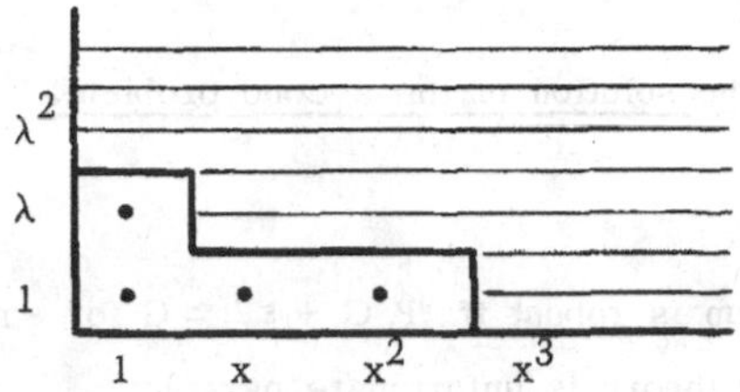


Figure 6.

We can get all of the monomials outside the "L" shaped region so we forget them. Inside, $3x^2 - \lambda$ is a relation (given either $x^2$ or $\lambda$ we can obtain the other). Including the ignored term $x\Lambda(\lambda)$ means we can get $x, \lambda x, \lambda^2 x$ . So we make a choice of 1 and $x^2$, and can get all other monomials in terms of these. That is, given $P(x,\lambda)$, $\exists \alpha, \beta \in \mathbb{R}$ s.t.

$$P(x,\lambda) = \alpha + \beta x^2 + \tau(x,\lambda)(x^3-\lambda x) + (3x^2-\lambda).\rho(x) - x\Lambda(\lambda).$$

Therefore a universal unfolding of G is

$$x^3 - \lambda x + \alpha + \beta x^2 .$$

The general machine tells us to put the extra terms into a universal unfolding, although the proof that this works is fairly long. See [6].

General theorem.

Given a bifurcation problem $G : \mathbb{R} \times \mathbb{R} \to \mathbb{R}$. If

$$\mathcal{E}_{x,\lambda} = \mathbb{R}\{p_1,\dots,p_k\} + \mathcal{E}_{x,\lambda}\{G, G_\lambda\} + \mathcal{E}_\lambda G_\lambda$$

then $F(x,\lambda,\alpha_1,\dots,\alpha_k) = G(x,\lambda) + \Sigma\alpha_j.p_j(x,\lambda)$ is a universal unfolding of G.

In this theorem $\mathcal{E}_{x,\lambda}$ refers to the set of germs of smooth functions of x and $\lambda$, $\mathcal{E}_\lambda$ to functions of $\lambda$ alone.

**Remarks.**

(i) The full theorem does not require the unfolding parameters to enter linearly.

(ii) The theorem in the vector valued case is analogous. We compute componentwise.

(iii) The result is readily computable in cases of practical interest.

## LECTURE 2.

**We begin with ideas for the solution of the second problem.**

**Definition.**

A bifurcation problem is robust if $\forall P, G + \varepsilon P \cong G$ for small $\varepsilon$. (The use of the term stable in singularity theory is unfortunate here.)

Consider $F(x,\lambda,\alpha,\beta) = x^3 - \lambda x + \alpha + \beta x^2$; for which values of $\alpha, \beta$ is $F(.,.,\alpha,\beta)$ robust?

There are three obvious ways in which the problem can fail to be robust:

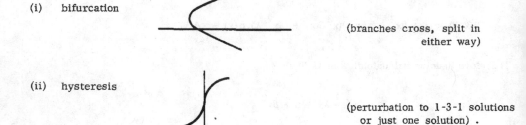

(i) bifurcation

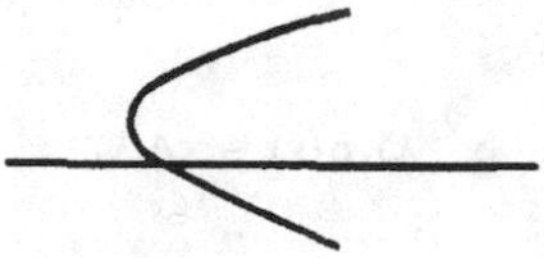

(branches cross, split in either way)

(ii) hysteresis

(perturbation to 1-3-1 solutions or just one solution) .

(iii) double limit

Figure 7.

**Remark.**

The double limit doesn't occur for $x^3 - \lambda x$ .

**Theorem.**

A bifurcation problem $G(x,\lambda) = 0$ is robust unless there is a point where one of the above three degeneracies occurs.

Proof.

See [6] .

Equation for bifurcation.

$$F = F_x = F_\lambda = 0 .$$

(At the bifurcation point $F(x,\lambda) = 0$ is not a manifold.)

Equation for hysteresis.

$$F = F_x = F_{xx} = 0 .$$

(At the hysteresis point $\lambda \sim x^3$.)

Computation of the bifurcation point for $x^3 - \lambda x + \alpha + \beta x^2$ .

$$F = x^3 - \lambda x + \alpha + \beta x^2 = 0 \qquad \alpha = 0$$

$$F_x = 3x^2 - \lambda + 2\beta x = 0 \qquad \lambda = 0$$

$$F_\lambda = -z = 0 .$$

Computation for the hysteresis points.

$$F = x^3 - \lambda x + \alpha + \beta x^2 = 0$$

$$F_x = 3x^2 - \lambda + 2\beta x = 0 \qquad \alpha = \frac{\beta^3}{27}$$

$$F_{xx} = 6x + 2\beta = 0 .$$

Relate this to the result of Theorem 2.

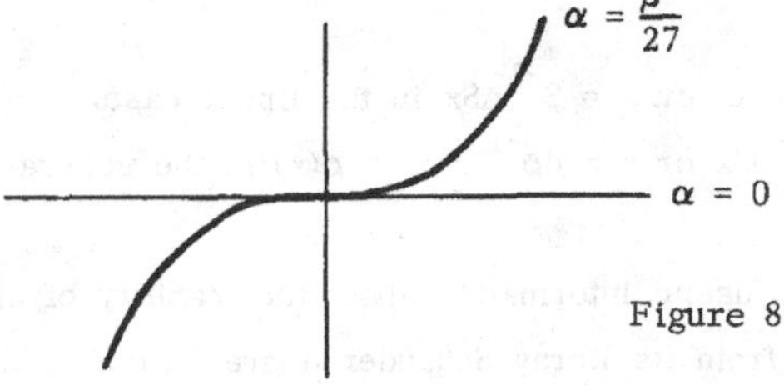


Figure 8.

In the general situation of $F(x,\lambda,\alpha_1,\ldots,\alpha_k)$ we define

$B = \{\alpha \in \mathbb{R}^k : \exists\ x,\lambda$ s.t. F has a bifurcation point at $(x,\lambda,\alpha)\}$.

$H = \{\alpha \in \mathbb{R}^k : \exists\ x,\lambda$ s.t. F has a hysteresis point at $(x,\lambda,\alpha)\}$.

$D = \{\alpha \in \mathbb{R}^k : \exists\ x,\lambda$ s.t. F has a double limit point at $(x,\lambda,\alpha)\}$.

B, H and D are real semi-algebraic surfaces in $\mathbb{R}^k$. $\mathbb{R}^k - B \cup H \cup D$ can be decomposed into connected components. The enumeration of the components is difficult and significant theoretical problems remain, but this construction leads to an enumeration of the possible perturbed diagrams, one for each component.

Remarks on the Leray-Schauder degree.

Warning. The fact that $G(x,\lambda)$ and $H(x,\lambda)$ are contact equivalent does not imply any relation between the phase portraits of the differential equations

$$\dot{x} = G(x,\lambda) \text{ and } \dot{x} = H(x,\lambda) .$$

Consider the following examples :

$$\begin{pmatrix}\dot{x}_1\\ \dot{x}_2\end{pmatrix} = \begin{pmatrix}1 & 0\\ 0 & -1\end{pmatrix}\begin{pmatrix}x_1\\ x_2\end{pmatrix} \qquad G(x,\lambda) = \begin{pmatrix}x_1\\ -x_2\end{pmatrix}$$

a saddle — independent of $\lambda$ .

$$\begin{pmatrix}\dot{x}_1\\ \dot{x}_2\end{pmatrix} = \begin{pmatrix}0 & -1\\ 1 & 0\end{pmatrix}\begin{pmatrix}x_1\\ x_2\end{pmatrix} \qquad H(x,\lambda) = \begin{pmatrix}-x_2\\ x_1\end{pmatrix}$$

a centre — independent of $\lambda$ .

The bifurcation problems G and H are contact equivalent. (Specifically chose $\tau = \begin{pmatrix}0 & 1\\ 1 & 0\end{pmatrix}$, $(\rho,\Lambda) = (x,\lambda)$), but of course the phase portraits are completely different. A different equivalence relation, namely similarity, is the one required for phase portraits to be related. That is,

$$x = Ax \text{ or } x = S^{-1}ASx \text{ in the linear case.}$$
$$x = Gx \text{ or } x = d\rho^{-1} \circ G \circ \rho(x) \text{ in the general case.}$$

Nevertheless, useful information about the stability of an equilibrium solution can often be extracted from its Leray-Schauder degree. Let $(x_0,\lambda_0)$ be a solution of $G(x,\lambda) = 0$, where $G : \mathbb{R}^n \times \mathbb{R} \to \mathbb{R}^m$. Let $L = (-dG)_{(x_0,\lambda_0)}$ (the derivative of G with respect to x only, so it is an n×n matrix). Suppose L has eigenvalues $\mu_1,\ldots,\mu_n$, and let k be the number of eigenvalues with negative real part.

Definition.

The Leray-Schauder degree $= (-1)^k$.

To understand the reason for this definition consider the equation

$$\dot{x} + G(x,\lambda) = 0 \ .$$

Suppose $\lambda = \lambda_0$, $x = x_0$ is an equilibrium solution. Linearise the equation near $x_0$, say $x = x_0 + \varepsilon y$; then

$$y + Ly = 0 \text{ to lowest order.}$$

The solutions of the equation are given by

$$y = e^{-\mu_j t} v_j \text{ where } v_j \text{ is an eigenvector .}$$

If Real $\mu_j > 0$ the solution tends to 0. If Real $\mu_j < 0$ the solution tends to $\infty$. Let k be the number of unstable directions. $k = 0 \Longleftrightarrow$ the solution is stable. Therefore, a necessary condition that the solution be stable is that the degree = +1.

Remark.

The degree is the sign of the determinant of L and is invariant under contact equivalence, provided we strengthen the definition of contact equivalence to require that $T$ and $d\phi$ have positive determinant.

If n = 1 the degree gives complete information about stability.

Example.

Stability and degree assignments for the pitchfork "s" stands for stable, "u" for unstable. The sign indicates the degree. It is also possible for all assignments to be reversed.

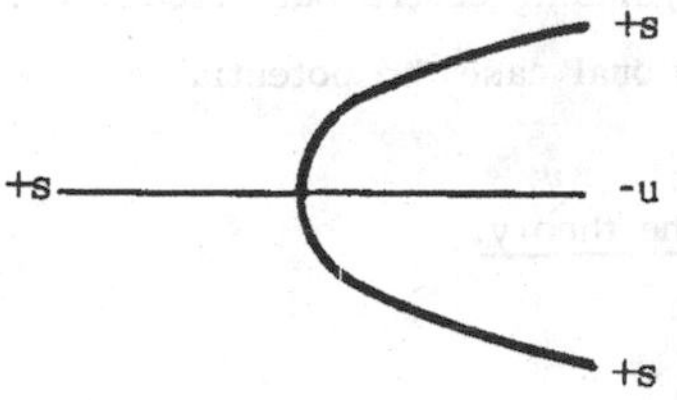


Figure 9.

Example with n = 1.

$$G(x,y,\lambda) = \begin{pmatrix} -x^2 - y^2 - \lambda x \\ -4xy - \lambda y \end{pmatrix}$$

There are three classes of solutions :

1. The trivial solution x = y = 0. Here $dG = \begin{pmatrix} -\lambda & 0 \\ 0 & -\lambda \end{pmatrix}$. Stable for $\lambda < 0$, unstable for $\lambda > 0$. Positive degree throughout.
2. The symmetric solution y = 0, x = $-\lambda$. Here $dG = \begin{pmatrix} \lambda & 0 \\ 0 & 3\lambda \end{pmatrix}$. Unstable for $\lambda < 0$, stable for $\lambda > 0$; positive degree throughout.
3. The symmetric solutions $x = \frac{-\lambda}{4}$, $y = \pm \frac{\sqrt{3}.\lambda}{4}$. Here

$$dG = \begin{pmatrix} \frac{-\lambda}{2} & \frac{\pm\sqrt{3}.\lambda}{2} \\ \pm\sqrt{3}.\lambda & 0 \end{pmatrix}$$

Note that det dG < 0, so the eigenvalues have opposite signs. Thus the degree is negative, unstable for all $\lambda$.

Remark.

This all carries over to the general case

$$u_t + \Phi(u,\lambda) = 0$$

where we consider the spectrum of $d\Phi$, given appropriate compactness hypotheses on $(d\Phi)^{-1}$ .

Shortcomings of the theory.

There are no dynamics or even a classification of stability, except for the degree. The general $C^\infty$ hypothesis can now be replaced by $C^k$ in particular problems. The results are local at the organising centre but already there is a large range of applications. In the variational case the potential has been ignored.

Comparison with catastrophe theory.

Consider the example

$$x^3 - \lambda x = 0 .$$

The Golubitsky and Schaeffer theory obtains a two parameter unfolding $x^3 - \lambda x + \alpha + \beta x^2 = 0$.

Catastrophe theory considers the potential $V(x,\lambda) = \frac{x^4}{4} - \frac{\lambda x^2}{2}$ and the equation $\frac{\partial V}{\partial x} = 0$. The basic singularity is $\frac{x^4}{4}$ with a universal unfolding $\frac{x^4}{4} + ax + bx^2$. One of the unfolding parameters is taken as $\lambda$ and it is argued that one additional parameter suffices to complete the picture.

The difference between the two point of view lies in the equivalence relations, i.e. in the changes of coordinate that are allowed. Consider the perturbed problem $x^3 - \lambda x + \beta x^2 = 0$, which has the bifurcation diagram

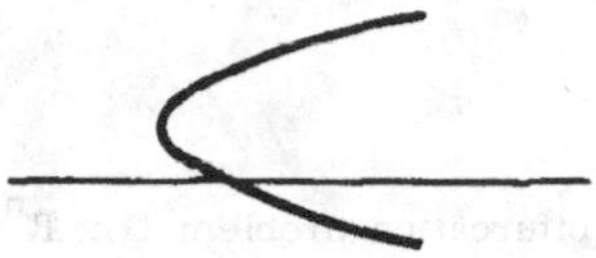

Figure 10.

Making the substitution $y = x + \frac{\beta}{3}$ and putting into potential form we rewrite the equation as

$$\frac{\partial}{\partial y}\left(\frac{y^4}{4} - \left(\lambda + \frac{\beta^2}{3}\right)\frac{y^2}{2} + \left(\frac{\lambda\beta}{3} + \frac{2}{27}\beta^3\right)y\right) = 0$$

Now suppose we define $\begin{cases} \bar{\lambda} = \lambda + \frac{\beta^3}{3} \\ a = \frac{\lambda\beta}{3} + \frac{2}{27}\beta^3 \end{cases}$

to obtain $\frac{\partial}{\partial y}\left(\frac{y^4}{4} - \lambda y^2 + ay\right) = 0$ .

This has a bifurcation diagram in $x, \lambda$ space of the form

Figure 11.

By making a $\lambda$-dependant change of coordinate in the unfolding parameter a, allowed by the catastrophe theory approach, we have changed the character of the bifurcation diagram. This supports our claim that contact equivalence is more suited to the study of bifurcation problems.

Wasserman considered a preferred parameter within the context of potential functions, but there seem to be difficulties in applications to bifurcation theory. In particular many problems have infinite codimension.

## LECTURE 3.

### Bifurcation with symmetry.

The codimension of a bifurcation problem $G : \mathbb{R}^n \times \mathbb{R} \to \mathbb{R}^n$ is the minimum number of parameters required for a universal unfolding. A problem is robust if it has codimension zero.

### Conceptual difficulty.

The pitchfork $x^3 - \lambda x$ has codimension two and is therefore extremely nonrobust and nongeneric. However, the pitchfork occurs frequently in mathematical theories. Why?

### Discussion.

Recall the equation for the Euler beam $u'' + \lambda \sin u = 0$. The equation is odd with respect to the reflection $u \to -u$. That is, we have a $\mathbb{Z}_2$ action. The oddness is preserved in the Lyapunov-Schmidt reduction $G(x,\lambda) \sim x^3 - \lambda x$.

In the mathematical theory we may restrict our attention to perturbations which preserve this symmetry (although perturbations of the physical problem need not do so). It turns out that the pitchfork is robust within the restricted class of symmetry preserving problems.

### The general context for bifurcation with symmetry.

Let $\Gamma$ be a compact Lie group (this is already of interest when $\Gamma$ is finite). We consider an action of $\Gamma$ on $\mathbb{R}^n$.

**Definition.**

An equivariant bifurcation problem. $G : \mathbb{R}^n \times \mathbb{R} \to \mathbb{R}^n$ is a bifurcation problem such that

$$G(\gamma . x, \lambda) = \gamma . G(x, \lambda) \text{ for } \gamma \in \Gamma .$$

A special case of the above group action is when $\Gamma = \mathbb{Z}_2$ and the action is $\gamma . x = -x$, where $\gamma$ is the non-trivial element of $\mathbb{Z}_2$ .

The whole theory can be repeated for equivariant problems and we again have two problems to consider .

1. Construct a universal unfolding to describe all equivariant perturbations (up to equivariant contact equivalence).
2. Enumeration of the perturbed diagram.

**Example.**

$n = 2$

$$G(x, y, \lambda) = \begin{pmatrix} ax^3 + bxy^3 - p\lambda x \\ cx^2 y + dy^3 - q\lambda y \end{pmatrix}$$

Consider the symmetry group $\Gamma = \mathbb{Z}_2 \times \mathbb{Z}_2 = \{(1,1), (1, 1), ( 1,1), ( 1, 1)\}$ acting on $(x, y)$. G is equivariant with respect to $\Gamma$; i.e. the first equation is odd in x and even in y, and the second equation is even in x and odd in y.

In the case of the single pitchfork the coefficients could be scaled away but in this case we have the following lemma.

**Lemma.**

G can be scaled so that $a = d = p = q = 1$. (Assuming all initially non-zero.)

**Proof.**

Consider the following equivariant contact equivalence transformation. Let $x' = \alpha_1 x$, $y' = \alpha_2 y$ and multiply the first equation by $\alpha_3$ and the second by $\alpha_4$. We have four parameters and can eliminate four coefficients. One choice gives $a = d = p = q = 1$.

The coefficients b and c cannot be scaled away and later we will see that they are essential. We are then left with considering the solution set of the following equations

$$x^3 + bxy^2 - \lambda x = 0$$
$$cx^2 y + y^3 - \lambda y = 0 \quad .$$

The equations factorise and we explicitly obtain the following solutions

$$x = 0,\ y = 0\ ;\ x = 0,\ y = \pm\sqrt{\lambda}\ ;\ y = 0,\ x = \pm\sqrt{\lambda}\ ;$$

$$x = \pm\left(\frac{b-1}{bc-1}\lambda\right)^{\frac{1}{2}}, \quad y = \pm\left(\frac{c-1}{bc-1}\lambda\right)^{\frac{1}{2}} \quad .$$

So we have the trivial solution, two bifurcating branches in the coordinate planes, and maybe other branches if the square roots are ever real. The possibilities are enumerated in the figure and the lemma below.

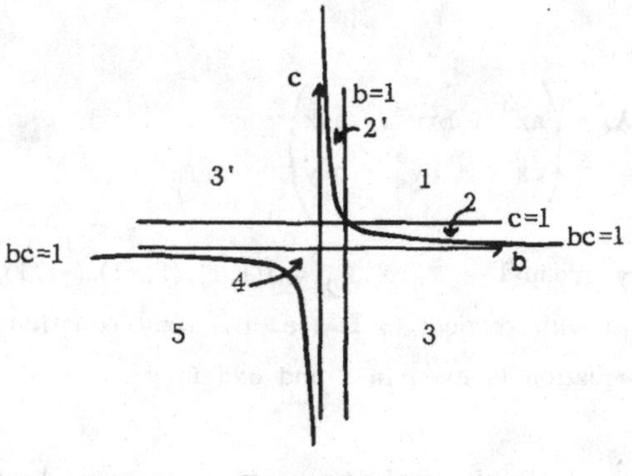


Figure 12.

Lemma.

The non-trivial solution branches have the following properties according to which region b and c lie in.

| Region | $y = 0$, $x \neq 0$ | $x = 0$, $y \neq 0$ | $x$ & $y \neq 0$ Existence | Stability |
|---|---|---|---|---|
| 1 | stable | stable | $\lambda > 0$ | unstable |
| 2 | stable | unstable | never real | - |
| 3 | stable | unstable | never real | - |
| 4 | unstable | unstable | $\lambda > 0$ | stable |
| 5 | unstable | unstable | $\lambda < 0$ | unstable |

Proof. See [7].

Remark.

The differences between region 2 and 3 only appear when considering perturbations.

Bifurcation diagrams for regions 1 and 2. (Dashed lines for negative degree.)

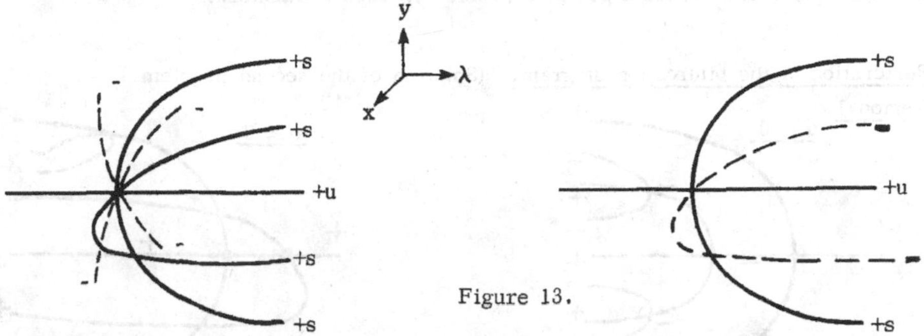


Figure 13.

The solutions with x and y both nonzero are only shown near the bifurcation point, to simplify the figure.

Proposition.

Provided that $b \neq 1$, $c \neq 1$ and $bc \neq 1$.

$$G = G + \text{higher order terms}.$$

Proof. See [7].

Proposition.

Provided that $b \neq 1$, $c \neq 1$ and $bc \neq 1$ then a universal unfolding is given by

$$\begin{pmatrix} x^3 + \beta y^2 - (\lambda+\alpha)x \\ \gamma x^2 y + y^3 - (\lambda-\alpha)y \end{pmatrix}$$

Remarks.

$\beta$ is near b, so we could write $\beta = b + \tilde{\beta}$ with $\tilde{\beta}$ near zero. Similarly for $\gamma$. The other parameter $\alpha$ is near zero. There are three parameters and so the problem has codimension three, but in a certain sense it should be called codimension one. There is a bifurcation from a double eigenvalue if $\alpha = 0$, independent of $\beta$ and $\gamma$. In other words, $\beta, \gamma$ do not effect the bifurcation diagram much provided they stay within a fixed region.

Linearisation of the problem.

At x = y = 0,

$$dG = \begin{pmatrix} \lambda+\alpha & 0 \\ 0 & \lambda-\alpha \end{pmatrix} .$$

There is a bifurcation from the trivial solution when $\lambda = \pm\alpha$. $\alpha$ splits the double eigenvalue into two simple eigenvalue problems, each a pitchfork.

Perturation of the bifurcation diagram. (Solution of the second problem.)

Region 1

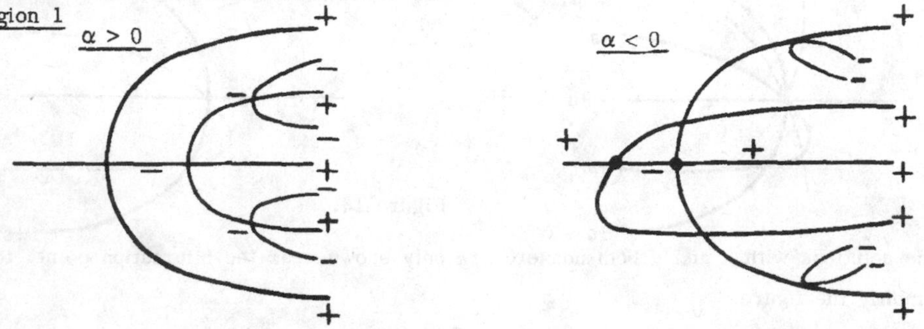


There is a secondary bifurcation to get the number of solution branches and the degrees right

Region 2, 3.

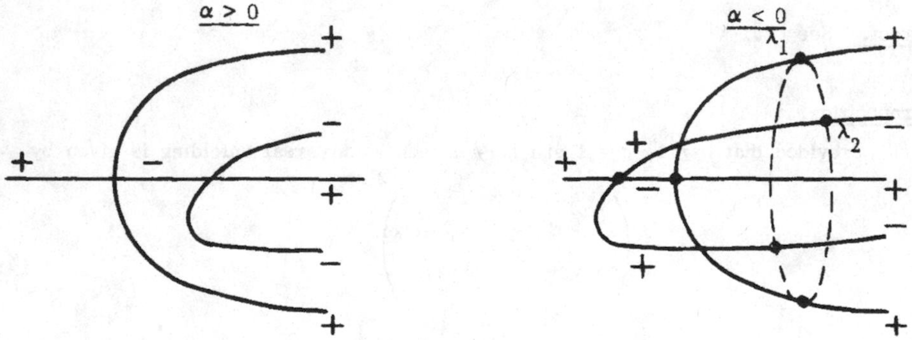


Figure 14.

Remarks on regions 2, 3; $\alpha < 0$.

In region 2, $\lambda_1 > \lambda_2$ and the dotted solution curve has negative degree. Thus for the range $\lambda_2 < \lambda < \lambda_1$ there are two competing stable solution possibilities in the two coordinate planes. In region 3, $\lambda_1 < \lambda_2$, and the dotted solutions are stable; these are the only stable solutions in the range $\lambda_1 < \lambda < \lambda_2$.

In region 1, when $\alpha \neq 0$ the evolution of the system under quasistatic variations of $\lambda$ will be completely smooth. But in regions 2 or 3, provided $\alpha < 0$, the system will first bifurcate into one mode and then at a larger value of $\lambda$ undergo a transition to the other mode, either smoothly along the dotted curve (region 3) or with a jump (region 2). More jumping is observed experimentally in the buckling of plates [15]. Indeed diagrams of the above type were obtained by Bauer, Keller & Reiss [2] for a related spring model.

Buckling of a rectangular plate.

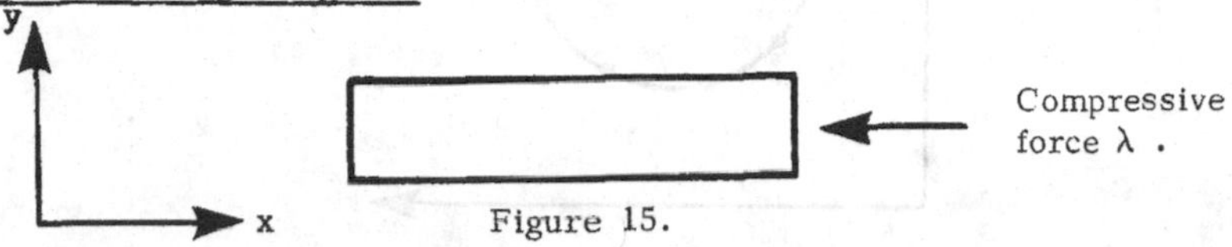


Figure 15.

When a plate buckles the displacement has the approximate form sin kx where $k \approx 1/\text{width}$. Thus for a plate with high aspect ratio there will be many buckles. Contrast this with a rod, which buckles as a single arch. The intuitive explanation for this different behaviour lies in the fact that the plate is supported all along its lateral edges, the rod only at the ends. Moreover, the number of buckles increases as the load is increased, typically with an abrupt and violent change in the mode. We propose to analyse this behaviour using singularity theory.

The mathematical model for the plate is as follows. Let the plate have dimensions $\ell\pi \times \pi$. Let $\Omega = (0,\ell\pi) \times (0,\pi)$. The von Karman equations for the plate are.

$$\Delta^2\omega = -\lambda\omega_{xx} + C(\omega) \text{ in } \Omega$$

with boundary conditions

$$\omega = \Delta\omega = 0 \text{ on } \partial\Omega .$$

$C(\omega)$ is a nonlocal cubic term. The boundary conditions assume that the plate is simply supported, or free to rotate on all sides. (One might wonder how this can be achieved in practice.)

Bifurcation analysis.

The linearised problem is $\Delta^2\omega = -\lambda\omega_{xx}$. The eigenfunctions are

$\omega_k(x,y) = \sin \frac{kx}{\ell} \cdot \sin y$ and $\omega_k$ is a solution $\Leftrightarrow \lambda = 2 + \frac{k^2}{\ell} + \frac{\ell^2}{k}$.

We graph $\lambda$ as a function of k, momentarily ignoring the fact that k is an integer variable.

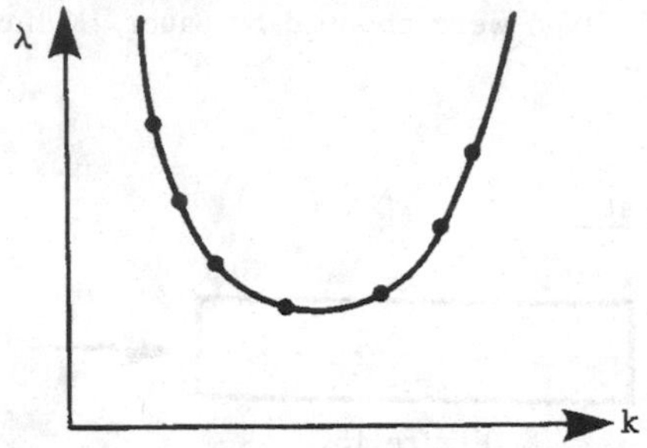


Figure 16.

Let $\lambda_k$ be the bifurcation load for buckling into the sin kx mode; i.e. the load for which the trivial solution becomes unstable with respect to perturbations of the form sin kx. As $\lambda$ is increased, the plate will buckle into the first mode it encounters, represented by the lowest dot on the above graph. Generically there will be a unique lowest dot, although it may happen that two dots are equal competitors for the lowest position. Specifically if $\ell = \sqrt{\ell(\ell+1)}$ then $\lambda_k = \lambda_{k+1}$. We analyse this case and then perturb $\ell$ (to change the relative heights), in order to understand the competition between the sin kx and sin (k+1)x modes.

The Lyapunov-Schmidt reduction leads to a 2×2 system of equations. These equations have $\mathbb{Z}_2 \times \mathbb{Z}_2$ symmetry, corresponding to up-down symmetry and to reflection about the midline of the plate. The equations have been computed to third order to evaluate the modal parameters b and c. Chow, Hall & Mallet-Paret [4], Magnus and Poston [10a] and Matkowsky-Reiss [11] all found region 1 and no mode jumping.

At first it was unclear what was wrong. Were there imperfections or were the equations at fault? The Von Karman equations are suspect but stability theorems cover equations nearby. Eventually the boundary conditions were suspected and mixed boundary conditions were considered, simply supported on the unloaded edges and clamped on the loaded ends.

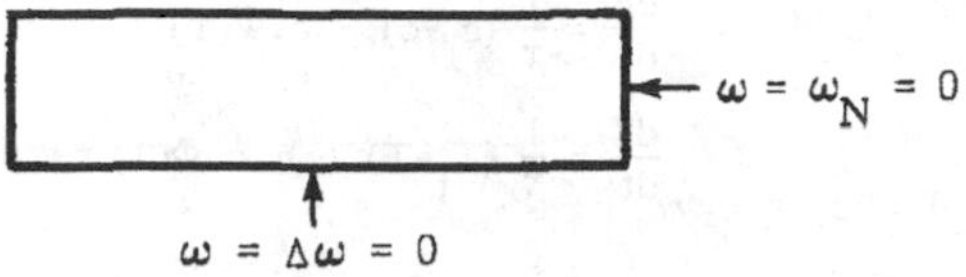


Figure 17.

With these boundary conditions it turns out that the computed modal parameters do lie in region 2, leading to a prediction of mode jumping. (Experimentally clamped boundary conditions on the ends are considered most realistic.)

The main contribution of singularity theory to this analysis was a psychological one; the classification above of bifurcation problems with $\mathbb{Z}_2 \times \mathbb{Z}_2$ symmetry is known to be complete. The classification closed loopholes of the form "Maybe another term in the expansion will change the result" and forced one either to work with the theory at hand or search for a radically different explanation.

Bifurcation in the presence of a continuous group of symmetries.

In [14] we study the Bernard problem in a spherical geometry from the singularity theory point of view. This problem is of special interest because it commutes with the action of the orthogonal group O(3).

Bifurcation problems in chemical reactors.

We give self-contained discussion on the continuous flow stirred tank reactor. This is taken from Golubitsky and Keyfitz [5].

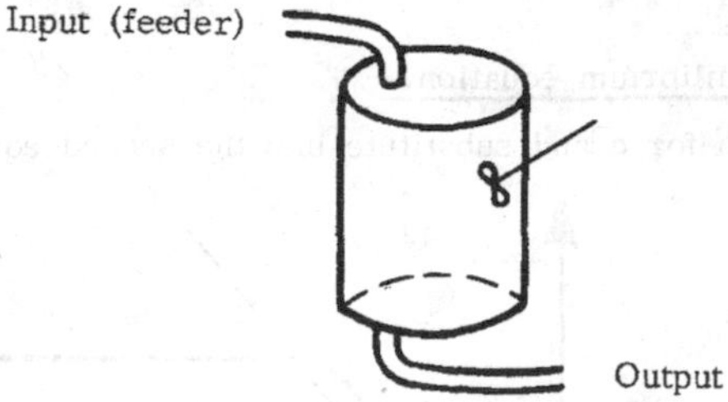


Figure 18.

The tank contains a chemical with a potential for reacting. The state is described by the temperature T(t) and the concentration c(t), both independent of position since the tank is assumed well stirred. The input rate equals the output rate. Let $c_f$ and $T_f$ denote the concentration and temperature at the feeder. The state is governed by the following differential equations.

$$\frac{dc}{dt} = \frac{1}{\tau}(c_f - c) - c.\Phi(T)$$

$$\frac{dT}{dt} = \frac{1}{\tau}(T_f - T) + h.c.\Phi(T) \quad .$$

The first term of the equations is the relaxation to conditions of the feeder at time scale $\tau$, which measures the flow rate. The second term involves the reaction rate; the reaction depletes the chemical and adds heat. For Airhenius kinetics $\Phi(T) = Ze^{-\gamma/T}$. Typically $\gamma \gg 0$ and the reaction rate is highly sensitive to temperature. We consider only steady state solutions of the equations. Intutitively as $\tau \to 0$ the flow is fast so that at steady state $c \approx c_f$ and $T \approx T_f$; effectively there is no time for the reaction to occur. As $\tau \to \infty$ the fow is very slow $c \approx 0$ and $T \approx T_f + hc_f$. The reaction continues until the reactant is depleted with an adiabatic temperature rise. For an intermediate flow rate we have high and low temperature states corresponding to fast and slow reactions. This will be proved below.

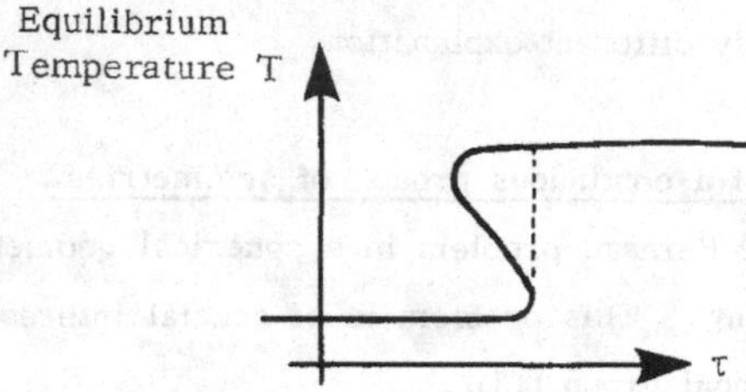


Figure 19.

Note there is no trivial solution. This difficulty is typical of problems of chemical origin.

Graphical solution of the equilibrium equation.

Solve the first equation for c and substitute into the second equation to obtain

$$T - T_f = \frac{h.c_f \tau \Phi(T)}{1+\tau.\Phi(T)}$$

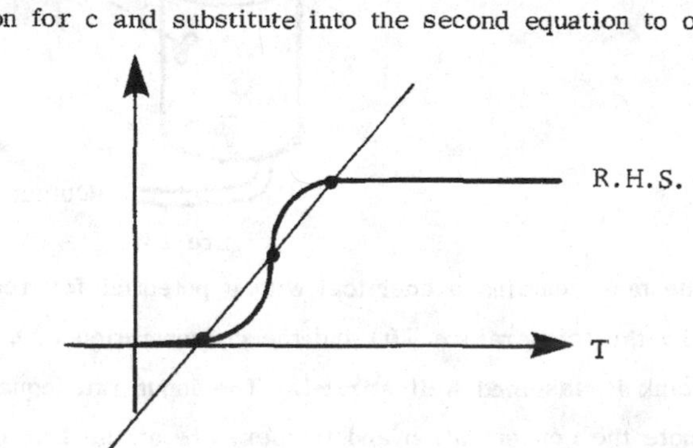


Figure 20.

The intersection of the graph gives the possible equilibrium temperature. As $\tau \to 0$ or $\infty$ we obtain the appropriate limits. The possible equilibrium temperature arise from the balance of two effects. The linear term is related to the heat exchange from the flow and the non linear term is related to the heat produced by the reaction. There is a problem with the limit as $\tau \to \infty$ since the heat loss to the exterior has been neglected. To compensate we add a heat loss term $-\ell(T-T_b)$ to the second equation. Now as $\tau \to \infty$, $c \to 0$ and $T \to T_b$. This additional term complicates the behaviour considerably.

Uppal, Ray & Poore ([16] page 967) found the following five bifurcation diagrams by a primarily numerical study in the limiting case $\gamma \to \infty$.

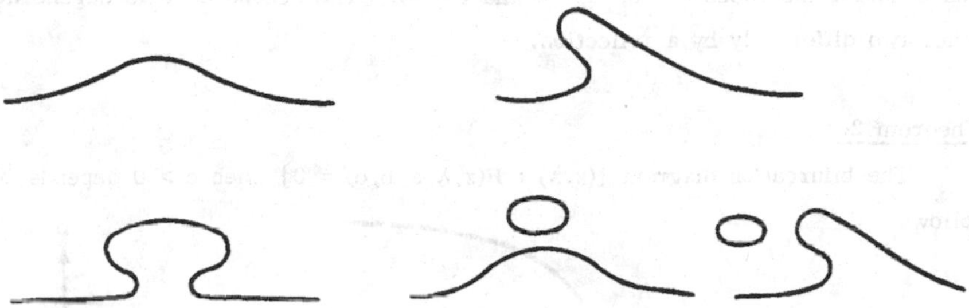

Figure 21.

Golubitsky and Keyfitz [5 ] using analytical methods with $\gamma < \infty$ found in addition two more diagrams.

Figure 22.

Russian workers have found the sixth diagram and numerical people are still looking for the seventh.

Golubitsky and Keyfitz study the organising centre $x^3 + \lambda^2 = 0$ named the winged cusp. An organising centre is the most singular problem that can occur. For certain special values of the parameters in the physical problem a bifurcation contact equivalent to $x^3 + \lambda^2$ occurs.

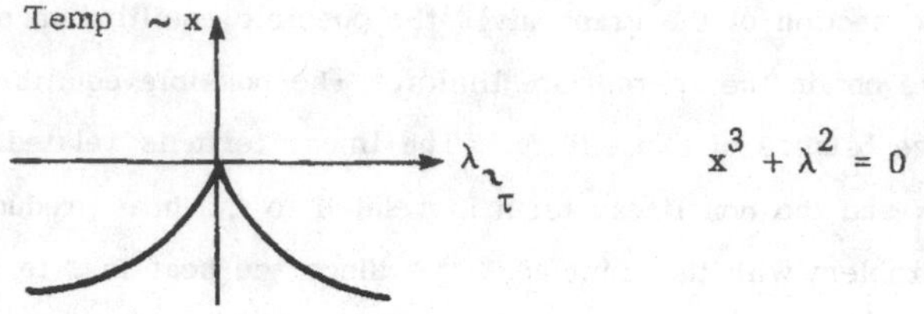


Figure 23.

<u>Theorem 1.</u>

$F(x, \lambda, a, b, c) = x^3 + \lambda^2 + a + bx + c\lambda x$ is a universal unfolding.

The solution of the second problem is difficult to plot in three dimensions. Fix c and consider the cases $c > 0$, $c = 0$ and $c < 0$. The second case is degenerate and the other two differ only by a reflection.

<u>Theorem 2.</u>

The bifurcation diagram $\{(x, \lambda) : F(x, \lambda, a, b, c) = 0\}$ when $c > 0$ depends on $(a, b)$ as follows

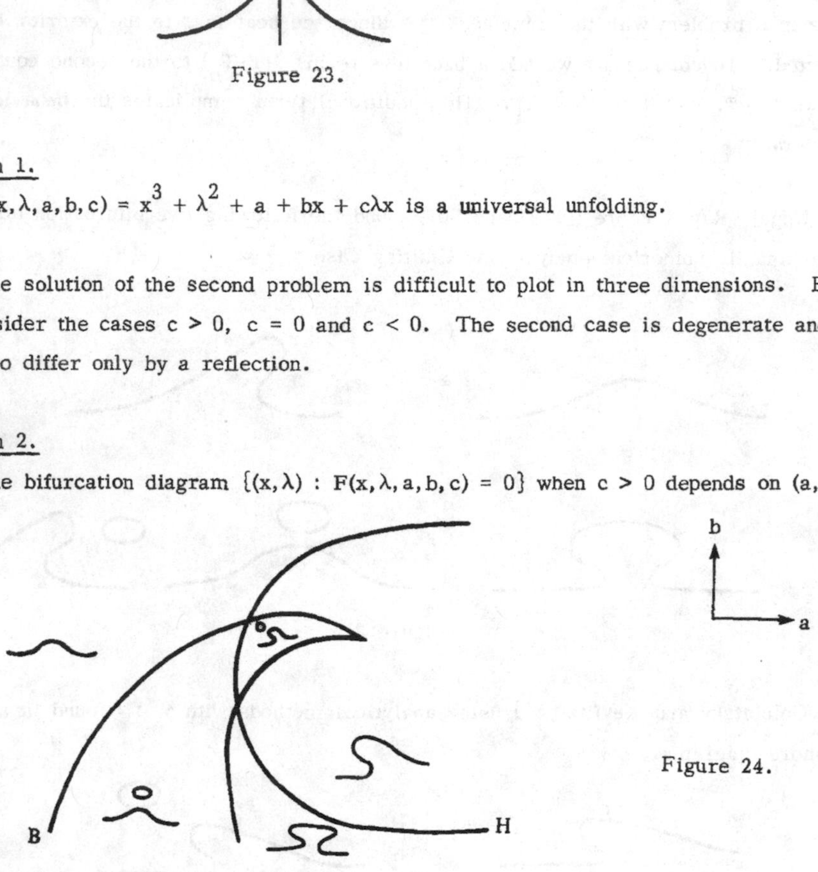


Figure 24.

For $c < 0$ we obtain the mirror image. For $c = 0$ the diagram collapses to

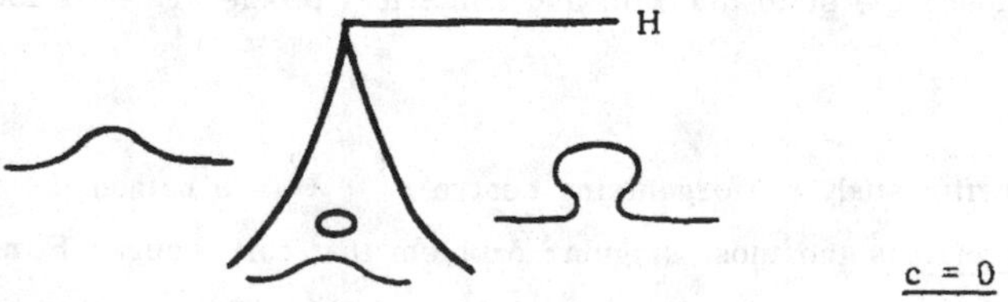


Figure 25.

Conjecture.

We conjecture that these results are quite global. In general if an organising centre is chosen and parameters adjusted to the worst case, everything of interest seems to happen after a perturbation in a small neighbourhood, since all solution branches are there competing competing against each other. These local models can often be used to give quasi-global results.

## LECTURE 4.

### Finite effects in the Taylor problem.

We discuss this application in greater detail. We give a case study of what the theory can contribute and how it has led to a better understanding of the experimental data. Numerical conjectures have been generated which appear to be verified by computations of P. Hall (no reference available yet).

We begin with a review of the experimental data. Two cylinders confine a viscous fluid, and the inner cylinder rotates at speed R. There is circulation in the (r, z) plane for R sufficiently large. We are interested in the steady state solutions.

The traditional analysis is for a cylinder of infinite length. The fundamental problem in a finite cylinder is that the number of cells is an integer which depends in some way on the length of the cylinder, a real parameter. We consider the following results of Benjamin's experiment [3] in a short cylinder, i.e. one having only 2 or 4 Taylor cells.

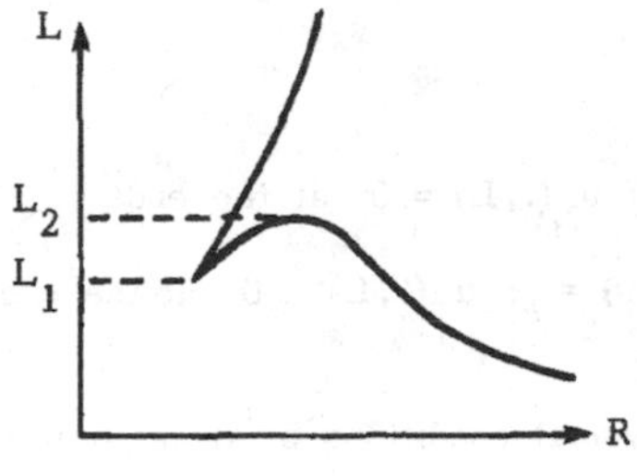


Figure 26.

If $L > L_2$, $L_1 < L < L_2$ or $L < L_1$ the primary mode is 4, 2 or 2 cells respectively. However if R is increased when $L_1 < L < L_2$ there is a jump as R passes the middle of the three intersections of the horizontal line with the cusped curve. Also there is hysteresis in the range $L_1 < L < L_2$. To the right of the curve there are two states, both 2 cell and 4 cell modes.

Our aim is to understand the 2 cell/4 cell competition pictured above, or more generally the 2k/2k+2 cell competition.

We begin the discussion with the traditional mathematical formulation of the problem.

Let $\Omega = \{(r,\theta,z) : r_1 < r < r_2 \ , \ 0 < z < L\}$ .
The Navier-Stokes equations are

$$\left.\begin{aligned} \frac{\partial u}{\partial t} + (u.\nabla)u &= \Delta u - \nabla p \\ \operatorname{div} u &= 0 \ . \end{aligned}\right\} \quad \text{in } \Omega \ .$$

We consider two sets of boundary conditions :

$$u(r_1,.) = R\hat{\theta}, \ u(r_2,.) = 0 \quad \text{at the sides}$$

and either

$$\text{(i)} \quad \begin{aligned} u_N(.,0) &= u_N(.,L) = 0 \quad \text{at the ends} \\ u_T(.,0) &= u_T(.,L) = 0 \quad \text{at the ends} \end{aligned}$$

or

$$\text{(ii)} \quad \begin{aligned} u_N(.,0) &= u_N(.,L) = 0 \quad \text{at the ends} \\ \frac{\partial}{\partial N} u_T(.,0) &= \frac{\partial}{\partial N} u_T(.,L) = 0 \quad \text{at the ends} \ . \end{aligned}$$

We refer to these as the physical conditions or quasi-periodic conditions respectively. Quasi-periodic conditions are a slight change from the traditional problem with an infinite cylinder. These boundary conditions imply that z = 0 is a plane of symmetry;

thus for $z < 0$ we can define

$$u_N(\cdot, z) = -u_N(\cdot, -z)$$
$$u_T(\cdot, z) = u_T(\cdot, -z)$$

and thereby obtain a continuous solution in a larger region. Repeated such reflections lead to a periodic solution.

These boundary conditions select only periodic solutions from the infinite problem whose period is $\frac{2L}{n}$ for some integer n. This is technically convenient. The physical boundary condition is of course $u = 0$ on the end faces, while $\frac{\partial u_T}{\partial N} = 0$ is analogous to a stress free surface.

<u>Bifurcation analysis in the traditional case.</u>

The trivial solution, Couette flow is given in cylindrical coordinates by $g = (0, ar+\frac{b}{r}, 0)$ , where a and b are adjusted to satisfy the boundary conditions. The trivial solution loses stability as R increases and other solutions bifurcate from it. The standard analysis is to linearise the Navier-Stokes equations around Couette flow and look to see whether the linearisation is invertible. It is easier to find the eigenfunctions than the eigenvalues

$$u = \begin{pmatrix} u_r(r).\cos\frac{k\pi z}{L} \\ u_\theta(r).\cos\frac{k\pi z}{L} \\ u_z(r).\sin\frac{k\pi z}{K} \end{pmatrix}$$

(We are only considering axisymmetric solutions since there is experimental justification for this.)

We pose the question : At what value of R does Couette flow become unstable with respect to perturbations of this form. This question was answered numerically by Kirchgassner [9] in 1961. The minimum of f occurs approximately at $3/(r_2-r_1)$ .

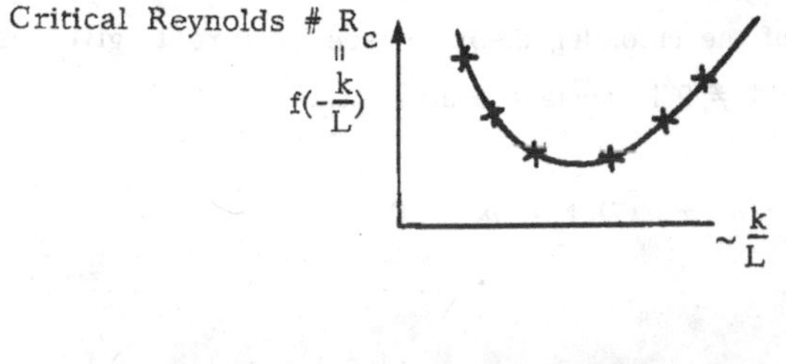


Figure 27.

As R is increased the system bifurcates into the first mode available to it. Apart from exceptional values of L there is a lowest integer point on the above graph. A unique k* can then be selected. The Lyapunov-Schmidt reduction gives

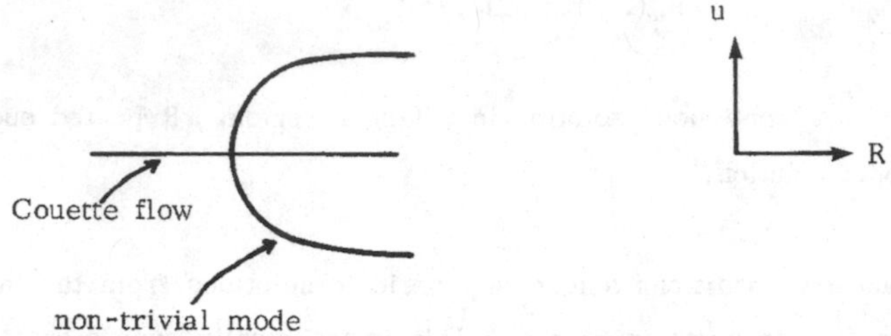


Figure 28.

The bifurcation diagram is a pitchfork, but this is not due to the symmetry you might expect i.e. reflection above the midpoint of the cylinder. Rather it is because

kth derivative of the reduced equation at the origin $\sim$ (k+1) st power of the eigenfunction,

and the integral of an odd power of a cosine is zero.

The symmetry is inherited from the translational invariance of the original problem.

<u>Modifications induced by the true boundary conditions.</u>

We introduce a homotopy parameter $0 \leq \tau \leq 1$. For each $\tau$ consider the boundary conditions

$$u_n = 0, \qquad (1-\tau)\frac{\partial u_T}{\partial N} + \tau u_T = 0 \quad \text{on the ends} .$$

When $\tau = 0$ we have the boundary conditions for the quasi-periodic solutions, and when $\tau = 1$ we have the boundary conditions for the physical problem. We analyse the behaviour for small $\tau$ as an unfolding of the pitchfork at $\tau = 0$ and hope that the discussion for small $\tau$ continues to apply out to $\tau = 1$. There are no proofs but as working hypothesis this seems to lead to agreement with experiments.

A non-trivial application of the unfolding theorem (see Lecture 1) gives us that the reduced bifurcation equation for $\tau \neq 0$ is contact equivalent to

$$x^3 - \lambda x + \alpha(\tau) + \beta(\tau)x^2 = 0$$

for some smooth functions $\alpha(\tau)$ and $\beta(\tau)$ . (The derivatives of $\alpha$ and $\beta$ at $\tau = 0$ are computable by numerical methods.)

We refer back to Lecture 1 to understand the bifurcation problem for $\tau \neq 0$.

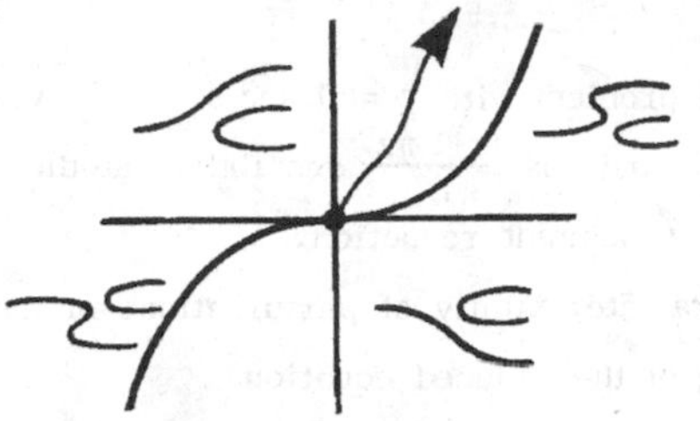

Figure 29.

We obtain a curve $(\alpha(\tau), \beta(\tau))$ starting at the origin. Assuming $\alpha'(0) \neq 0$ we find that a unique picture sets in for all small positive $\tau$. (The assumption $\alpha'(0) \neq 0$ is generic. Although $\alpha'(0) \neq 0$ ought to follow from the fact that Couette flow does not satisfy the boundary conditions, this has not been proved yet.) Thus for $\tau \neq 0$ we obtain the bifurcation diagram.

Figure 30.

Two comments on the interpretation of this diagram. (i) The appearance of circulation is not abrupt. (ii) There is a preferred direction of bifurcation. Both predictions are verified experimentally. Indeed, as regards (ii), there is always an inward flow on the end faces in a gradual spin up experiment. (A reversed flow is possible if the apparatus is started with a jerk.) It should be recalled that on the upper branch $u \simeq q + u_1$ while on the lower branch $u \simeq q - u_1$, where $u_1$ is a solution of the linearised equation. $u_1$ gives the circulation, so choosing between the upper and lower branches specifies the circulation.

We have just presented the traditional picture although phrased in the language of singularity theory, but the above discussion has been well understood before. However, the discussion is inadequate to explain Benjamin's experiment which indicates that even after the original bifurcation occurs the system can bifurcate a second time in jumping to a new mode. There is competition between two modes which must be analysed as a double eigenvalue. This leads to the programme below. An alternative point of view is that Benjamin's experiment performs a one parameter scan (in L), and in that one parameter family there will be values of L for which the generic hypothesis of s simple eigenvalue fails.

Programme.

1. Define the idealised problem with $\tau = 0$ and $L = L^*$, where $L^*$ is chosen so that the eigenfunctions $\cos \frac{k\pi z}{L}$ and $\cos \frac{(k+1)\pi z}{L}$ destabilise at the same value of R.

2. Perform the Lyapunov-Schmidt reduction.

3. Discuss the two parameter family of perturbations of the idealised problem in terms of the universal unfolding of the reduced equation.

1. The mathematical analysis is not yet complete but we outline our philosophy of approach. We refer back to the experiment to decide certain questions which sufficient computation could answer. The main question posed is, have these simple ideas any chance of explaining the data?

The first instance of this philosophy is to restrict attention to an even number of cells. We redefine $L^*$ as the special value where the bottom two modes $\cos \frac{2k\pi z}{L}$ and $\cos \frac{2(k+1)\pi z}{L}$ are on the same level.

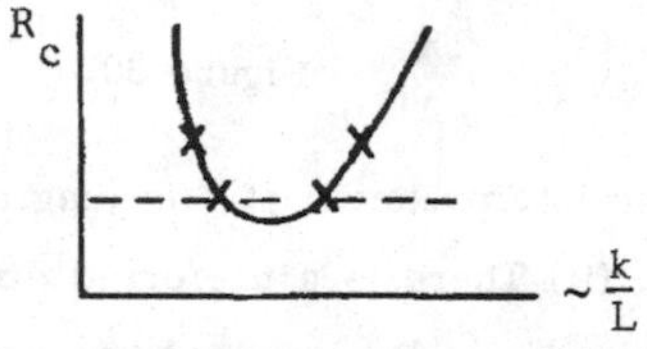


Figure 31.

2. We look for a solution of the Navier-Stokes equation of the form $u = q + x.\psi_k + y.\psi_{k+1} + \omega$, where x and y are unknown coefficients, not space coordinates and $\langle \omega, \psi_j \rangle = 0$. We can solve for $\omega$ using all but two dimensions of the equation. We are left with two equations for the idealised problem where $\lambda = R - R_c$

$$G_1(x, y, \lambda) = 0$$

$$G_2(x, y, \lambda) = 0 \, .$$

By symmetry and after scaling coefficients these can be rewritten as

$$G(x,y,\lambda) = \begin{pmatrix} x^3 + bxy^3 - \lambda x \\ cx^2y + y^3 - \lambda y \end{pmatrix} + \text{higher order terms} .$$

Remark.

There is symmetry on the cross terms only to sufficiently low order since

$$\int_0^{2\pi} (\cos kz)^{k+1} (\cos (k+1)z)^k dz \neq 0.$$

The higher order terms can be transformed away provided $b \neq 1$, $c \neq 1$ and $bc \neq 1$.

No calculations for b and c are available yet, but we can ask which regions in the b, c plane fit the experimental data. A fundamental experimental fact is that for large R both the 2k and (2k+2) cell modes are stable. (They become time dependent, but there is no change in the number of cells.) If b, c belong to region 1 (Fig. 12) the pure modes (i.e. either $x = 0$ or $y = 0$) are both stable as $R \to \infty$. In no other region is this true - in regions 2 and 3 only one of the pure modes is stable, and in region 4 only mixed modes (both x and y nonzero) are stable. Thus only if b, c belong to region 1 does a fit with the experiment seem possible. Therefore we conjectured [12] that $b > 1$, $c > 1$, and calculation [8] seems to verify this.

Remark.

The above argument could be criticised on the grounds that experimental data is used to make predictions about the idealised model. However quantitatively the physical boundary conditions seem to perturb averaged measures of the flow less than 10%.

3. We consider the perturbation parameters $h, \tau$ where $h = L - L^*$.

Remark.

The symmetry of the problem is not destroyed if $h \neq 0$. The symmetry comes from properties of trigonometric functions over a period interval. Therefore there is a unique way to insert h into the unfolded equation - see the universal unfolding result of Lecture 3.

The full unfolding requires sixteen parameters, and since $\tau$ destroys the symmetry it appears that all are needed for a complete analysis. However the zero'th order terms control the situation for $\tau$ small but non-zero, and we retain only those. Justifications are given in [12].

Therefore from the reduced bifurcation equation we form the unfolding

$$x^3 + bxy^2 - (\lambda+h)x + \alpha\tau = 0$$
$$cx^2y + y^3 - (\lambda-h)y + \beta\tau = 0 \ .$$

<u>Bifurcation diagrams.</u>

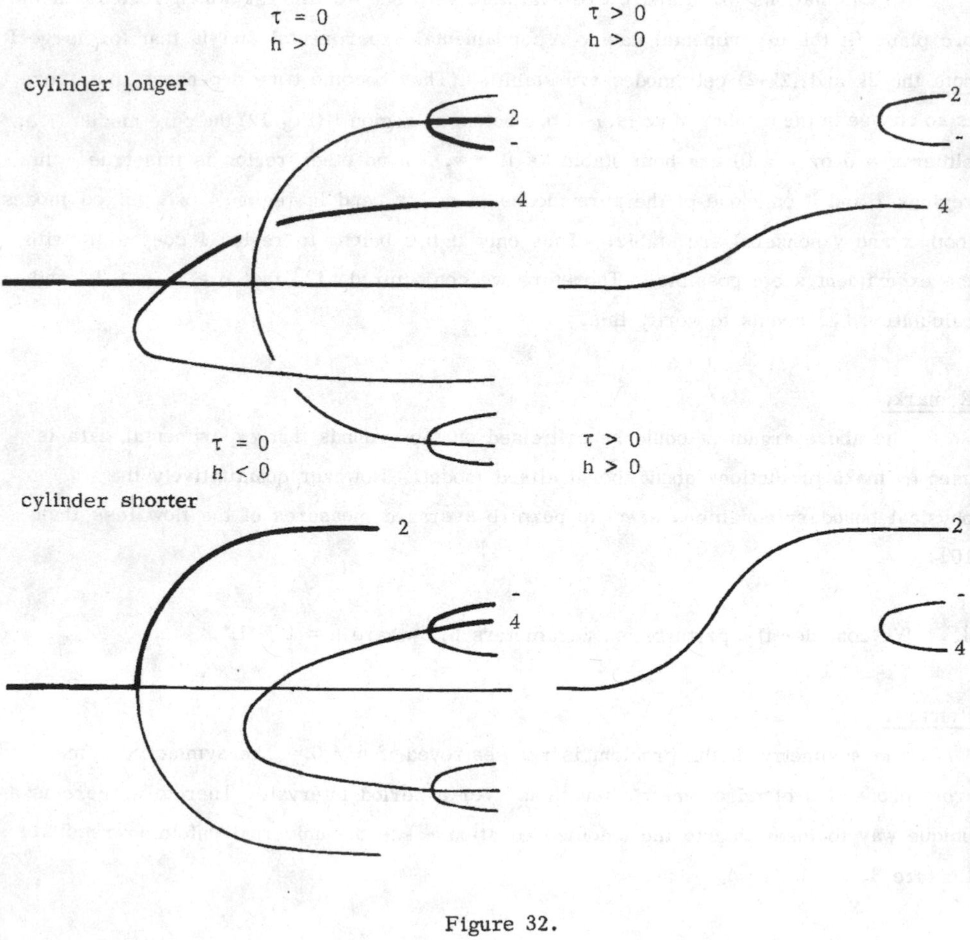


Figure 32.

The diagrams are first drawn for $\tau = 0$. Then the bold portions (the only part relevant for the experiment) are redrawn with $\tau \neq 0$ (but still small). Think of the labels "2", "-", "4" and the "trivial solution at small Reynolds number" as the vertices of a graph, and note that the connections are different for h large positive and h large negative. The transitions are effected by a bifurcation

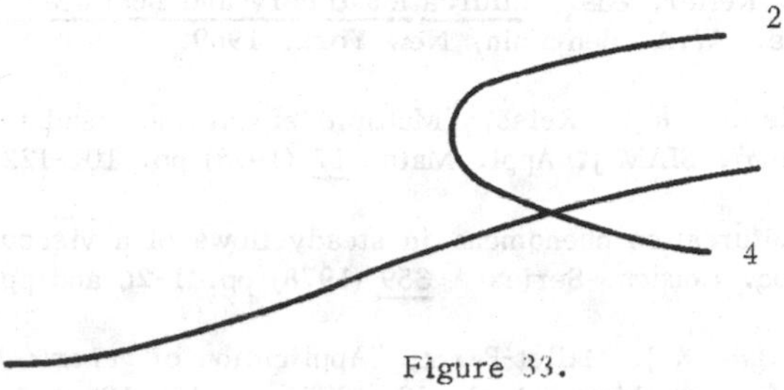


Figure 33.

Generically (an explicit condition is given in [12]) the bifurcation is transcritical, and the parameter h provides a universal unfolding.

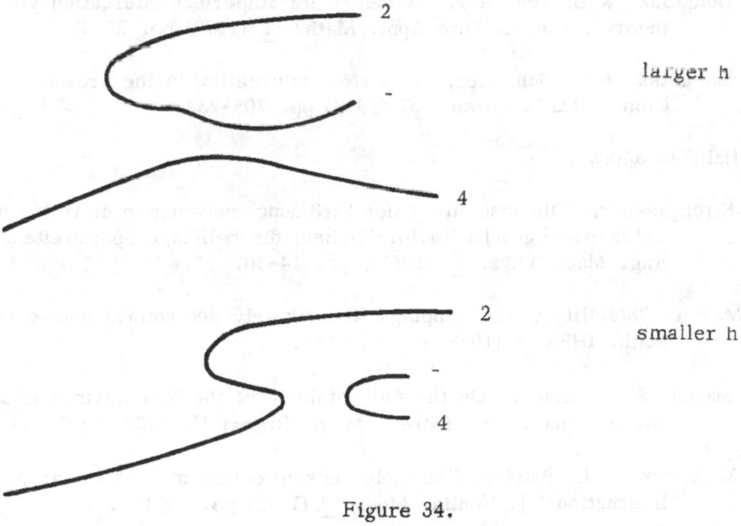


Figure 34.

Hysteresis appears naturally just after the bifurcation, as seen in the experiment. Indeed precisely these bifurcation diagrams were proposed by Benjamin [3]. Our analysis provides a possible explanation for their occurence.

Remark.

It turns out that the case k = 1 is exceptional-there is a resonance between sin z and sin 2z which leads to nonzero quadratic terms in the bifurcation equations. Unfortunately the experiments for k > 1 are not complete yet, so there is a slightly embarassing mismatch between theory and experiment.

References.

1. S. Antman & H.B. Keller, eds., Bifurcation theory and nonlinear eigenvalue problems. W.A. Benjamin, New York, 1969.

2. L. Bauer, H.B. Keller, & E. Reiss, "Multiple eigenvalues lead to secondary bifurcation", SIAM J. Appl. Math. 17 (1975) pp. 101-122.

3. T.B. Benjamin, "Bifurcation phenomena in steady flows of a viscous fluid", Proc. Royal Soc. London, Series A 359 (1978) pp. 1-26 and pp. 27-43.

4. S.N. Chow, J.K. Hale & J. Mallet-Paret, "Application of generic bifurcation. I and II", Arch. Rat. Mech. Anal. 59 (1975) pp. 159-188 and 62 (1976) pp. 209-235.

5. M. Golubitsky & B. Keyfitz, "A qualitative study of the steady state solutions for a continuous flow stirred tank chemical reactor", SIAM J. Math. Anal. 11 (1980) pp. 316-339.

6. M. Golubitsky & D. Schaeffer, "A theory for imperfect bifurcation via singularity theory", Comm. Pure Appl. Math. 32 (1979) pp. 21-98.

7. M. Golubitsky & D. Schaeffer, "Imperfect bifurcation in the presence of symmetry", Comm. Math. Physics 67 (1979) pp. 205-232.

8. P. Hall, to appear.

9. K. Kirchgassner, "Die Instabilität der Strömung zwischen zwei rotierenden Zylindern gegenuba Taylor-Wirbeln für beliebige Spaltbreiten," Z. fur Ang. Math. Phys. 12 (1961), pp. 14-30.

10. J. Mather, "Stability of $C^\infty$ mappings III : Finitely determined map germs", Publ. Math. IHES 35 (1968) pp. 127-156.

10a. R. Magnus & T. Poston, "On the full unfolding of the Von Karman equations at a double eigenvalue", Battelle Math. Report No. 109 (1977), Geneva.

11. B. Matkowsky & L. Putnick, "Multiple buckled states of rectangular plates", International J. Nonlin. Mech. 9 (1973) pp. 89-103.

12. D. Schaeffer, "Qualitative analysis of a model for boundary effects in the Taylor problem", Math. Proc. Camb. Phil. Soc. 87 (1980) pp. 307-337.

13. D. Schaeffer & M. Golubitsky, "Boundary conditions and mode jumping in the buckling of a rectangular plate", Comm. Math. Phys.

14. D. Schaeffer & M. Golubitsky, "Bifurcation with 0(3) symmetry including applications to the Benard problem", to appear.

15. M. Stein, "The phenomenon of change in buckle pattern in elastic structures", NASA Technical Report R-39 (1959).

16. A. Uppal, W.H. Ray & A.B. Poore, "The classification of the dynamic behaviour of continuous stirred tank reactors - influence of reactor residence time", Chem. Eng. Sci. 31 (1976).

## Acknowledgement.

I am exceedingly grateful to John Hayden for his very competent work in writing up these notes - a very frustrating task indeed, given the casualness of the lectures.

D. Schaeffer : Department of Mathematics, Duke University, Durham, North Carolina, U.S.A.

# Anosov diffeomorphisms with pinched spectrum.

M. Brin and A. Manning.

An important open problem in smooth dynamical systems is to classify Anosov diffeomorphisms (see [1,16] for the definition). The only known examples act on infranilmanifolds [8,16,7] and the natural question is whether there are any others. Our main result is that for a certain class of Anosov diffeomorphisms (those with "pinched spectrum") the manifold must be infranil.

Every diffeomorphism f of a compact Riemannian manifold $M^n$ induces a bounded linear operator $f_*$ in the Banach space of continuous vector fields defined by

$$(f_* v)(x) = df_{f^{-1}(x)} \cdot v(f^{-1}x) \quad .$$

According to Mather [14], Anosov diffeomorphisms are characterised by the condition that the spectrum $\sigma$ of the complexification of $f_*$ does not meet the unit circle. Thus, $\sigma$ is contained in the interiors of two annuli with radii $0 < \lambda_1 < \lambda_2 < 1$ and $1 < \mu_2 < \mu_1 < \infty$ .

Theorem. Suppose the numbers $\lambda_1, \lambda_2, \mu_2, \mu_1$ associated with an Anosov diffeomorphism $f: M \to M$ satisfy either

$$1 + \frac{\log \mu_2}{\log \mu_1} > \frac{\log \lambda_1}{\log \lambda_2} \tag{1}$$

or

$$1 + \frac{\log \lambda_2}{\log \lambda_1} > \frac{\log \mu_1}{\log \mu_2} \tag{2}$$

Then M is homeomorphic to an infranilmanifold and f is topologically conjugate to a hyperbolic infranilmanifold automorphism.

Remark. Generally speaking the smooth structure on M can be different from the algebraic one [7].

The authors are grateful to W. Neuman for useful discussions. The first author thanks the Mathematics Institute of the University of Warwick for inviting him to the Symposium on Dynamical Systems and Turbulence during which this paper was written.

Proof of Theorem. Step 1. In [4, 5, 6] it was shown that either of the pinching conditions (1) and (2) implies that the universal cover $\tilde{M}$ of M is homeomorphic to $\mathbb{R}^n$, $\pi_1(M)$ has polynomial growth, all points of M are non-wandering for f and f has a fixed point. (The reason is basically that (1) or (2) makes it possible to extend the local product structure of the foliations of f to a global product structure.)

Thus, M is of type $K(\pi,1)$. Gromov's theorem on groups of polynomial growth states that $\pi_1(M)$ has a nilpotent subgroup of finite index [11]. L. Auslander and Schenkman showed (in Theorem 3 of [3]) that there is a compact infranilmanifold P whose fundamental group is isomorphic to $\pi_1(M)$. See also Shub [15]. The automorphism $f_*$ of $\pi_1(M)$ (we use the fixed point of f as base point) determines an infranilmanifold automorphism $g: P \to P$ induced by an automorphism A of the nilpotent Lie algebra (see Theorem 2 of [2]).

Step 2. We now show that A is hyperbolic by counting periodic orbits of f. By taking double covers of M and P if necessary we can assume that the bundle $E^u$ for f is orientable. Then all fixed points of $f^m$ have the same index ±1 and so $\# \mathrm{Fix}(f^m)$ is the modulus of the Lefschetz number $|L(f^m)|$. Since M and P are of type $K(\pi,1)$ they have the same cohomology and $L(f^m) = L(g^m)$. Taking powers of f and g if necessary we may assume that the automorphism induced by g in the finite group of covering transformations of the nilmanifold which covers P (with multiplicity p ) is the identity. Therefore, the calculation in [12] gives

$$\# \mathrm{Fix}(f^m) = |L(f^m)| = |L(g^m)| = \frac{1}{p}\prod_{i=1}^{n} |1-\lambda_i^m|$$

where $\lambda_i$ are the eigenvalues of A. We now show that $|\lambda_i| \neq 1$ for all i. No $\lambda_i$ can be a root of unity for then $\# \mathrm{Fix}(f^m) = 0$ for some large m. We follow an argument of Anosov [17, p.102] to show that there exists k such that

$$\# \mathrm{Fix}(f^m) \leq \# \mathrm{Fix}(f^{m+k}) \quad \forall m \geq 1 .$$

This is sufficient to show that $|\lambda_j| \neq 1$. Let $\delta$ be an expansive constant for f and choose $\varepsilon$ so that every closed $\varepsilon$-orbit $x_0, x_1, \ldots, x_i = x_0$ is $\frac{1}{3}\delta$-shadowed by a genuine closed orbit $f^i x = x$. Since all points are non-wandering f is topologically mixing. Choose k so that some finite cover by $\frac{1}{2}\varepsilon$-balls $V_j$ has $f^k V_j \cap V_{j'} \neq \emptyset$ $\forall j, j'$. Every $\varepsilon$-ball contains one of these $\frac{1}{2}\varepsilon$-balls so $f^k U \cap U' \neq \emptyset$ for any $\varepsilon$-balls $U, U'$. For any $x \in \mathrm{Fix}(f^m)$ take y such that $\rho(x,y) < \varepsilon$ and $\rho(f^k y, x) < \varepsilon$. The closed $\varepsilon$-orbit $x, fx, \ldots, f^{m-1}x, y, fy, \ldots, f^{k-1}y, x$ is $\frac{1}{3}\delta$-shadowed by a genuine periodic orbit $f^{m+k}z = z$. By expansiveness, distinct points x give rise to distinct points z. Thus, $\# \mathrm{Fix}(f^m) \leq \# \mathrm{Fix}(f^{m+k})$ and A is hyperbolic.

<u>Step 3.</u> Since the infranilmanifold P is of type $K(\pi, 1)$ the isomorphism $\pi_1(M) \to \pi_1(P)$ that conjugates $f_*$ and $g_*$ is induced by some continuous based map $M \to P$. According to Franks (Theorem 2.2 of [8]) there is actually a continuous map $j: M \to P$ (sending some fixed point of f to the identity coset fixed point of g) making the diagram

$$\begin{array}{ccc} M & \xrightarrow{\;f\;} & M \\ {\scriptstyle j}\big\downarrow & & \big\downarrow{\scriptstyle j} \\ P & \xrightarrow[\;g\;]{} & P \end{array}$$

commute.

We must show that j is a homeomorphism. It is of degree $\pm 1$ and hence surjective because $j_*: H_n(M) \to H_n(P)$ is an isomorphism.

We follow Franks [9] and show that j is injective. Lift j, f, g to maps $\tilde{j}, \tilde{f}, \tilde{g}$ in the universal covers. Then $\tilde{g}\tilde{j} = \tilde{j}\tilde{f}$ so $\tilde{j}$ maps stable and unstable manifolds for $\tilde{f}$ to those of $\tilde{g}$. Theorem 5 of [4] shows that a (and therefore the) universal cover $\tilde{M}$ of M is a product of a stable and an unstable leaf so that, as in the case of $\tilde{g}$, every stable leaf for $\tilde{f}$ meets every unstable leaf just once. Suppose $\tilde{j}(x) = \tilde{j}(y)$ and $x \neq y$. If $z = W^s_{\tilde{f}}(x) \cap W^u_{\tilde{f}}(y) \neq y$ then $\tilde{j}$ identifies two points y and z on the same unstable leaf. This will lead to a contradiction. Let K be a compact fundamental domain in the nilpotent Lie group that is the universal cover of P. Its preimage $\tilde{j}^{-1}(K)$ is compact in $\tilde{M}$. Let $W^s(v,R)$, $W^u(v,R)$ be the balls of radius R in $W^s(v)$, $W^u(v)$ with centre v. For some large R and some $v \in \tilde{j}^{-1}(K)$

$$\bigcup_{w \in W^s(v,R)} W^u(w,R) \supset \tilde{j}^{-1}(K) \quad .$$

However, for appropriately large k, the distance between the points $\tilde{f}^k y$ and $\tilde{f}^k z$ measured along $W^u(\tilde{f}^k y)$ is greater than 2R although $\tilde{j}$ identifies these two points. This contradiction shows that j is a homeomorphism. Thus f and g are conjugate and the theorem is proved.

Corollary. If the numbers $\lambda_1, \lambda_2, \mu_2, \mu_1$ associated with the spectrum of the Anosov diffeomorphism $f:M \to M$ satisfy both the conditions (1) and (2) of the theorem then M is homeomorphic to a torus or a flat manifold.

Remark. It was shown in [10] that a nilmanifold which is not a torus can sometimes admit an Anosov diffeomorphism satisfying one of conditions (1) and (2).

Proof. The topological conjugacy $j:M \to P$ given by the theorem lifts to the universal cover to give a commutative square

$$\begin{array}{ccc} \tilde{M} & \xrightarrow{\tilde{f}} & \tilde{M} \\ \tilde{j}\downarrow & & \downarrow\tilde{j} \\ N & \xrightarrow[A]{} & N \end{array}$$

By compactness of P we see from the corresponding property in P that for any positive $\varepsilon$ there is K such that, whenever x, y in N satisfy $d(x,y) > \varepsilon$, we have

$$K^{-1}d'(\tilde{j}^{-1}x, \tilde{j}^{-1}y) \le d(x,y) \le Kd'(\tilde{j}^{-1}x, \tilde{j}^{-1}y) \tag{3}$$

where d' is the metric in $\tilde{M}$ arising from some Riemannian metric in M. If $1 < |\nu_2| \le |\nu_1|$ and $\nu_2$ and $\nu_1$ are the smallest and largest expanding eigenvalues of $dA_e$ then any unstable vector in P grows at a rate between $|\nu_2|$ and $|\nu_1|$. Thus the expanding part of the spectrum of $g:P \to P$ is between the circles of radii $|\nu_2|$ and $|\nu_1|$. The geodesic one parameter subgroups of N defined by the corresponding eigenvalues grow in length under powers of A at the rates $|\nu_2|$ and $|\nu_1|$. Now, by (3), the image under $\tilde{j}^{-1}$ of one of these paths grows at the same rate under powers of $\tilde{f}$. The numbers $1 < \mu_2 < \mu_1$ bound the expanding part of the spectum of f so paths in unstable manifolds in $\tilde{M}$ grow in length faster than $\mu_2$ and slower than $\mu_1$, by [14]. Thus $1 < \mu_2 \le |\nu_2| \le |\nu_1| \le \mu_1$ so that the expanding spectrum of g is no wider than the expanding spectrum of f. The same follows for the contracting parts of the spectra by consideration of $\tilde{f}^{-1}$ and $A^{-1}$.

Since the spectrum of f satisfies (1) and (2) do does that of g. In [4] it was proved in Proposition 2 that the nilpotent group N is abelian provided either condition (1) holds and $\lambda_2\mu_1 \geq 1$ or condition (2) holds and $\lambda_1\mu_2 \leq 1$. Now at least one of $\lambda_2\mu_1 \geq 1$ and $\lambda_1\mu_2 \leq 1$ always holds so in our case N is abelian and M is a torus or flat manifold.

## Refcrences.


1. D.V. Anosov, Geodesic flows on closed Riemannian manifolds of negative curvature, Proc. Steklov Inst., 90 (1969).

2. L. Auslander, Bieberbach's theorems on space groups and discrete uniform subgroups of Lie groups, Ann. Math., 71 (1960), 579-590.

3. L. Auslander & E. Schenkman, Free groups, Hirsch-Plotkin radicals, and applications to geometry, Proc. Amer. Math. Soc., 16 (1965), 784-788.

4. M.I. Brin, Nonwandering points of Anosov diffeomorphisms, Astérisque, 49 (1977), 11-18.

5. M.I. Brin, On the fundamental group of a manifold admitting a U-diffeomorphism, Sov. Math. Dokl., 19 (1978), 497-500.

6. M.I. Brin, On the spectrum of Anosov diffeomorphisms, Isr.Jour. Math., to appear.

7. F.T. Farrell & L.E. Jones, Anosov diffeomorphisms constructed from $\pi_1\mathrm{Diff}(S^n)$, Topology, 17 (1978), 273-282.

8. J. Franks, Anosov diffeomorphisms, Global Analysis, Proc. Symp. Pure Math., 14 (1970), 61-93.

9. J. Franks, Anosov diffeomorphisms on tori, Trans. Amer. Math. Soc., 145 (1969), 117-124.

10. D. Fried, Some non-toral pinched Anosov maps, to appear in Proc. Amer. Math. Soc.

11. M. Gromov, Groups of polynomial growth and expanding maps, Preprint, IHES, 1980.

12. A. Manning, Anosov diffeomorphisms on nilmanifolds, Proc. Amer. Math. Soc., 38 (1973), 423-426.

13. A. Manning, There are no new Anosov diffeomorphisms on tori, Amer. J. Math., 96 (1974), 422-429.

14. J. Mather, Characterisation of Anosov diffeomorphisms, Indag. Math., 30 (1968), 479-483.

15. M. Shub, Expanding Maps, Global Analysis, Proc. Symp. Pure Math., 14 (1970), 273-276.

16. S. Smale, Differentiable dynamical systems, Bull. Amer. Math. Soc., 73 (1967), 747-817.

17. Smooth dynamical systems (Russian), ed. D.V. Anosov, MIR, Moscow, 1977.

M.I. Brin, Department of Mathematics, University of Maryland, College Park, Md. 20742, U.S.A.

A. Manning, Mathematics Institute, University of Warwick, Coventry, CV4 7AL, England.

# Formal Normal Form Theorems for Vector Fields and some Consequences for Bifurcations in the Volume Preserving Case*.

Henk Broer.

In the following we shall treat a formal normal form theorem for vector fields which generalises e.g. Takens [28, 29]. The results of [28, 29], where the ∞-jet of a singularity of a vector field by a change of coordinates is put into a normal form, are extended to a rather general situation where some structure must be respected and where parameters are involved. One may think of a volume or a symplectic structure. A natural language for this appears to be that of completely filtered Lie algebras, see Gérard and Levelt [12]. In fact we shall study subalgebras of the Lie algebra of all the ∞-jets of singularities of vector fields : an unfolding of a vector field then is interpreted as a vertical vector field in a higher dimension. A form of Takens' [29], theorem (2.1) -as is mentioned there - was known before, e.g. see Poincaré [20]. Similar theorems can be found for instance in Birkhoff [3] for a hamiltonian case, in Sternberg [25] for a volume preserving case and in Moser [16] for a reversible case. See below.

In the second part of this paper we shall apply this result to investigate bifurcations of singularities in one-parameter families of volume preserving (divergence free) vector fields. We try to prove results analogous to Arnold [2], Sotomayor [24] and Brunovsky [7] in the case of ordinary (i.e. non volume preserving) vector fields. The only new phenomena of interest are two bifurcations, one in the dimension 3 and one in the dimension 4. Analogous to [29] we use the normal form theorem to symmetrise the ∞-jets of the bifurcations.

At the end of this contribution we try to interpret some of our results in terms of fluid mechanics and we raise a question concerning turbulence. I wish to express my gratitude to Floris Takens for introducing me into these problems and his stimulative help in getting them solved. Also I thank Hans Duistermaat, Freddie Dumortier and Wim Hesselink with whom I had helpful discussions.

---

*Partly this publication was made possible by the Netherlands Organisation for the Advancement of Pure Research (ZWO).

§1. Formal Normal Forms.

Let $\mathfrak{X}_0(\mathbb{R}^n)$ denote the Lie algebra of all (germs of) $C^\infty$ vector fields on $\mathbb{R}^n$ which vanish in the origin. We consider a Lie subalgebra $\mathfrak{X}_0'(\mathbb{R}^n)$ of $\mathfrak{X}_0(\mathbb{R}^n)$. One may think e.g. of the set of all r-parameter families of volume preserving vector fields on $\mathbb{R}^m$ where $m + r = n$. More examples will be given below. In the usual way for $k = 1,2,\ldots$ define the singular k-jet space $\mathscr{J}_k(\mathbb{R}^n)$ and $\mathscr{J}_k'(\mathbb{R}^n)$ respectively, corresponding with the sets of truncated Taylor series of vector fields in $\mathfrak{X}_0^{(\prime)}(\mathbb{R}^n)$. For $k = \infty$ we define the k-jet spaces by taking the inverse limits : now the coefficient functions are formal power series. We recall that in a natural way for $1 \leq k < \infty$ the sets $\mathscr{J}_k(\mathbb{R}^n)$ and $\mathscr{J}_k'(\mathbb{R}^n)$ become Lie algebras, the second being a subalgebra of the first. We also consider the corresponding Lie groups, to be denoted by $J_k(\mathbb{R}^n)$ and $J_k'(\mathbb{R}^n)$, now the latter being a Lie subgroup of the former. $J_k(\mathbb{R}^n)$ is the group of k-jets of fixed points of diffeomorphisms. In terms of the above example think of $J_k'(\mathbb{R}^n)$ as being the k-jets of r-parameter families of volume preserving diffeomorphisms of $\mathbb{R}^m$. In all cases the canonical k-jet projection is denoted by $j_k$. For $X \in \mathfrak{X}_0(\mathbb{R}^n)$ the time t flow is denoted by $X_t$. The exponential map $\mathrm{Exp} : \mathscr{J}_k^{(\prime)}(\mathbb{R}^n) \to J_k^{(\prime)}(\mathbb{R}^n)$ is given by $j_k(X) \to j_k(X_1)$. Finally for $k \geq 0$ we introduce the set $H_k(\mathbb{R}^n)$, containing the vector fields in $\mathfrak{X}_0(\mathbb{R}^n)$ which have as coefficient functions homogeneous polynomials of degree k. Also define $H_k'(\mathbb{R}^n) = H_k(\mathbb{R}^n) \cap \mathfrak{X}_0'(\mathbb{R}^n)$. Observe that $\mathscr{J}_k(\mathbb{R}^n) \cong \bigoplus_{i=1}^{k} H_i(\mathbb{R}^n)$, see above, and that by E. Borel's theorem (see Narasimhan [17]) the jet projection $j_\infty : \mathfrak{X}_0(\mathbb{R}^n) \to \mathscr{J}_\infty(\mathbb{R}^n)$ is surjective.

In order to formulate our generalisation of [29] we give the following definitions (see [12]) :

1.1 Definition.

A Lie algebra L is called filtered if a sequence $F_0 \supseteq F_1 \supseteq F_2 \ldots$ of linear subspaces of L is given such that

(i) $L = F_0$ and $\bigcap_k F_k = 0$ and
(ii) $[F_k, F_\ell] \subseteq F_{k+\ell}$.

Since $[F_k, L] \subseteq F_k$ the $F_k$ are ideals of L and we may define quotient algebras $L \twoheadrightarrow L/_{F_k} = L_k$. Apparently natural projections $L_k \twoheadrightarrow L_\ell$ exist for $k \geq \ell$.

1.2 **Definition.**

We say that the filtration $F_0 \supseteq F_1 \supseteq F_2 \ldots$ is complete if L is the inverse limit of the system $\{L_k\}_{k \geq 0}$.

This property of completeness means that for each sequence $\{X_{(k)}\}_{k \geq 0}$ in L, such that for all m we have $X_{(m)} \in F_m$, there exists a unique $X \in L$ such that for all m $X \equiv \sum_{k<m} X_{(k)} (\text{mod } F_m)$. We write formally $X = \sum_{k=0}^{\infty} X_{(k)}$.

One should think of L and $L_k$ as being $\mathscr{J}'_{\infty}(\mathbb{R}^n)$ and $\mathscr{J}'_k(\mathbb{R}^n)$ respectively. A practically very important special case of a filtration is the case where the filtration is given by a formal gradation :

1.3 **Definition.**

The filtration $F_0 \supseteq F_1 \supseteq \ldots$ in 1.1 given by a formal gradation of L if for $k \geq 0$ linear subspaces $Gr_k$ of L are given such that

(i) $L = \prod_{k=0}^{\infty} Gr_k$, $F_k = \prod_{j=k}^{\infty} Gr_j$ and

(ii) $[Gr_k, Gr_\ell] \subseteq Gr_{k+\ell}$.

Note that 1.3 (ii) is stronger than 1.1 (ii). Here we also have that $L_k = \bigoplus_{j=0}^{k-1} Gr_j$. In many cases - see all our examples below - the Lie algebra $L = \mathscr{J}'_{\infty}(\mathbb{R}^n)$ can be graded by a convenient choice of the coordinates. E. g. consider the above example where $\mathfrak{X}'_0(\mathbb{R}^n)$ consists of the r-parameter families of volume preserving vector fields on $\mathbb{R}^m$, where the standard volume is used and define

$$(*) \ldots\ldots\ldots\ldots \quad Gr_j = H'_{j+1}(\mathbb{R}^n) \; .$$

Here and in many other cases the fact that a vector field $X \in \mathfrak{X}_0(\mathbb{R}^n)$ is an element of the Lie algebra $\mathfrak{X}'_0(\mathbb{R}^n)$ means that X has to preserve some structure which means that X should satisfy a differential equation that can be made homogeneous by a proper choice of coordinates (e.g."div (X) = 0"). In all these cases (*) yields a formal gradation of $\mathscr{J}'_{\infty}(\mathbb{R}^n)$.
Now let L be a completely filtered Lie algebra. Consider $X \in L$ and $Y \in F_1$. The formal sum $e^{\text{ad}\, Y} X = \sum_0^{\infty} \frac{1}{i!} (\text{ad}\, Y)^i X$ then is a well defined element of L because of the completeness. Again consider $X \in L$ and write $X = X_{(0)} + X'$ where $X \equiv X_{(0)} (\text{mod } F_1)$. Note that for all k we have : $\text{ad}\, X_{(0)}(F_k) \subseteq F_k$. In our considerations the following lemma is crucial :

1.4 **Lemma.**

Let $C \subseteq F_1$ be a linear subspace such that

(i) $\cap_k (C+F_k) = C$ and

(ii) for all $k$ : $F_k \subseteq \mathrm{ad}\, X_{(0)}(F_k) + C$

Then $Y \in F_1$ exists such that $e^{\mathrm{ad}\, Y} X \in X_{(0)} + C$ .

**Proof.**

We prove that it is sufficient to show that for all $k \geq 1$ there exists $Y_{(k)} \in F_k$ such that

$$e^{\mathrm{ad} \sum_{\ell=1}^{k} Y_{(\ell)}} X \in X_{(0)} + C + F_{k+1} .$$

For then $Y = \sum_{\ell=1}^{\infty} Y_{(\ell)}$ and $e^{\mathrm{ad}\, Y} X$ are well defined elements of $F_1$ and

$$e^{\mathrm{ad}\, Y} X - e^{\mathrm{ad} \sum_{\ell=1}^{k} Y_{(\ell)}} X \in F_{k+1},$$

which means that for all $k$

$$e^{\mathrm{ad}\, Y} X \in X_{(0)} + C + F_{k+1} .$$

This by (i) yields the conclusion of the lemma. The existence of $Y_{(k)}$ can be established using induction on $k$. Q.E.D.

1.5 **Remark.**

Condition (i) of 1.4 is fulfilled if e.g. $C$ is a complete Lie subalgebra of $F_1$. Condition (ii) is equivalent with $F_k = \mathrm{ad}\, X_{(0)}(F_k) + (C \cap F_k)$. Moreover it is easy to reformulate 1.4 in terms of formally graded Lie algebras. Then condition (i) is fulfilled if $C = \prod_{k=1}^{\infty} C_k$ for linear subspaces $C_k \subseteq Gr_k$, while in that case condition (ii) means that $Gr_k = \mathrm{ad}\, X_{(0)}(Gr_k) + C_k$ .

We apply 1.4 to $L = \mathcal{J}'_\infty(\mathbb{R}^n)$, the $\infty$-jets belonging to a Lie subalgebra $\mathfrak{X}'_0(\mathbb{R}^n)$ of $\mathfrak{X}_0(\mathbb{R}^n)$. Define $F_k = \{X \in L \mid p_k^\infty(X) = 0\}$, where $p_k^\infty : \mathcal{J}'_\infty(\mathbb{R}^n) \twoheadrightarrow \mathcal{J}'_k(\mathbb{R}^n)$ is the natural projection. Of course now $F_0 \supseteq F_1 \supseteq \ldots$ is a complete filtration of $L$ and $L_k = L/_{F_k} = \mathcal{J}'_k(\mathbb{R}^n)$. We have :

1.6 Theorem.

Let $X \in \mathfrak{X}_0'(\mathbb{R}^n)$ and $X_{(0)} \in \mathscr{J}_\infty'(\mathbb{R}^n)$ such that $j_1(X) \equiv X_{(0)} (\mathrm{mod}\ F_1)$. Suppose that $C \subseteq F_1$ is a linear subspace such that

(i) $\cap_k (C+F_k) = C$ and

(ii) for all $k$ : $F_k = \mathrm{ad}\ X_{(0)}(F_k) + C \cap F_k$ .

Then $\hat{Y} \in F_1$ exists such that

$$d(\mathrm{Exp}\ \hat{Y})\ (j_\infty(X)) \in X_{(0)} + C .$$

Proof.

Consider $\mathscr{J}_k(\mathbb{R}^n) \xrightarrow{\mathrm{Exp}} J_k(\mathbb{R}^n)$ . We know that $e^{\mathrm{ad}\ Y} = \mathrm{Ad}(\mathrm{Exp}\ Y)$ and that $(\mathrm{Ad}\ \psi)(X) = j_k((d\psi)(X))$. Now apply 1.4. Q.E.D.

Theorem 1.6 yields a formal change of coordinates Exp $\hat{Y}$ which puts the $\infty$-jet of X in $X_{(0)} + C$. Here in the analytic case problems of convergence arise. E.g. see Siegel [23]. Note that the 1-jet of this formal diffeomorphism equals that of the identity map. It is rather convenient to have a "real" germ $Y \in \mathfrak{X}_{(0)}'(\mathbb{R}^n)$ such that $j_\infty(Y) = \hat{Y}$.

1.7 Definition.

We say that $\mathfrak{X}_{(0)}'(\mathbb{R}^n)$ has the Borel property if $j_\infty : \mathfrak{X}_{(0)}'(\mathbb{R}^n) \to \mathscr{J}_\infty'(\mathbb{R}^n)$ is surjective.

1.8 Remarks.

(i) It is not vacuous to require the Borel property since it is easy to provide a set of functions which do not have it : think of the harmonic functions.

(ii) Clearly in the case where $\mathfrak{X}_0'(\mathbb{R}^n)$ has the Borel property, in the conclusion of 1.6 we have a real change of coordinates. Then this conclusion is that $Y \in \mathfrak{X}_0'(\mathbb{R}^n)$ exists such that for its time 1 flow $Y_1$ we have : $j_\infty(dY_1(X)) \in X_{(0)} + C$. If this conclusion holds we say that $\mathfrak{X}_0'(\mathbb{R}^n)$ admits the N(ormal) F(orm) T(heorem).

1.9 List of Lie subalgebras of $\mathfrak{X}_0(\mathbb{R}^n)$ that admit the NFT.

(Always $n = m + r$).

(i) The set $\mathfrak{X}_0(\mathbb{R}^m | \mathbb{R}^r)$ of all r-parameter families of vector fields on $\mathbb{R}^m$ that vanish in 0 admits the NFT. For r = 0 this result reduces to Takens [29] (2,1).

(ii) If $\Omega$ is a volume form on $\mathbb{R}^m$ then $\mathfrak{X}_0^{\Omega}(\mathbb{R}^m|\mathbb{R}^r) = \{X \in \mathfrak{X}_0(\mathbb{R}^m|\mathbb{R}^r) | \mathcal{L}_X\Omega = 0\}$ admits the NFT.

(iii) If $m = 2p$ and $\omega$ is a symplectic 2-form on $\mathbb{R}^m$ then define $\mathfrak{X}_0^{\omega}(\mathbb{R}^m|\mathbb{R}^r)$ to be the set of the corresponding unfoldings of hamiltonian vector fields that vanish in 0. $\mathfrak{X}_0^{\omega}(\mathbb{R}^m|\mathbb{R}^r)$ admits the NFT.

(iv) If $m = 2p + 1$ and $\alpha$ is a contact 1-form on $\mathbb{R}^m$, consider $\mathfrak{X}_0^{\alpha}(\mathbb{R}^m|\mathbb{R}^r)$ being the set of unfoldings of infinitesimal contact transformations that vanish in 0, which admits the NFT.

(v) Let $\tau:\mathbb{R}^m \to \mathbb{R}^m$ be a linear involution. The corresponding set of unfoldings of reversible vector fields ($\tau_* X = -X$) is denoted by $\mathfrak{X}_0^{\tau}(\mathbb{R}^m|\mathbb{R}^r)$ and also admits the NFT. (See Moser [16].)

Proof.

We only have to check the Borel property.

(ii) Suppose that r = 0. Consider $U \in \mathcal{J}_\infty^{\Omega}(\mathbb{R}^m)$ and use Borel's theorem ([17]) to find $X \in \mathfrak{X}_0(\mathbb{R}^m)$ with $j_\infty(X) = U$. Then $d\iota_X\Omega = \mathcal{L}_X\Omega$ is flat in 0. Let H be a right inverse of d (as in the Poincaré lemma) carrying flat m forms to flat m - 1 forms. Then define Y by $\iota_Y\Omega = H(\mathcal{L}_X\Omega)$ and verify that $j_\infty(X-Y) = U$ while $\mathcal{L}_{X-Y}\Omega = 0$. For $r \geq 1$ proceed as before, now use vertical differential forms.

(iii) Use Borel's theorem for the hamiltonian functions.

(iv) The correspondence $X \to \alpha(X)$ is a bijection from $\mathfrak{X}^{\alpha}(\mathbb{R}^m)$ to functions on $\mathbb{R}^m$. This yields the Borel property. Q.E.D.

From now on restrict to the case where the Lie algebra $\mathcal{J}'_\infty(\mathbb{R}^n)$ is formally graded by (*):$Gr_k = H'_{k+1}(\mathbb{R}^n)$. As is said before we may identify $\mathcal{J}'_k(\mathbb{R}^n) = \bigoplus_{j=1}^{k} H'_j(\mathbb{R}^n)$. Consider $X \in \mathfrak{X}'_0(\mathbb{R}^n)$ as above and take $X_{(0)} \in \mathcal{J}'_1(\mathbb{R}^n) = H'_1(\mathbb{R}^n)$ such that $j_1(X) = X_{(0)}$. The endomorphism ad $X_{(0)}: \mathcal{J}'_\infty(\mathbb{R}^n) \to \mathcal{J}'_\infty(\mathbb{R}^n)$ splits nicely over the gradings $H'_k(\mathbb{R}^n)$, $k \geq 1$. Identifying $\mathcal{J}_1(\mathbb{R}^n) \cong gl(n, \mathbb{R})$ in the usual way, suppose that $X_{(0)} = S + N$ is the Jordan decomposition in $gl(n, \mathbb{R})$. We consider the corresponding subgroup $J'_1(\mathbb{R}^n)$ of $J_1(\mathbb{R}^n) \cong Gl(n, \mathbb{R})$ generated from $\mathcal{J}'_1(\mathbb{R}^n)$ by the exponential map. By a result of A. Borel [4] we know that if $J'_1(\mathbb{R}^n)$ is an algebraic subgroup of $Gl(n,\mathbb{R})$ both S and N are elements of $\mathcal{J}'_i(\mathbb{R}^n)$. Observe that all examples in 1.9 have this property of algebraicness. In this case the Jordan decomposition of ad $X_{(0)}$ is ad $X_{(0)}$ = ad S + ad N, which also splits over the gradings. Now consider C = Ker (ad S).

If $C_k = \ker(\text{ad } S) \cap H'_{k+1}(\mathbb{R}^n)$ then $C = \prod_{k=1}^{\infty} C_k$. See above. If $X_{(0)}$ were semisimple

(N=0) then for all $k \geq 1$

$$C_k \oplus \mathrm{ad}\, X_{(0)}(H'_{k+1}(\mathbb{R}^n)) = H'_{k+1}(\mathbb{R}^n) .$$

The next theorem states that this choice of C also suffices for $N \neq 0$. See Varadarajan [30], 3.1.4.

1.10 Theorem.

Let $\mathfrak{X}'_0(\mathbb{R}^n)$, $X_{(0)}$, S and N be as above. Suppose that $J'_1(\mathbb{R}^n)$ is an algebraic subgroup of $Gl(n, \mathbb{R})$. Then for $C_k = \ker(\mathrm{ad}\, S) \cap H'_{k+1}(\mathbb{R}^n)$ we have

$$C_k + \mathrm{ad}\, X_{(0)}(H'_{k+1}(\mathbb{R}^n)) = H'_{k+1}(\mathbb{R}^n), \quad k \geq 1 \quad .$$

1.11 Remark.

By 1.10 the NFT obtains a geometric meaning (see [28,29]) : In the normal form the vector field X can be written as

$$X = X_{(0)} + c_{(2)} + \ldots + c_{(\ell)} + \text{Higher Order Terms},$$

where $c_{(i)} \in H'_i(\mathbb{R}^n) \cap \mathrm{Ker}\, (\mathrm{ad}\, S)$, implying that $\mathcal{L}_S(c_{(i)}) = 0$, $2 \leq i \leq \ell$. In other words: the $c_{(i)}$ are invariant under the flow of S.

We proceed in giving two examples of germs of a 1-parameter family of divergence free vector fields : one in dimension 3 and one in dimension 4. In the next section it will be shown that these examples are the only bifurcations which are essentially different from the theory in e.g. [2]. For analogous results see [29], proposition 2.5.

1.12 Example in dimension 3.

Let $X \in \mathfrak{X}_0^{\Omega}(\mathbb{R}^3|\mathbb{R})$ be a 1-parameter family parametrised by $\mu$ where $\Omega = dx_1 \wedge dx_2 \wedge dx_3$ ( the standard volume). See 1.9. Assume that $X_{(0)} = -x_2 \frac{\partial}{\partial x_1} + x_1 \frac{\partial}{\partial x_2} + \mu \frac{\partial}{\partial x_3}$ is the 1-jet of X. Clearly $S = -x_2 \frac{\partial}{\partial x_1} + x_1 \frac{\partial}{\partial x_2}$ and $N = \mu \frac{\partial}{\partial x_3}$ . As an immediate consequence of the NFT and 1.10 it follows that modulo a $\mu$-dependent, volume preserving change of coordinates $X = \tilde{X} + p$ where

(i) $j_\infty(p) = 0$ : $p(x,\mu)$ is flat in $(x,\mu) = (0,0) \in \mathbb{R}^3 \times \mathbb{R}$ ,

(ii) Both $\tilde{X}$ and p are divergence free.

(iii) $\tilde{X} = f(r^2, x_3, \mu)\frac{\partial}{\partial\varphi} + g(r^2, x_3, \mu)\, r\frac{\partial}{\partial r} + h(r^2, x_3, \mu)\frac{\partial}{\partial x_3}$ $((\varphi, r, x_3)$ being cylinder coordinates), with $f(0,0,0) = \frac{\partial h}{\partial \mu}(0,0,0) = 1$ and $g(0,0,0) = \frac{\partial h}{\partial x_3}(0,0,0) = 0$ .

1.13 <u>Example in dimension 4.</u>

Consider $X \in \mathfrak{X}_0^{\Omega}(\mathbb{R}^4|\mathbb{R})$, also parametrised by $\mu$ and where again $\Omega = dx_1 \wedge dx_2 \wedge dx_3 \wedge dx_4$ is the standard volume. Suppose that X has a 1-jet $X_{(0)} = \lambda(-x_2\frac{\partial}{\partial x_1} + x_1\frac{\partial}{\partial x_2}) + (-x_4\frac{\partial}{\partial x_3} + x_3\frac{\partial}{\partial x_4})$ where $\lambda \in \mathbb{R}\setminus\mathbb{Q}$ ❀. Now $S = X_{(0)}$ and $N = 0$. Then modulo a $\mu$-dependent, volume preserving change of coordinates $X = \tilde{X} + p$ where

(i) $j_\infty(p) = 0$ ,

(ii) Both $\tilde{X}$ and p are divergence free.

(iii) $\tilde{X} = f_1(r_1^2, r_2^2, \mu)\frac{\partial}{\partial\varphi_1} + f_2(r_1^2, r_2^2, \mu)\frac{\partial}{\partial\varphi_2} + g_1(r_1^2, r_2^2, \mu)\, r_1\frac{\partial}{\partial r_1} + g_2(r_1^2, r_2^2, \mu)\, r_2\frac{\partial}{\partial r_2}$

$((r_1, \varphi_1)$ and $(r_2, \varphi_2)$ are polar coordinates in the planes $x_3 = x_4 = 0$ and $x_1 = x_2 = 0$ respectively), with $f_1(0,0,0) = \lambda$, $f_2(0,0,0) = 1$, $g_1(0,0,0) = g_2(0,0,0) = 0$ . Condition ❀ can be weakened if one wishes to "normalise" up to a finite order. (See [29].)

1.14 <u>Remark.</u>

In general one may not hope to be able to normalise without a flat term p which destroys the symmetry. See Takens [27] and the results below.

§2. <u>Catalogue of Codimension One Singularities in Divergence Free Vector Fields.</u>

From now on we study the singularities of generic 1-parameter families of divergence free vector fields on $\mathbb{R}^m$. We try to develop a theory analogous to [2, 7, 24]. Therefore we consider the same equivalence relation as used in e.g. [2] : the local $C^0$-equivalence of 1-parameter families. (I.e. the local homeomorphisms are of the form $H(x,\mu) = (\bar{H}^{\mu}(x), h(\mu))$, $x \in \mathbb{R}^m$, $\mu \in \mathbb{R}$.) Note that we do not require the equivalences to preserve the volume. We shall see that using this equivalence relation (structural) stability is not generic.

Consider a $C^\infty$ 1-parameter family $X = X(x,\mu)$, $x \in \mathbb{R}^m$, $\mu \in \mathbb{R}$, of divergence free vector fields. Define $N(X) = \{(x,\mu) \in \mathbb{R}^m \times \mathbb{R} | X(x,\mu) = 0\}$, the set of zero's of X. It follows from the implicit function theorem that for generic X the set

N(X) is an embedded 1-dimensional $C^\infty$ submanifold of $\mathbb{R}^m \times \mathbb{R}$. For $\mu \in \mathbb{R}$ we write $X^\mu = X(-,\mu)$. If $(x,\mu) \in N(X)$, i.e. $X^\mu(x) = 0$, then our first interest is for the 1-jet of $X^\mu$ in x. For the set of such singular 1-jets we have $\mathcal{J}_1^\Omega(\mathbb{R}^m) \cong s\ell(m,\mathbb{R})$, see §1. This defines a map $\varphi_X: N(X) \to s\ell(m,\mathbb{R})$. According to Mather [15] for generic X this map $\varphi_X$ is transversal on any given semialgebraic subset $\Sigma$ of $s\ell(m,\mathbb{R})$.

So we look for a semialgebraic decomposition of $s\ell(m,\mathbb{R})$ related to whether or not the eigenvalues are purely imaginary. Especially the search is for codimension 0 and 1 subsets in such a decomposition : the codimension 1 singularities occur as bifurcations in a discrete subset of the curve N(X), the codimension 0 singularities are seen in the open pieces of arc in between. Without proof we state :

2.1 <u>Theorem.</u>

In the following list for m = 2, 3, ... in the left column an open codimension 0 subset of $s\ell(m,\mathbb{R})$ is given. In the right column a semialgebraic codimension 1 subset of $s\ell(m,\mathbb{R})$ is indicated, which moreover is a $C^\infty$-manifold. The complement of the union of these two sets is a closed semialgebraic subset of $s\ell(m,\mathbb{R})$ of codimension ≥2.

| m | codimension 0 | codimension 1 | |
|---|---|---|---|
| 2 | Eigenvalues $\pm\lambda$ (hyperbolic), or $\pm i\lambda$ (elliptic), $\lambda \in \mathbb{R}\backslash\{0\}$ . | Both eigenvalues 0, without the matrix being 0. | |
| 3 | Hyperbolic eigenvalues. | Eigenvalues | |
| | | 0 and $\pm\lambda$ | case (3,1) |
| | | or 0 and $\pm i\lambda$ | case (3,2) |
| | | $\lambda \in \mathbb{R}\backslash\{0\}$ . | |
| 4 | Hyperbolic eigenvalues. | Eigenvalues | |
| | | 0 and rest hyperbolic | case (4,1) |
| | | or $\pm i\lambda_1$ and rest hyperbolic | case (4,2) |
| | | or $\pm i\lambda_1$ and $\pm i\lambda_2$ | case (4,3) |
| | | $\lambda_1, \lambda_2 \in \mathbb{R}\backslash\{0\}, \lambda_1 \neq \lambda_2$. | |
| ≥5 | Hyperbolic eigenvalues. | Eigenvalues | |
| | | 0 and rest hyperbolic | case (m,1) |
| | | or $\pm i\lambda$ and rest hyperbolic | case (m,2) |
| | | $\lambda \in \mathbb{R}\backslash\{0\}$. | |

The codimension 0 singularities are all structurally stable in the above sense, see e.g. Hartman [13] and Andronov et al. [1].

2.2 In the hamiltonian case (m = 2), which is well known, also the codimension 1 singularity is stable, even when we use smooth equivalences. It is equivalent to $x_2 \frac{\partial}{\partial x_1} + (3x_1^2 + \mu) \frac{\partial}{\partial x_2}$ in $(x_1, x_2, \mu) = (0,0,0)$. See Andronov et al. [1].

2.3 The codimension 1 cases (m, 1) for m ≥ 3 and (m, 2) for m ≥ 4 are also stable. They are the bifurcations that occur in [2, 24, 7]. Up to considerations involving normally hyperbolic invariant manifolds (see Palis & Takens [18])

* the cases (m, 1) are $C^0$-equivalent to the SADDLE NODE $(x^2+\mu)\frac{\partial}{\partial x}$ in $(x, \mu) = (0, 0)$,

* the cases (m, 2) are $C^0$-equivalent to the HOPF BIFURCATION

$$-x_2 \frac{\partial}{\partial x_1} + x_1 \frac{\partial}{\partial x_2} + \{(x_1^2+x_2^2) - \mu\}(x_1 \frac{\partial}{\partial x_1} + x_2 \frac{\partial}{\partial x_2}) \text{ in } (x_1, x_2, \mu) = (0,0,0) .$$

2.4 The remaining cases are the bifurcations (3, 2) and (4, 3), both having a strong non hyperbolicity. Up to some linear algebra they are exactly the examples 1.12 and 1.13 above. In the following section these cases will be studied. It will appear that they are not stable. First, however, we give a heuristic proof of the fact that the bifurcations (3, 2) and (4, 3) do occur and the fact that such strong non-hyperbolicities do not appear in higher dimensions :

Consider - for generic X - the curve $N(X) \xrightarrow{\varphi_X} s\ell(m, \mathbb{R})$. Suppose that for some $(x, \mu) \in N(X)$ the matrix $\varphi_X(x, \mu)$ has a conjugated pair of eigenvalues. As we move along the curve N(X) this conjugated pair varies continuously and now and again it may pass through the imaginary axis. But trace $\varphi_X(x, \mu) = 0$, so if the considered pair of eigenvalues is on the imaginary axis, then the sum of the remaining eigenvalues also must be 0. For m = 3 this means that the 3rd eigenvalue is 0, for m = 4 it implies that generically we are in the cases (4, 2) or (4, 3). One easily sees that in this way the constraint trace $\varphi_X(x, \mu) = 0$ is not felt for m ≥ 5.

## §3. If the Germ is Symmetric.

First we consider the bifurcation (3, 2), the 4-dimensional case (4, 3) is dealt with later on. We assume to have a 1-parameter family X of divergence free vector fields on $\mathbb{R}^3$ which in $(x_1, x_2, x_3, \mu) = (0,0,0,0)$ has a bifurcation of type (3,2). According to example 1.12 we may assume that X has the following normal form decomposition :

$$X = \tilde{X} + p$$

where

(i) p is flat in $(x,\mu) = (0,0) \in \mathbb{R}^3 \times \mathbb{R}$,

(ii) $\tilde{X}$ and p are divergence free,

(iii) in cylinder coordinates $r, \varphi, z$ : $X = f(r^2, z, \mu)\frac{\partial}{\partial\varphi} + g(r^2, z, \mu)r\frac{\partial}{\partial r} + h(r^2, z, \mu)\frac{\partial}{\partial z}$

where $f(0,0,0) = \frac{\partial h}{\partial \mu}(0,0,0) = 1$ and $g(0,0,0) = h(0,0,0) = \frac{\partial h}{\partial z}(0,0,0) = 0$.

In this section we investigate the topological type of the rotationally symmetric vector field $\tilde{X}$ by reducing it to a 2-dimensional situation : we forget the $\frac{\partial}{\partial\varphi}$-component (see [29] Afterwards we may attempt to evaluate the effect of a small perturbation p to the phase portrait of $\tilde{X}$. It is easy to see that in our context it is no loss of generality to assume that $f(r^2, z, \mu) \equiv 1$. The reduced vector field on the (r, z)-plane is $\bar{X} = g(r^2, z, \mu)\, r\frac{\partial}{\partial r} + h(r^2, z, \mu)\frac{\partial}{\partial z}$.

3.1 Remark.

If $T:\mathbb{R}^2 \to \mathbb{R}^2$ is defined by $(r,z) \mapsto (-r,z)$ then $T_*\bar{X} = \bar{X}$. Considerations in this section concerning the case (3,2) are required to be equivariant with respect to the $\mathbb{Z}_2$-action induced by T. Also observe that $\bar{X}$ respects the 2-form $\omega = r dr \wedge dz$ which is only degenerated on the line $r = 0$. This fact enables us to introduce technically convenient "hamiltonian" functions $H(r^2, z)$.

We treat the cases $\mu \neq 0$ and $\mu = 0$ separately. Firstly suppose that $\mu \neq 0$. Define a $\mu$-dependent change of coordinates $D:\mathbb{R}^2 \times \{\mu\neq 0\} \to \mathbb{R}^2 \times \{\mu \neq 0\}$ by $(r,z,\mu) \mapsto (r\sqrt{|\mu|}, z\sqrt{|\mu|}, \mu)$. D is the so called blowing down map. For $\mu$ fixed we write $D^\mu$ and we consider the blown up vector field

$$Z^\mu = \frac{1}{\sqrt{|\mu|}}(D^\mu)_*^{-1}\bar{X}^\mu \quad .$$

One checks easily that if one defines $Z^+(r,z) = \lim_{\mu \downarrow 0} Z^\mu(r,z)$ and $Z^-(r,z) = \lim_{\mu\uparrow 0} Z^\mu(r,z)$ then for some $a, b \in \mathbb{R}$ we have

$$Z^{\pm}(r,z) = azr\frac{\partial}{\partial r} + (br^2 - az^2 \pm 1)\frac{\partial}{\partial z} .$$

First we present pictures of the phase portraits of $Z^{\pm}$. Note that it is no loss of generality to assume that $b \geq 0$, otherwise replace z by -z. Under the generic assumption $a \neq 0$ and $b \neq 0$ we find the following two cases :

$$\text{I} : a < 0,\ b > 0 \quad \text{and II} : a > 0,\ b > 0 .$$

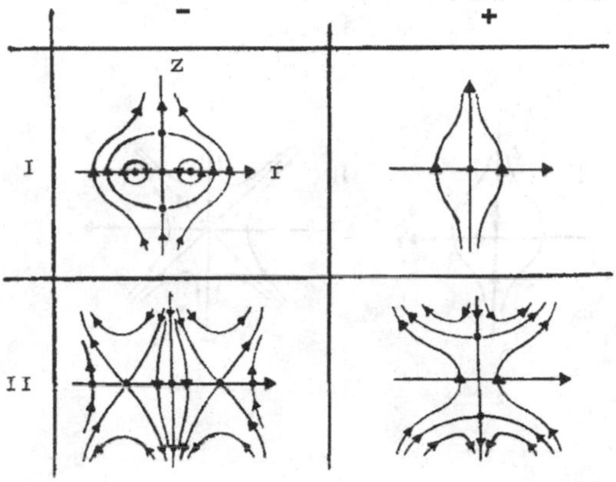


Figure 1.

Observe that $\{Z^{\mu}\}_{\mu>0} \cup \{Z^+\}$ and $\{Z^{\mu}\}_{\mu>0} \cup \{Z^-\}$ form $C^{\infty}$ 1-parameter families in $\sqrt{|\mu|}$. We want to prove that the vector fields $Z^{\pm}$ are stable for equivariant perturbations which preserve $\omega$. Since we have blown up and since we aim to draw conclusions concerning the $\overline{X}^{\mu}$ for $\mu \neq 0$, it is not useful to prove that the $Z^{\pm}$ are locally stable. Therefore let us choose a compact neighbourhood K of (0,0) in $\mathbb{R}^2$ such that T(K) = K, large enough to contain the interesting details of the phase portraits of $Z^{\pm}$. (For instance let $\partial K$ be a circle with (0,0) as centre and a radius larger than max $\{\sqrt{\frac{2}{b}}, \sqrt{\frac{1}{|a|}}\}$). The equivalence relation that underlies the concept of stability that we need here can be defined in terms of equivariant homeomorphisms on neighbourhoods of K. See Poènaru [19]. Then

3.2 Theorem.

For $a \neq 0$ the $Z^{\pm}$ within the class of equivariant $\omega$-preserving vector fields are $C^{\infty}$-equivariantly stable in the above sense. There are two $C^{\infty}$-equivalence classes, corresponding to the situations I and II. (See fig. 1.)

We do not prove this theorem : see [5]. It can be done by proving that the corresponding hamiltonians are infinitesimally stable in the equivariant sense. Secondly we come to the situation $\mu = 0$. The vector field $\bar{X}^0$ has a 2-jet $azr\frac{\partial}{\partial r} + (br^2 - az^2)\frac{\partial}{\partial z}$ for $a, b \in \mathbb{R}$. Again we assume $b \geq 0$. We now quote [29] (3, 8) which states that $\bar{X}^0$ is 2-determined :

3.3 Theorem.

Let Y be a $C^3$ vector field on $\mathbb{R}^2$ with $j_2(Y) = j_2(\bar{X}^0)$. Suppose that $a \neq 0 \neq b$. Then for any $C^3$ vector field Z with $j_1(Z) = 0$ and $j_2(Z)$ close enough to $j_2(Y)$ is locally $C^0$-equivalent to Y. Moreover there are two $C^0$-equivalence classes, again corresponding to the cases I and II.

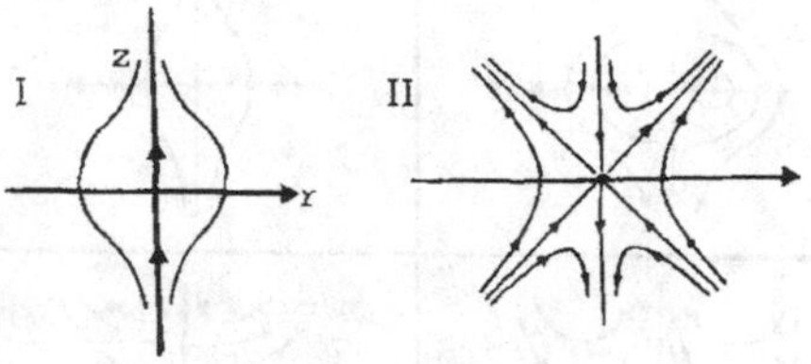


Figure 2.

3.4 Remarks.

(i) According to Dumortier [8] 3.3 can be restated in terms of conjugations instead of equivalences.

(ii) In our case we like the equivalences (conjugations) to be equivariant. This is no problem.

(iii) The perturbations need not be $\omega$-preserving!

A completely similar elaboration can be given of the 4-dimensional case (4, 3), now using example 1.13. On $\mathbb{R}^4$ we use polar coordinates $(r_1, \varphi_1)$ and $(r_2, \varphi_2)$ in the planes $x_3 = x_4 = 0$ and $x_1 = x_2 = 0$ respectively. The symmetric vector field $\tilde{X}$ is in this case $\tilde{X} = f_1(r_1^2, r_2^2, \mu)\frac{\partial}{\partial\varphi_1} + f_2(r_1^2, r_2^2, \mu)\frac{\partial}{\partial\varphi_2} + g_1(r_1^2, r_2^2, \mu)r_1\frac{\partial}{\partial r_1} + g_2(r_1^2, r_2^2, \mu)r_2\frac{\partial}{\partial r_2}$, which is reduced to $\bar{X} = g_1(r_1^2, r_2^2, \mu)r_1\frac{\partial}{\partial r_1} + g_2(r_1^2, r_2^2, \mu)\frac{\partial}{\partial r_2}$ on the $(r_1, r_2)$-plane. Observe that here two evolutions have to be respected : $(r_1, r_2) \mapsto (-r_1, r_2)$ and $(r_1, r_2) \mapsto (r_1, -r_2)$ and a 2-form $\omega = r_1 r_2 dr_1 \wedge dr_2$. See 3.1. Firstly restrict to $\mu \neq 0$. As in the 3-dimensional case we blow up with $\sqrt{|\mu|}$ and obtain limits

$$Z(r_1, r_2) = (ar_1^2 + 2br_2^2 \pm 1)r_1\frac{\partial}{\partial r_1} - (2ar_1^2 + br_2^2 \pm 1)r_2\frac{\partial}{\partial r_2} ,$$

for some $a, b \in \mathbb{R}$. Up to permuting $r_1$ and $r_2$ or changing the sign of $Z^{\pm}$ we may restrict ourselves to the situation where $a \geq |b|$. Under the generic assumption $a \neq |b| > 0$ we depict the phase portraits. We find the following cases :

I : $a > b > 0$ and II : $a > -b > 0$

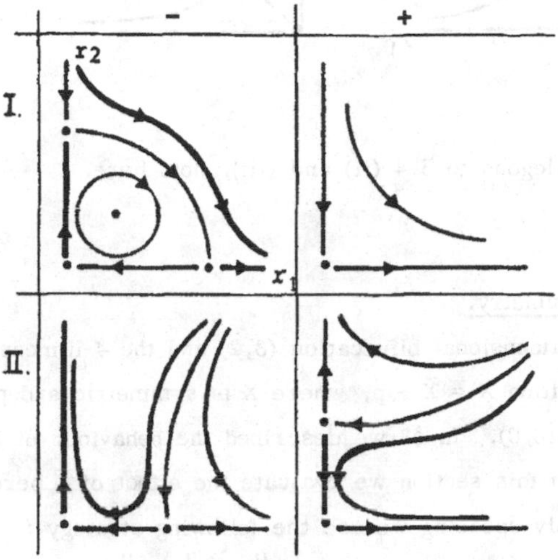


Figure 3.

Analogous to 3.2 we have

3.5 Theorem.

Suppose that $a \neq |b| < 0$. Then the vector fields $Z^{\pm}$ are $C^{\infty}$-equivariantly stable within the class of equivariant $\omega$-preserving vector fields. Moreover there are two $C^{\infty}$-equivalence classes, corresponding to the situations I and II. (See fig. 3.)

Secondly consider the case $\mu = 0$. $\bar{X}^0$ has a 3-jet $(ar_1^2 + 2br_2^2)r_1\frac{\partial}{\partial r_1} - (2ar_1^2 + br_2^2)r_2\frac{\partial}{\partial r_2}$ for $a, b \in \mathbb{R}$. Again assume $a \geq |b|$. We quote [29] (3.10), which states that $\bar{X}^0$ is 3-determined.

3.6 Theorem.

Let Y be a $C^4$ vector field on $\mathbb{R}^2$ with $j_3(Y) = j_3(\bar{X}^0)$. Suppose that $a \neq |b| > 0$. Then for any $C^4$ vector field Z on $\mathbb{R}^2$ with $j_2(Z) = 0$ and $j_3(Z)$ close enough to $j_3(Y)$, Z is locally $C^o$-equivalent to Y. There are two $C^o$-equivalence classes, corresponding to I and II.

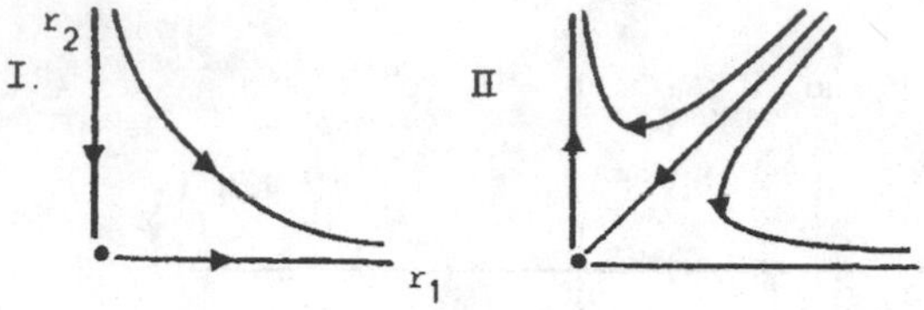


Figure 4.

Remarks, completely analogous to 3.4 (ii) and (iii), hold here.

§4. Close to Symmetry.

Both the 3-dimensional bifurcation (3,2) and the 4-dimensional case (4,3) have a normal form decomposition $X = \tilde{X} + p$, where $\tilde{X}$ is symmetric and p very small in a neighbourhood of $(x, \mu) = (0, 0)$. In §3 we described the behaviour of $\tilde{X}^\mu$ if $\mu$ varies over a neighbourhood of 0. In this section we evaluate the effect of a perturbation p in a number of cases. Roughly speaking we use the following strategy : First we keep $\mu$ fixed in a small neighbourhood of 0 and study $X^\mu = \tilde{X}^\mu + p^\mu$ as a divergence free vector field on $\mathbb{R}^3$ ($\mathbb{R}^4$). In some cases we are able to establish structural stability or something very close to this. Only in these cases there is any hope to be able to prove stability of the bifurcation-singularity as a 1-parameter family. (See the introduction to §2.)

4.1 Quasi periodic flows.

Consider in the 3-dimensional bifurcation the case I,- (fig. 1). The symmetric vector field $\tilde{X}^\mu$($\mu < 0$, $|\mu|$ small) foliates an open piece of $\mathbb{R}^3$ with invariant 2-tori which enclose an elliptic orbit. In Broer [6] it is established that the elliptic orbit is stable and that for sufficiently small $|\mu|$ the vector field $X^\mu = \tilde{X}^\mu + p^\mu$ also has many invariant tori with a quasi periodic flow. Problems of small denominators are involved. It follows immediately that there is no structural stability.

A very similar situation appears in the 4-dimensional case I,- (fig. 3). We conjecture that for each invariant 3-torus of $\tilde{X}^\mu$ with "sufficiently independent" frequencies and $|\mu|$ small enough, an appropriate $\mu'$ close to $\mu$ exists, such that $X^{\mu'}$ has a nearby invariant 3-torus with a quasi periodic flow. Also here we have no structural stability.

4.2 Generically the vector field $X^0$ is 2-determined (dimension 3 only).

Recall 3.3 and 3.4. A similar result holds for the 3-dimensional vector field $X^0$. In part 4.2 we abandon the volume preservingness which makes our results more general : the only importance is that $X^0$ has a good 2-jet, not that it is divergence free. We emphasise that our method is not easily carried over to a 4-dimensional analogue of 3.6. Let $X^0 = \tilde{X}^0 + p^0$ be as above. From §3 recall that in appropriate cylinder coordinates $r, \varphi, z$ for some $a, b \in \mathbb{R}$

$$j_2(X^0) = \frac{\partial}{\partial\varphi} + azr\frac{\partial}{\partial r} + (br^2 - az^2)\frac{\partial}{\partial z} .$$

Again assume $b \geq 0$. Consider a $C^\infty$ vector field Y on $\mathbb{R}^3$ with $Y(0) = 0$ and $j_1(Y)$ possessing exactly one eigenvalue zero and two non-zero purely imaginary eigenvalues. Then

4.2.1 Theorem.

Suppose that $a \neq 0 \neq b$. If $j_2(Y)$ is close enough to $j_2(X^0)$ then $X^0$ and Y are locally $C^0$-equivalent. There are two $C^0$-classes : I and II. (See 3.3 and fig. 2.)

If $X^0$ and Y are both symmetric then 4.2.1 is a consequence of 3.4 : first reparametrise Y to a rotational velocity 1, then conjugate the reductions $\bar{X}^0$ and $\bar{Y}$. Note that the methods from [29] - see e.g. the proof of 3.3 - do not work directly in our 3-dimensional case! Our proof of 4.2.1 consists of constructing a $C^0$-equivalence $X^0 \simeq \tilde{X}^0$, this is sufficient by the normal form theorem. We only present a rough outline of this proof. Observe that case I is trivial : use the Liapunov function $(x_1, x_2, x_3) \mapsto x_3$. So from now on restrict to case II. The major problem in making an equivalence $X^0 \simeq \tilde{X}^0$ is to show that the Poincaré map $\tilde{\Phi}$ of $\tilde{X}^0$, from the plane $x_2 = 0$ onto itself, is 2-determined. This is done in Dumortier, Rodrigues & Roussarie [9]. A merely technical problem is that the Poincaré map $\Phi$ of $X^0$ is not well-defined near the $x_3$-axis. We therefore replace $X^0$ by X', obtained after a flat perturbation. X' coincides with $\tilde{X}^0$ near the $x_3$-axis and with $X^0$ further from this axis. X' has a well defined Poincaré map and the theory of [9] yields an equivalence $X' \simeq \tilde{X}^0$. For the proof of 4.2.1 we have to adjust this equivalence near the $x_3$-axis. This is straightforward.

4.3 The (almost) stable $\tilde{X}^\mu (\mu \neq 0)$.

In a number of cases the vector field $\tilde{X}^\mu$ is Morse-Smale and therefore stable, even within the class of all vector fields. In dimension 3 these are the cases

I,+ and II,- Case II,+ has a saddle connection which can be broken with an arbitrarily small divergence free perturbation, see Robinson [21]. After that it is Morse-Smale. In dimension 4 case I,+ is stable.

### 4.4 As a 1-parameter family (dimension 3 only).

In case (3,2) I by 4.1 there is no stability. However, using the Liapunov function $x_3$ one can prove the following partial result :

#### 4.4.1 Proposition.

Let X be a 1-parameter family of divergence free vector fields on $\mathbb{R}^3$ which in $(x,\mu) = (0,0)$ has a type (3,2) I-bifurcation. Then any 1-parameter family Y, close enough to X, has a type (3,2) I-bifurcation in some $(x_0,\mu_0)$ close to (0,0). Moreover there exist neighbourhoods U of (0,0) in $\mathbb{R}^3 \times [0,\infty)$ and V of $(x_0,\mu_0)$ in $\mathbb{R}^3 \times [\mu_0,\infty)$ as well as a homeomorphism $H:U \to V$ of shape $H(x,\mu) = (\bar{H}^\mu(x), h(\mu))$ which makes the restrictions of X and Y equivalent.

In case (3,2) II one might hope to prove stability after the breaking of the saddle connection. We expect however that the broken "saddle tails" $(\mu > 0)$ cause difficulties in spiralling around each other and coming close to the "opposite" saddles. Observe that the topology is rather strange since by the normal form theorem the symmetric specimen are dense.

In the complement of this dense set the "generic" bifurcations are open and dense. Therefore we conjecture that, analogous to 4.4.1, we may find local $\bar{H}^\mu$ for $\mu$ in a full neighbourhood of 0, which can not always be chosen continuous in $\mu$ for $\mu \geq 0$. For $\mu \leq 0$ we do not expect such difficulties.

### 4.5 Some "global" results in the 3-dimensional case I,-.

In 4.1 we mentioned the fact that a great number of the invariant 2-tori of $\tilde{X}^\mu(\mu < 0, |\mu|$ small) survive a flat perturbation. Here we make some observations concerning the saddles and their connecting manifolds. If a small, divergence free perturbation is carried out, then

a. The hyperbolic saddles survive.

b. The 1-dimensional connection may break, see [21].

c. Also the 2-dimensional invariant saddle manifolds generally do not coincide any more.

Since $X^{\mu}$ is divergence free it cannot happen that the "upper manifold" fully encloses the "lower" or v.v. Generically there is transversal intersection, visualised in fig. 5 (helpful suggestions were made by Duistermaat).

d. It is possible to have non transversal homoclinic points. How often this will happen is not yet clear. If the bifurcation I occurs in a global family on a compact manifold, by Takens [26], we suspect that such homoclinic points appear in a countable, dense set of parameter values.

What happens during such a homoclinic intersection is described by Šil'nikov [11]. The dynamics on a part of the non wandering set involves a shift on an alphabet with infinitely many symbols.

4.6 Some remarks on fluid mechanics.

A 1-parameter family $X^{\mu}(x)$ of divergence free vector fields can be regarded as the velocity field of an incompressible fluid by saying e.g. that $\mu = t$ (time). One may ask whether the bifurcations (3,1), (3,2)I and (3,2)II do occur in solutions of the Euler equations or the Navier Stokes equation. For the Euler equations the answer is positive, see [5]. It is a theorem of Ebin & Marsden [10] that the solutions of the Navier Stokes equation converge to the solutions of the Euler equations if the viscosity $\nu \to 0$. We speak of a strong limit in the Sobolev-$H^s$-topology, where $s > 1$. By taking s sufficiently large it follows that our bifurcations can also occur in solutions of the Navier-Stokes equation. Now consider the bifurcation (3,2)I which looks rather wild. If it occurs in a solution of the Navier Stokes equation there also will be dissipation of kinetic energy into heat. In this case one wonders whether the adjective "turbulent" would be appropriate. Note that such a geometric approximation of turbulence does not fit into the Hopf, Landau & Lifshitz -, nor in the Ruelle & Takens-picture on the nature of turbulence. See [14] and [22]. Nevertheless we like to point out that there is some resemblance between the phenomena described in 4.1 and smoke rings.

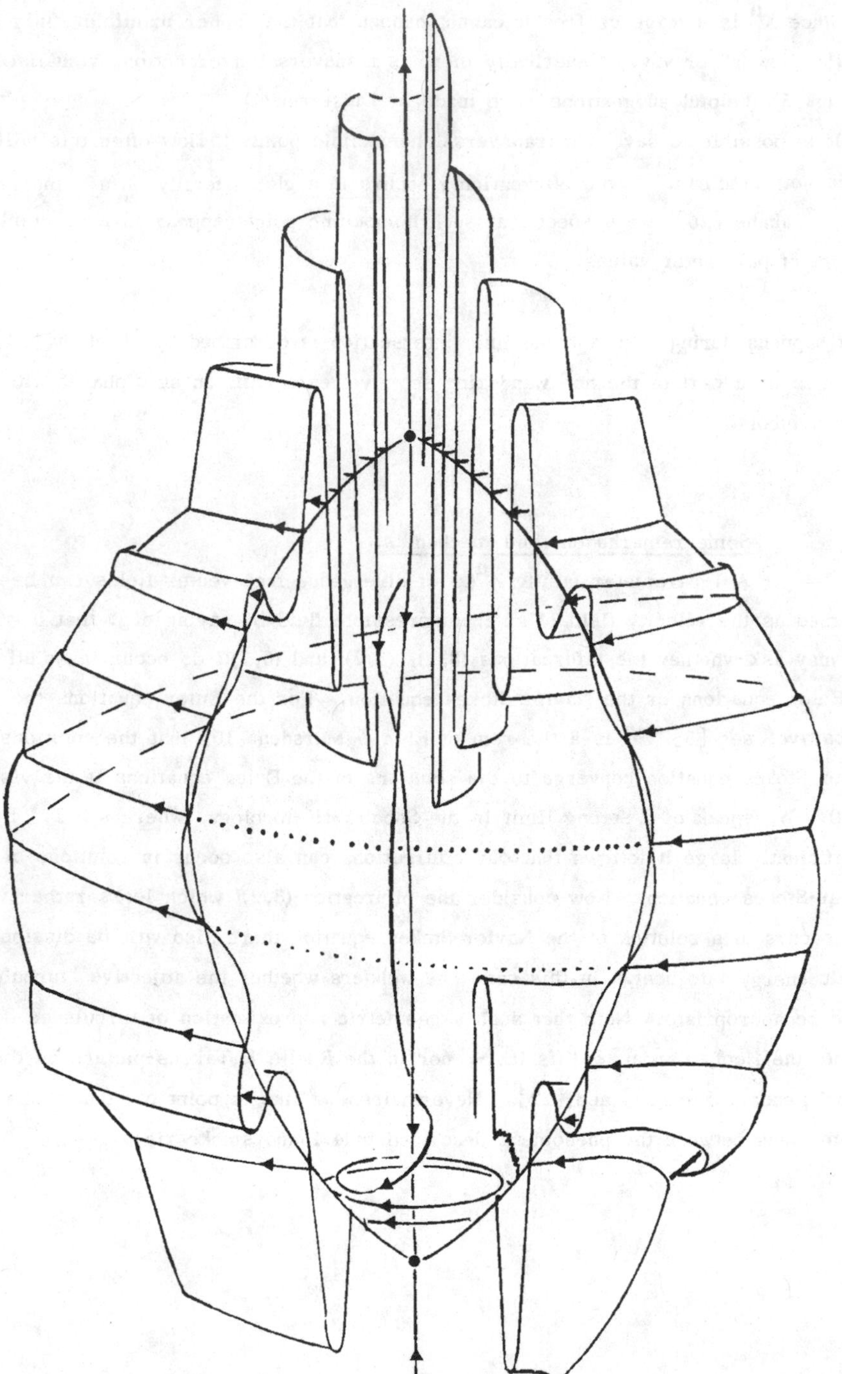

Figure 5.

## References.

1. Andronov, A.A., Leontovich, E.A., Gordon, I.I., & Maier, A.G. : Theory of bifurcations of dynamical systems on a plane. Israel Progr. for Sc. Transl. Ltd. 1971.

2. Arnold, V.I. : Lectures on bifurcations and versal families. In : Russ. Math. Surv. 27, 54-123 (1972).

3. Birkhoff, G.D. : Dynamical systems. In : AMS. Coll. Publ. 9 New York : Amer. Math. Soc. 1927. Reprinted 1966.

4. Borel, A. : Linear algebraic groups. (Notes by H. Bass). Benjamin 1979.

5. Broer, H.W. : Bifurcations of singularities in volume preserving vector fields. Ph.D. thesis, Groningen 1979.

6. Broer, H.W. : Quasi periodic flow near a codimension one singularity of a divergence free vector field in dimension three. This volume.

7. Brunovksy, P. : One parameter families of diffeomorphisms. In : Symposium on Differential Equations and Dynamical Systems, pp 29-33. Springer 1971.

8. Dumortier, F. : Singularities of vector fields. In : Monografias de Mathemàtica 32, IMPA.

9. Dumortier, F., Rodrigues, P.R., & Roussarie, R. : Germs of diffeomorphisms in the plane. To be published.

10. Ebin, D., & Marsden, J.E. : Groups of diffeomorphisms and the motion of an incompressible fluid. In : Ann. of Math. 92, 102-163 (1970).

11. Šil'nikov, L.P. : A contribution to the problem of the structure of an extended neighbourhood of a rough equilibrium state of saddle focus type. In : Math. USSR Sbornik, 10, 91-102 (1970).

12. Gérard, R., & Levelt, A.H.M. : Sur les connexions a singularités régulières dans le cas de plusiers variables. In : Funkcialaj Elevacioj 19, 149-173 (1976).

13. Hartman, P. : Ordinary differential equations. Wiley & Sons 1964.

14. Landau L.D., & Lifshitz, E.M. : Fluid mechanics, Pergamon 1959.

15. Mather, J.N. : Stratifications and mappings. In : Dynamical Systems, ed. M.M. Peixoto, pp. 195-232. Acad. Press 1973.

16. J. Moser. : Stable and random motions in dynamical systems. Princeton Univ. Press, 1973.

17. Narasimhan, R. : Analysis on real and complex manifolds. North Holland 1968.

18. Palis, J. & Takens, F. : Topological equivalence in normally hyperbolic dynamical systems. In : Topology 16, 335-345 (1977).

19. Poènaru, V. : Singularités $C^{\infty}$ en présence de symmétrie. Springer 1976.

20. Poincaré, H. : Thèse. In : Oeuvres I, pp. LIX-CXXIX (1879). Gauthiers Villars 1928.

21. Robinson, R.C. : Generic properties of conservative systems I, II. In : Amer. J. Math., 92, 562-603, 879-906 (1970).

22. Ruelle, D., & Takens, F. : On the nature of turbulence. In : Comm. Math. Phys. 20, 167-192 (1971).

23. Siegel, C.L. : Über die analytische Normalform analytischer Differentialgleichungen in der Nähe einer Gleichgewichtslösung. In : Nachr. Akad. Wiss. Göttingen, Math. Phys. Kl. 21-30 (1952).

24. Sotomayor, J. : Generic one-parameter families of vector fields on two-dimensional manifolds. In : Publ. Math. IHES 43, 5-46 (1973).

25. Sternberg, S. : On the structure of local homeomorphisms of Euclidean n-space III. In : Amer. J. Math. 81, 578-604 (1959).

26. Takens, F. : Homoclinic points in conservative systems. In : Inventiones Math. 18, 267-292 (1972).

27. Takens, F. : A nonstabilisable jet of a singularity of a vector field. In : Dynamica Systems, ed. M.M Peixoto, pp. 583-597, Acad. Press 1973.

28. Takens, F. : Forced oscillations and bifurcations. In : Applications of Global Analysis I, Comm. of the Math. Inst. Rijksuniversiteit Utrecht (1974).

29. Takens, F. : Singularities of vector fields. In : Publ. Math. IHES 43, 48-100 (1974).

30. Varadarajan, V.S. : Lie groups, Lie algebras and their representations. Prentice Hall 1974.

H.W. Broer : Department of Mathematics, Groningen University, P.O. Box 800, Groninger The Netherlands.

# Quasi Periodic Flow near a Codimension One Singularity of a Divergence Free Vector Field in Dimension Three*.

Henk Broer

It is the aim of this paper to show that within the class of smooth one parameter families of divergence free (or volume preserving) vector fields on $\mathbb{R}^3$, the phenomenon of invariant two-dimensional tori with a quasi-periodic flow has open occurence.

For this purpose we study a specific codimension one singularity in this class : a bifurcation that may appear in generic one parameter families of divergence free vector fields.

We shall consider such a generic unfolding and show that for the parameter in a neighbourhood of the bifurcation value, many such invariant 2-tori with a quasi-periodic flow come into existence.

For the proof of this fact we shall make use of Moser's twist mapping theorem, see e.g. [1,6] and of Rüssman [7]. For extra details we also refer to [4,5].

In [4] we made a general study of bifurcations of singularities in volume preserving vector fields, by investigating generic one parameter families of such vector fields. It appeared that only in the dimensions three and four the results are different from those in the bifurcation theory of ordinary - i.e. not necessarily volume preserving vector fields. See e.g. Arnold [2a]. The present study treats one of the bifurcations occuring in dimension three. It illustrates the fact that in this divergence free case topological stability is not a generic property.

In dimension four we meet a similar bifurcation. Our conjecture is, that analogous results can be obtained for this siutation. (See [4].)

I wish to express my gratitude to Floris Takens and Boele Braaksma : during the preparation of this paper discussions with them were very helpful.

---

* Partly this publication was made possible by the Netherlands Organisation for the Advancement of Pure Research (Z.W.O.).

## §1. Introduction.

Consider a divergence free vector field on $\mathbb{R}^3$, which has the origin as a singular point, where the eigenvalues of the linearised vector field are 0, $i\alpha$ and $-i\alpha$, for some $\alpha > 0$. One easily sees that such singularities have codimension one, i.e. that they may occur in generic one parameter families of divergence free vector fields : in the divergence free case the trace of the linearised part in a singularity must be zero. So let us consider such a generic one parameter family $X = X^{\mu}(x)$, where $\mu$ is a real parameter, which unfolds the above singularity $x = 0 \in \mathbb{R}^3$ of the vector field $X^0$. In [4] we proved a normal form theorem, which states that up to a volume preserving, $\mu$-dependent change of coordinates and some rescaling of time, we may write $X^{\mu} = \tilde{X}^{\mu} + p^{\mu}$, where

i. Both $\tilde{X}^{\mu}$ and $p^{\mu}$ are divergence free,

ii. $p = p^{\mu}(x)$ is flat in $(x,\mu) = (0,0) \in \mathbb{R}^3 \times \mathbb{R}$,

iii. In cylindrical coordinates $r, \varphi$ and $z$ the vector field $\tilde{X}^{\mu}$ has the form

$$\begin{cases} \dot{\varphi} = 1 \\ \dot{r} = r.g(r^2, z, \mu) \\ \dot{z} = h(r^2, z, \mu) , \end{cases}$$

which expresses rotational symmetry.

Moreover :

$$\frac{\partial h}{\partial \mu}(0,0,0) = 1 \text{ and } g(0,0,0) = h(0,0,0) = \frac{\partial h}{\partial z}(0,0,0) = 0 \ .$$

Forgetting the angle $\varphi$, from $\tilde{X}^{\mu}$ we obtain a reduced vector field $\bar{X}^{\mu}$, defined on the $(r,z)$-plane. In this plane we rescale, or blow up, with $\sqrt{|\mu|}$, i.e. for $\mu \neq 0$ we introduce new variables $\bar{r}$ and $\bar{z}$, defined by

$$r = \bar{r}\sqrt{|\mu|} \text{ and } z = \bar{z}\sqrt{|\mu|}$$

and consider

$$Z^{\mu}(\bar{r},\bar{z}) = \frac{1}{\sqrt{|\mu|}} \bar{X}^{\mu}(\bar{r}\sqrt{|\mu|}, \bar{z}\sqrt{|\mu|}) \ .$$

Replacing $\bar{r}$ and $\bar{z}$ by $r$ and $z$ respectively, we write for $Z^{\mu}$ $(\mu \neq 0)$ :

$$\begin{cases} \dot{r} = arz + O(\sqrt{|\mu|}) \\ \dot{z} = br^2 - az^2 - \text{sgn}\{\mu\} + O(\sqrt{|\mu|}), \end{cases}$$

an evident correspondence between closed orbits of the vector fields and periodic points of the Poincaré maps. Let $\tilde{\Phi}^\mu$ be the Poincaré map belonging to $\tilde{X}^\mu$, then

$$\tilde{\Phi}^\mu = \tilde{X}^\mu_{2\pi} = \bar{X}^\mu_{2\pi} \quad .$$

(Where for an arbitrary vector field Y we denote the time t flow by $Y_t$.) Also consider the blown up version of this map :

$$\tilde{\alpha}^\mu = Z^\mu_{2\pi\sqrt{\mu}} \quad .$$

Note that the right hand side of this equality is also defined for $\mu = 0$. Accordingly define $\tilde{\alpha}^0$ to be the identity map.

If one attempts to define a similar Poincaré map $\Phi$ for X, then problems arise from the lack of symmetry near the z-axis. These merely technical problems can be easily overcome by adjusting $X^\mu$ slightly in an arbitrarily small neighbourhood of the z-axis, there making it symmetric. This adjustment avoids as large a region of $\tilde{X}^\mu$-invariant tori, nested around the elliptic closed orbit, as one wishes. For more details see [5]. So from now on $\Phi^\mu$ is the Poincaré map for $X^\mu$ and $\alpha^\mu$ is its blown up version. Note that $\alpha - \tilde{\alpha}$ is flat for $\mu = 0$, uniformly for (r,z) in an arbitrarily large compact neighbourhood of (0,0).

We observe that in the region relevant to us, both $\tilde{\Phi}$ and $\Phi$ (and therefore also $\tilde{\alpha}$ and $\alpha$) preserve a 2-form. This is a consequence of the volume preservingness of $\tilde{X}$ and X, as can be seen using Stokes' theorem. If $\Omega$ denotes the standard volume 3-form on $\mathbb{R}^3$, then $\tilde{\Phi}$ and $\tilde{\alpha}$ both preserve $\iota_{\tilde{X}}\Omega|_{\{\varphi=0\}} = r dr \wedge dz$, while $\Phi$ and $\alpha$ both preserve $\iota_X\Omega_{\{\varphi=0\}}$, which is very close to $r dr \wedge dz$. In the region under consideration the 2-form $\omega = r dr \wedge dz$ is not degenerate and hence defines a volume (or rather area), which in our 2-dimensional case also gives a symplectic structure. It would be preferable to us, if $\Phi$ and $\tilde{\Phi}$ both preserved the same area. This can be arranged by adjusting $\Phi$ slightly with a change of coordinate $\Gamma$: $\Gamma$ is close to the identity map and $\Gamma \circ \Phi \circ \Gamma^{-1}$ also preserves $\omega$. See [5] for more details. Write again $\Phi$ instead of $\Gamma \circ \Phi \circ \Gamma^{-1}$. We end these technical preliminaries by passing to coordinates $(\xi, \eta)$ such that $\omega = d\xi \wedge d\eta$ : the standard area, and such that modulo reparametrisation $Z^0$ has the form

$$\begin{cases} \dot{\xi} = 2\eta + 2\xi\eta \\ \dot{\eta} = -2\xi - \eta^2 \end{cases} ,$$

uniformly on compacta, where a and b are real constants. We put a generic condition on the second order terms of $X^0$ by requiring that $a \neq 0$ and $b \neq 0$. It is no essential restriction to assume that $b \geq 0$, otherwise replace z by -z.

The present study deals with the case where $a < 0$ and $b > 0$ (which in [4] was labelled (3,2) I). Furthermore we restrict ourselves to the situation where $\mu > 0$ : the parameter $\mu$ varies in a right hand neighbourhood of 0.

Define $Z^0$ to be the limit for $\mu \downarrow 0$ and observe that the family $\{Z^\mu\}_{\mu \geq 0}$ is smoothl parametrised by $\sqrt{\mu}$.

According to [4], theorem 3.2, this vector field $Z^0$ is $C^\infty$-stable within the class of all such reductions of symmetric, divergence free vector fields. Figure 1 below depicts the phase portrait of $Z^0$. (Cf. [4], fig.1, case I,-.)

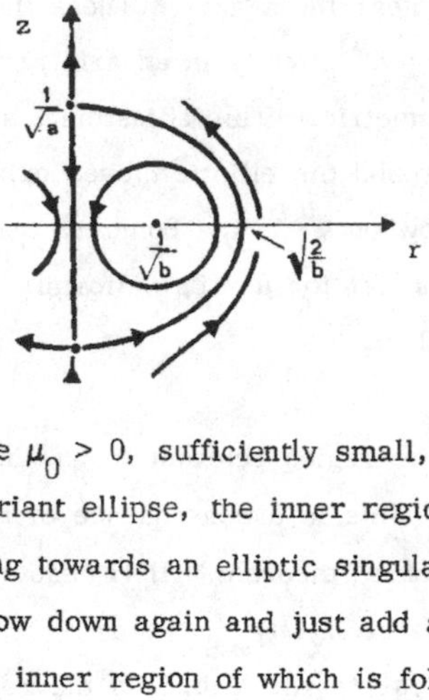


Figure 1.

This stability means that for some $\mu_0 > 0$, sufficiently small, and for all $0 \leq \mu < \mu_0$ the vector field $Z^\mu$ possesses an invariant ellipse, the inner region of which is foliated by a cylinder of closed orbits, shrinking towards an elliptic singularity. For the corresponding vector field $\tilde{X}^\mu$ (now $\mu \neq 0$) we blow down again and just add a rotational component. Thu we find an invariant ellipsoid, the inner region of which is foliated by a one parameter family of invariant 2-tori, shrinking towards an elliptic closed orbit. Note that, because of the blow down operation, all interesting phenomena occur asymptotically at a distance $\sqrt{\mu}$ from the origin. In the region under consideration the symmetric vector field $\tilde{X}^\mu$ defines a "completely integrable" system and the flat term p, for $\mu \geq 0$ and small, is a small perturbation. In general this perturbation destroys the symmetry.

We shall study the original system $X = \tilde{X} + p$ by investigating the first return Poincaré maps, belonging to $\tilde{X}$ and X, from the plane $\varphi = 0$ onto itself. Invariant tori of the vector fields correspond to invariant circles of these Poincaré maps. Also there is

with a Hamiltonian function $H_0(\xi,\eta) = \xi^2 + \eta^2 + \xi\eta^2$. See figure 2.

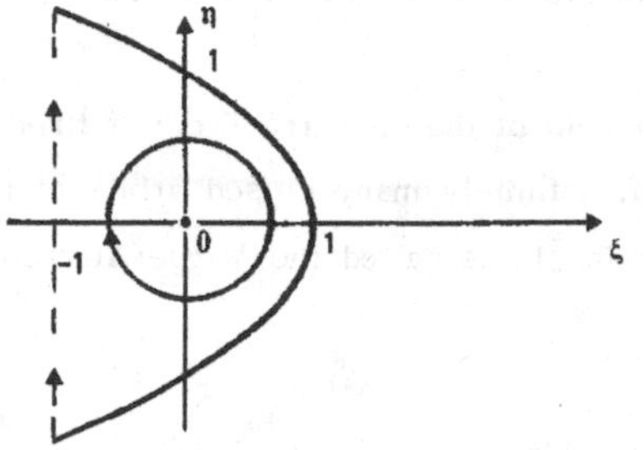


Figure 2.

For $\mu \geq 0$ the vector field $Z^\mu$ has a Hamiltonian $H_\mu$, which for $\mu \geq 0$ and sufficiently small, has a unique minimum in a point denoted by $\tilde{q}_\mu$ : the elliptic singularity of $Z^\mu$, for $\mu \neq 0$ also the elliptic fixed point of $\tilde{\alpha}^\mu$. We may and do assume that $H_\mu(\tilde{q}_\mu) = 0$. Note that $\tilde{q}_0 = (0,0)$ and also that for some $\varkappa < 0$ :

$$\tilde{\alpha}^\mu = Z^\mu_{\varkappa\sqrt{\mu}} .$$

We conclude this introduction by giving a brief sketch of what follows in the sections 2 and 3 .

In section 2 a neighbourhood of the elliptic fixed point $\tilde{q}_\mu$ will be treated. We shall prove that in a sufficiently small neighbourhood of $\tilde{q}_\mu$, the perturbed diffeomorphism $\alpha^\mu$ has a unique fixed point $q_\mu$, which again is elliptic. From the explicit form in which $H_0$ is given, we can compute the first two terms of the Birkhoff normal form for $\alpha^\mu$ in $q_\mu$, and hence conclude that for positive, but small $\mu$ $\alpha^\mu$ in $q_\mu$ is a twist mapping. Moser's twist mapping theorem then yields that $\alpha^\mu$ possesses very many invariant circles, with a rotation number which, divided by $2\pi$, is badly approximated by rational numbers. It is easy to translate these results back to the three dimensional situation for the vector field $X^\mu$.

In section 3 we study an arbitrary compact annulus, invariant for the unperturbed map $\tilde{\alpha}^\mu$. We shall prove that for $\mu > 0$ and sufficiently small, the frequency of the closed orbit of $Z^\mu$ decreases monotonically with the level of $H_\mu$, which provides us with a kind of action-angle variable. Then, on our annulus we apply Rüssman [7]. Here we have to take account of the presence of our parameter $\mu$, which is problematic. We have to repeat Rüssmann's results "with a parameter". With help of Liouville's theorem on the approximation of algebraic numbers by rationals, we construct for each small, positive $\mu$ a countable set of $\alpha^\mu$-invariant circles : the algebraic numbers (multiplied with $2\pi$) being the rotation numbers of these invariant circles.

Moreover, with the help of the Poincaré-Birkhoff fixed point theorem one finds, in between the invariant tori, infinitely many closed orbits of large period : the well known phenomena of what in [1] is called the Vague Attractor of Kolmogorov.

## §2. Close to the elliptic orbit.

In this section we restrict ourselves to a small neighbourhood of the elliptic fixed point $\tilde{q}_\mu$ ($\mu > 0$ and small). We begin by stating the results of our considerations. Let $\tilde{F} \subseteq \mathbb{R}^2 \times \mathbb{R}$ be defined as $\tilde{F} = \{(\tilde{q}_\mu, \mu) | \mu \geq 0 \text{ and small }\}$, then

**2.1 Theorem.**

For $\mu > 0$ and sufficiently small, a neighbourhood of $\tilde{q}_\mu$ in $\mathbb{R}^2$ exists, such that within this neighbourhood the diffeomorphism $\alpha^\mu$ has exactly one fixed point $q_\mu$. This fixed point is elliptic. Moreover, if $F = \{(q_\mu, \mu) | \mu > 0 \text{ and small}\} \cup \{(0,0,0)\}$, then $F$ and $\tilde{F}$ in $(\xi, \eta, \mu) = (0,0,0)$ have an infinitely high order on contact. Also

**2.2 Theorem.**

For $\mu > 0$ and sufficiently small, $\alpha^\mu$ in its elliptic fixed point $q_\mu$ is a twist mapping (see [1, 6]).

For our 1-parameter family X it follows

**2.3 Corollary.**

For $\mu > 0$ and sufficiently small

(i) There exists a thin solid torus which is a neighbourhood of the elliptic orbit of $\tilde{X}^\mu$ such that within this neighbourhood $X^\mu$ has exactly one closed orbit $\gamma_\mu$: the period of $\gamma_\mu$, which is elliptic, is approximately $2\pi$.

(ii) In every neighbourhood of $\gamma_\mu$ there is an $X^\mu$-invariant 2-torus on which the flow is quasi periodic.

(iii) For all $\varepsilon > 0$ there is a $\delta > 0$ such that the union of these invariant 2-tori in a $\delta$-neighbourhood of $\gamma_\mu$ has a measure greater than $(1-\varepsilon) \times$ volume of this neighbourhood.

(iv) Take any two of the invariant tori from (ii). Then there exists an interval $J \subseteq \mathbb{R}$

such that for each $p/q \in J$, in between these tori, there is a periodic orbit of $X^{\mu}$ with approximate period $2\pi q$.

The corollary follows from the theorems with help of Moser's twist mapping theorem and the Poincaré-Birkhoff fixed point theorem. See e.g. [1,6]. Theorem 2.1 can be proved using an index argument : note that $d\tilde{\alpha}^{\mu}(\tilde{q}_{\mu})$ is a rotation over an angle asymptotically proportional to $\sqrt{\mu}$, while $\alpha - \tilde{\alpha}$ is flat for $\mu = 0$, i.e. behaves like $\mu^{\infty} \ll \sqrt{\mu}$ for $0 < \mu \ll 1$. A proof of theorem 2.2 runs along the following lines : For $\mu \geq 0$ Birkhoff normal form coordinates x and y can be obtained by a $\mu$-dependent canonical transformation $(\xi,\eta) \to (x(\xi,\eta,\mu), y(\xi,\eta,\mu))$ such that, if $x = r\cos\theta$ and $y = r\sin\theta$, then

$$H_{\mu}(r,\theta) = \beta_1(\mu)r^2 + \beta_2(\mu)r^4 + O(r^6)$$

The functions $\beta_1(\mu)$ and $\beta_2(\mu)$ are smooth. Note that $\beta_1(\mu) = 1 + O(\sqrt{\mu})$ and that $x(\tilde{q}_{\mu},\mu) = y(\tilde{q}_{\mu},\mu) = 0$ . For $Z^{\mu}$ we have in the polar coordinates r and $\theta$ :

$$\begin{cases} \dot{r} = O(r^5) \\ \dot{\theta} = 2\beta_1(\mu) + 4\beta_2(\mu)r^2 + O(r^4) , \end{cases}$$

such that

$$\tilde{\alpha}^{\mu}(r,\theta) = (r, \theta + 2\varkappa\beta_1(\mu)\sqrt{\mu} + 4\varkappa\beta_2(\mu)r^2\sqrt{\mu}) + \sqrt{\mu}.O(r^4) \quad .$$

We prove theorem 2.2 by showing that $\tilde{\alpha}^{\mu}$, for $\mu > 0$ and small, is a twist mapping and again using that for $0 < \mu \ll 1$ also $\mu^{\infty} \ll \sqrt{\mu}$. In order to prove that $\tilde{\alpha}^{\mu}$ is a twist mapping we need to control $2\varkappa\beta_1(\mu)\sqrt{\mu}$ and the twist coefficient $4\varkappa\beta_2(\mu)\sqrt{\mu}$. First observe that for $\mu > 0$ and small enough, the first of these coefficients is not an integral multiple of $0, \frac{\pi}{2}$ or $\frac{2\pi}{3}$ . Then the following proposition yields that $\tilde{\alpha}^{\mu}$ ($\mu > 0$, small) is a twist mapping. See [1,6]. For more details see [5].

2.4 <u>Proposition.</u> $\qquad \beta_2(0) \neq 0$ .

<u>Proof</u> : For $z \in (0,1)$ the set $H_0^{-1}(z)$ contains a closed orbit of $Z^0$. Let $A(z)$ be the area of the disc $H_0^{-1}([0,z])$, then

$$A(z) = 2\int_{-\sqrt{z}}^{\sqrt{z}} \sqrt{\frac{z-\xi^2}{1+\xi}}\,.d\xi \quad .$$

Of course $A(0) = 0$. We differentiate $A$, realising that the integrand vanishes for $\xi = \pm\sqrt{z}$. Substitute $\xi = \eta\sqrt{z}$ :

$$A'(z) = \int_{-1}^{+1} \frac{d\eta}{\sqrt{(1+\eta\sqrt{z})(1-\eta^2)}} \quad .$$

It follows that $A'(0) = \pi$. Differentiating once more we find :

$$A''(z) = -\frac{1}{4\sqrt{z}} \int_{-1}^{+1} \frac{\eta d\eta}{(1+\eta\sqrt{z})^{\frac{3}{2}}(1-\eta^2)^{\frac{1}{2}}} \quad ,$$

such that $A''(0) = -\frac{1}{4}f'(0)$ where

$$f(z) = \int_{-1}^{+1} \frac{\eta d\eta}{(1+\eta z)^{\frac{3}{2}}(1-\eta^2)^{\frac{1}{2}}} \quad .$$

It is easy to see that

$$f'(z) = -\frac{3}{2}\int_{-1}^{+1} \frac{\eta^2 d\eta}{(1+\eta z)^{\frac{5}{2}}(1-\eta^2)^{\frac{1}{2}}}$$

such that

$$f'(0) = -\frac{3}{2}\int_{-1}^{+1} \frac{\eta^2 d\eta}{\sqrt{1-\eta^2}} = -\frac{3\pi}{4} \quad .$$

So $A''(0) = \frac{3}{16}\pi$ and $A(z) = \pi z + \frac{3}{32}\pi z^2 + O(z^3)$ . On the other hand, since $\beta_1(0) = 1$, for $(r,\theta) \in H_0^{-1}(z)$ we have

$$z = r^2 + \beta_2(0)r^4 + O(r^6), \qquad \text{such that}$$

$$r^2 = z - \beta_2(0)z^2 + O(z^3) \quad .$$

Then $$A(z) = \frac{1}{2}\int_{H_0(r,\theta) = z} r^2 d\theta = \pi z - \pi\beta_2(0)z^2 + O(z^3) \; .$$

Comparing the two expansions immediately yields that $\beta_2(0) = -\frac{3}{32}$ QED.

§3. Further from the elliptic orbit.

Define the real number $c_\mu$ by saying that $H_\mu^{-1}(c_\mu)$ contains the saddles of the vector field $Z^\mu$. Note that $c_0 = 1$ . In this section we consider the annulus $H_\mu^{-1}(0, c_\mu)$, containing all the invariant circles of $Z^\mu$, $\mu \geq 0$ and small. First we shall construct cylinder coordinates $(r, \theta, \mu)$ such that Z is of the form

$$(*) \begin{cases} \dot{r} = 0 \\ \dot{\theta} = 2\pi r \\ (\dot{\mu} = 0, \ \mu \text{ being the parameter}) \end{cases}$$

It will appear that a good choice for r is that, where $r(\xi, \eta, \mu)$ denotes the frequency of the closed orbit $Z^\mu$ through the point $(\xi, \eta)$. The polar angle $\theta$ then can be defined using the time parameterisation of these closed orbits. The only thing we have to show is that in this way we have defined coordinates. To see this for $0 < z < c_\mu$ let $P_\mu(z)$ denote the period of the closed orbit in $H_\mu^{-1}(z)$, then $r(\xi, \eta, \mu) = \frac{1}{P_\mu H_\mu(\xi, \eta)}$ . Note that

$\lim_{z \downarrow 0} P_\mu(z) = \frac{\pi}{\beta_1(\mu)}$ , see §2.

3.1 Proposition.

$P_0 : (0, 1) \to \mathbb{R}$ is strictly increasing, while $\lim_{z \uparrow 1} P_0(z) = \infty$ .

Proof: As in the proof of 2.4 let A(z) be the area of the disc $H_0^{-1}([0, z])$ then it follows from e.g. Arnold [2] that $P_0(z) = \frac{d}{dz} A_0(z)$, such that

$$P_0(z) = \int_{-1}^{+1} \frac{d\eta}{\sqrt{(1+\eta\sqrt{z})(1-\eta^2)}} .$$

From this the proof is straight forward. QED.

Now take any compact interval $I = [a, b] \subseteq (0, \frac{1}{\pi})$. The above proposition together with the implicit function theorem then yields a constant $\mu_I > 0$ such that the above construction of cylinder coordinates works on $C_I = \{(\xi, \eta, \mu) | 0 \leq \mu \leq \mu_I \text{ and } r(\xi, \eta, \mu) \in I\}$. In other words : for $0 \leq \mu \leq \mu_I$ on the annulus $a \leq r \leq b$ the vector field $Z^\mu$ is defined by (*). Note that $\tilde{\alpha}^\mu$ is given by $(r, \theta) \mapsto (r, \theta + 2\pi \varkappa r/\mu)$. We conclude the preliminaries of this section by blowing down again. Write r instead of $-2\pi\varkappa r/\mu$ etc. Then for $0 \le \mu \le \mu_I$ the diffeomorphisms $\tilde{\Phi}^\mu$ and $\Phi^\mu$ on the annulus $a/\mu \le r \le b/\mu$ have the form

$$\tilde{\Phi}^\mu(r,\theta) = (r, \theta + r) \text{ and } \Phi^\mu(r,\theta) = \tilde{\Phi}^\mu(r,\theta) + (g(r,\theta,\mu), f(r,\theta,\mu)) \quad ,$$

f and g having period $2\pi$ in $\theta$ and being flat for $r = 0$ and $\mu = 0$.

The rest of this paper consists of an application of Rüssmann [7] to this situation. This is problematic because of the $\mu$-dependency. Consider the following inequalities

$$\text{(A)} \qquad a/\mu + \gamma \le r \le b/\mu - \gamma$$

$$\text{(B)} \qquad \forall k \in \mathbb{N}.\ \forall \ell \in \mathbb{Z}\ .\ |rk - 2\pi\ell| \ge \gamma k^{-\frac{5}{4}} \quad ,$$

where $\gamma$ is a parameter, $0 < \gamma < 1$ . In [7] it is proved that $\delta_0 > 0$ exists, depending on $\gamma$, such that for

$$\text{(C)} \qquad |f(r,\theta,\mu)| + |g(r,\theta,\mu)| < \delta, \quad \delta < \delta_0$$

-up to some conditions - in the neighbourhood of every solution $r = r_0$ of (A) and (B), the diffeomorphism $\Phi^\mu$ has an invariant circle with rotation number $r_0$. We consider the following two questions :

(i) Can $\gamma = \gamma(\mu)$ be chosen in such a way that for every $\mu > 0$ (small) the inequalities (A) and (B) have solutions $r = r_0$ - depending on $\mu$ - ?

(ii) How in [7] does $\delta_0$ depend on $\gamma$ ?

We conclude this section by giving answers to these questions and deriving some consequences from these answers. Proofs will be presented in the next section. Question (i) can be answered using Liouville's theorem on algebraic numbers. See e.g. Schneider [8].

3.2 <u>Lemma.</u>

For each $R \ge 1$ a constant $\mu' = \mu'(I,R)$ exists, such that by choosing $\gamma(\mu) = \mu^R$, for all $\mu \in (0,\mu')$ there is a set of solutions of (A) and (B), which is dense in $[a/\mu + \gamma(\mu), b/\mu - \gamma(\mu)]$.

Question (ii) requires an exploration of [7]. We found

3.3 Lemma.

In Rüssmann [7], theorem 1, the conclusion holds for sufficiently small $\gamma > 0$ if one takes $\delta_0(\gamma) = \gamma^{5740}$ .

An immediate consequence from the above lemmas :

3.4 Theorem.

A constant $\mu'' = \mu''(I, R)$ exists, such that for all $0 < \mu \leq \mu''$ and each solution $r = r_0$, as in 3.2, the diffeomorphism $\Phi^{\mu}$ has an invariant circle with rotation number $r_0$.

Summarising we reformulate these results for the 1-parameter family X. Blow down $C_I$ and rotate around the z-axis. The area of revolution is denoted by $T_I$.

3.5 Corollary.

(i) For $0 < \mu \leq \mu''$ the vector field $X^{\mu}$ has at least countably many invariant 2-tori in $T_I$. The flow on these tori is quasi periodic.

(ii) For any two of the tori from (i) a real interval exists, such that for each p/q in this interval, $X^{\mu}$ has a closed orbit of approximate period $2\pi q$ in between these two tori.

The last statement again uses the Poincaré-Birkhoff fixed point theorem, see [6]. Compare corollary 2.3.

§4. Proofs of the lemmas 3.2 and 3.3, the theorems of Liouville and Rüssmann.

We refer to [5] for more details. First consider lemma 3.2. According to Liouville's theorem for an algebraic number $\alpha$ of degree s a constant $c_\alpha > 0$ exists such that for all $p \in \mathbb{Z}$ and $q \in \mathbb{N}$ :

$$|\alpha - \frac{p}{q}| \geq \frac{c_\alpha}{q^s} . \text{ See [8].}$$

Our interest is for the case $s = 2$ : we consider positive real numbers $\alpha$ of the form $\alpha = \frac{h}{n}\sqrt{p}\text{-}1$ ($h, n, p \in \mathbb{N}$, p prime and $(n, p) = 1$) and in the inequalities (A) and (B) from §3 we try solutions $r_0 = 2\pi\alpha$ .

Retracing the proof in [8] we find that for such $\alpha$ we may take $c_\alpha = \frac{1}{4hn\sqrt{p}}$. From this we conclude :

4.1 Proposition.

For $\mu > 0$ consider the inequalities (A) and (B) from §3. Take $\alpha = \frac{h}{h}\sqrt{p}-1$ as above. Suppose that $\mu$ is sufficiently small and that $n \geq \frac{6\pi\sqrt{p}}{b-a} \cdot \frac{1}{\sqrt{\mu}}$. Then $h \in \mathbb{N}$ exists such that for the corresponding $\alpha$ the inequalities (A) and (B) are satisfied by $r_0 = 2\pi\alpha$ where $\gamma = \frac{\pi}{4n^2}$.

Proof: Let $c_\alpha$ be as above.

If $|\alpha - \frac{p}{q}| \geq c_\alpha q^{-2}$ then for $r_0 = 2\pi\alpha$ we have $|r_0 q - 2\pi p| \geq 2\pi c_\alpha q^{-1} \geq 2\pi c_\alpha q^{-\frac{5}{4}}$.
Compare (B), §3. Our assumptions imply that $2\pi c_\alpha \geq \frac{\pi}{4n^2} = \gamma$ for $\mu$ sufficiently small. The rest of the proof is straightforward. (See [5].) QED.

The prime number p is kept fixed.

4.2 Remarks.

i) For $R \geq 1$ it is possible to choose $n \sim \mu^{-\frac{R}{2}}$. By the above proposition then $\gamma(\mu) \sim \mu^R$.

ii) Evidently, if $r_0$ satisfies (A) and (B), so does $r_0 + 2\pi\nu$ for every $\nu \in Q$, if only $r_0 + 2\pi\nu$ satisfies (A), i.e. if

$$a\sqrt{\mu} + \gamma(\mu) \leq r_0 + 2\pi\nu \leq b\sqrt{\mu} - \gamma(\mu) .$$

Now the proof of lemma 3.2 is straight forward.

Secondly we come to lemma 3.3. Since it is quite difficult to give a concise presentation of the considerations made in Rüssmann [7], writing this I have assumed the reader to have a copy of [7] in front of him. The pages in [7] will be referred to as R[i] if [i] is the bottom number of the page in question. A formula which in [7] is numbered i we refer to as R(i), etc. etc. The namegiving of the variables in §3 corresponds to that in [7], the only difference is the $\mu$-dependency : Rüssmann considers an annulus $a \leq r \leq b$ and perturbation functions $f(r,\theta)$, $g(r,\theta)$ not depending on $\mu$. Also, instead of $\tilde{\Phi}(r,\theta) = (r, \theta + r)$ his unperturbed map is $(r,\theta) \mapsto (r, \theta + h(r))$, where for some $K \geq 1$ we have : $K^{-1} \leq \frac{dh}{dr}(r) \leq K$ ................ R(4). We take $K = 1$. (A), (B) and (C) correspond to R(6), R(7) and R(3) respectively. As was indicated in §3, R(Satz 1) states that a constant $\delta_0 = \delta_0(\gamma, K) = \delta_0(\gamma)$ exists, such that R(3) with $\delta < \delta_0$ implies that for every $r_0$ satisfying R(6) and R(7), a closed curve R(8) is found. This curve by $\Phi$ is mapped to a radially translated curve R(9). Since in our case $\Phi$ has to preserve some volume the radial translation must be zero : the curve R(8) is invariant. We proceed by briefly describing the successive parts of Rüssmann's proof, stopping at the points which are of interest to us. For more details see [5].

In §R2 "Vereinfachende Überlegungen" are made. Constants $C = C(K)$ and $c = c(\gamma, K)$ are introduced. In R(Staz 2) the existence of $\delta_1 = \delta_1(\gamma, K, C, c)$ is claimed with certain properties. It is proved that R(Satz 2) implies R(Satz 1) by taking $\delta_0(\gamma, K) = \delta_1(\gamma, K, C(K), c(\gamma, K))$. Since $K = 1$ the constant $C(K)$ does not bother us and it appears that in our case we may choose $c(\gamma, 1) = 1$. The parameter $\gamma$ has to be taken sufficiently small.

In §§R3 - 7 technicalities are developed which are independent of the value of $\gamma$.

In §R8 "Eine Lineare Differenzengleichung" (for a technical detail see §R9) the important R(Satz 6) is proved. Here the small denominator problem is solved. Our special interest is for the estimate R(75).

In §R10 "Implizite Funktionen" once again the techniques are independent of $\gamma$.

In §R11 "Linearisierung" an iteration step is developed for the Newtonian process to be used in the proof of R(Satz 2). In R(Satz 8) it is proved that after such a step, which is a change of coordinates applied to a map like $\Phi$, we obtain a similar map with much smaller perturbation functions than $\Phi$. Here R(Satz 6) is used fundamentally. The constant $\mu, \beta, \sigma$ and $\varkappa$ obeying R(106) are chosen according to R(115) :

$$\mu = 4.9; \quad \beta = 1.25; \quad \sigma = 3.6 \quad \text{and} \quad \varkappa = 1 + 2^{-7} \quad \ldots\ldots \quad (*)$$

In R(Satz 8) the existence of $\varepsilon^* = \varepsilon^*(\gamma) \in (0,1)$ is claimed such that R(101-106) imply R(107-114) .
We state that, for the choice (*) of $\mu, \beta, \sigma$ and $\varkappa$, and for $\gamma$ sufficiently small, R(Satz 8) is valid for $\varepsilon^*(\gamma) = \gamma^{41}$ .

For a proof see [5].

In §§R12-13 "Der k-te Schritt des Newtonschen Iterationsverfahren" and "Der Grenzübergang" the proof of R(Satz 2) is given. R(Satz 8) is used essentially. Rüssmann has to find $\delta_1 = \delta_1(\gamma, K, C, c)$ such that for $0 < \delta < \delta_1$ the conclusions of R(Satz 2) are valid. Depending on $\gamma, K, C, c, \beta, \sigma, \mu$ and $\varkappa$ new constants $c_1, \ldots\ldots, c_{13}$ are introduced in §R12 and $c_{14} = c_{14}(K)$ in §R13 such that $\delta_1 = \min\{c_1, \ldots\ldots, c_{14}\}$ suffices.

The constants $\beta, \sigma, \mu$ and $\varkappa$ are determined in (*) and do not involve $\gamma$. The same holds for $C = C(K)$ and $c_{14} = c_{14}(K)$, since $K = 1$. Also recall that $c(\gamma, 1) = 1$. Remain the choices of $c_1, c_2, \ldots, c_{13}$. We state that for $\gamma > 0$, sufficiently small, the following choices work :

$$c_1 = \frac{1}{3}, \quad c_2 = (1 + 2 \times 3^8 \times C)^{-1}, \quad c_6 = \frac{1}{2} ,$$

$$c_4 = \left(\frac{c_2}{4}\right)^{\frac{20}{17}}, \quad c_5 = 8^{-2^7 \times \frac{10}{49}}, \quad c_7 = 14^{-2^7}, \quad c_9 = 7^{-2.55} ,$$

$$c_{10} = 2^{-2^7}, \quad c_{11} = 2^{-2^6}, \quad c_{12} = \left(\frac{4}{5}\right)^{2^7 \times \frac{5}{18}}, \quad c_{13} = 3^{-2^{14} \times \frac{10}{49}} ,$$

$$c_3 = \min\{\gamma^{41}, (4 \times 3^{16} \times C)^{-1}\} , \quad c_8 = c_3^{20} .$$

For a proof see [5].

Lemma 3.3 now finally follows from the fact that $41 \times 20 \times 7 = 5740$.

## References.


1. Abraham, R., & Marsden, J.E., Foundations of mechanics, Benjamin/Cummings 1978.

2. Arnold, V.I., Méthodes mathématiques de la mécanique classique, MIR 1976.

2a. Arnold, V.I., Lectures on bifurcations and versal families, Russian Math. Surveys 27, 54-123 (1972).

3. Arnold, V.I., & Avez, A., Problèmes ergodiques de la mécanique classique, Gauthier-Villars 1967.

4. Broer, H.W., Formal normal form theorems for vector fields and some consequences for bifurcations in the volume preserving case. This volume.

5. Broer, H.W., Bifurcations of singularities in volume preserving vector fields. Ph.D. thesis, Groningen 1979.

6. Moser, J., Lectures on Hamiltonian systems, Memoirs of the AMS. 81 (1968).

7. Rüssmann, H., Über invariante Kurven differentierbarer Abbildungen eines Kreisringes, Nachr. Akad. Wiss., Gottingen II, Math. Phys. Kl., 67-105 (1970).

8. Schneider, Th., Einführung in die transzendenten Zahlen, Springer 1957.

--------------------

Added in proof :

The conjecture concerning the four dimensional analogue has been proved by B.L.J. Braaksma and H.W. Broer. The results will be published elsewhere.

--------------------

H.W. Broer : Department of Mathematics, Groningen University, P.O. Box 800, 9700 AV Groningen, The Netherlands.

# A $C^2$ Kupka-Smale Diffeomorphism of the Disk With No Sources or Sinks.

John Franks & Lai-Sang Young.

In 1968 Smale [S] raised the question of whether or not there exists a Kupka-Smale diffeomorphism of $S^2$ with neither sources nor sinks. An example of a $C^1$ diffeomorphism which is Kupka-Smale and has no sources or sinks was given in [B-F]. This example has certain similarities to the Denjoy example of a diffeomorphism of $S^1$. For example, in both cases most orbits (all for the Denjoy diffeomorphism) have as $\omega$-limit set a Cantor set which is a minimal set of the map. This raises the question whether Kupka-Smale diffeomorphisms of the disk or $S^2$ with no sinks or sources are necessarily only $C^1$ like the Denjoy example or can be further smoothed.

In this note we give an example of a $C^2$ Kupka-Smale diffeomorphism of $D^2$ or $S^2$ with no sources or sinks. It can, in fact, be constructed to be topologically conjugate to the $C^1$ example of [B-F]. The example given here is definitely not $C^3$ (though it is $C^{2+\varepsilon}$ for some $\varepsilon > 0$), and it is the feeling of the authors that it is not topologically conjugate to a $C^3$ map.

Recall that a diffeomorphism $f:M \to M$ is called <u>Kupka-Smale</u> if every periodic point p is hyperbolic (i.e., $df^n_p : TM_p \to TM_p$, n = period of p has no eigenvalues of absolute value one) and for any periodic points p, q the stable manifold $W^s(p)$ intersects the unstable manifold $W^u(q)$ transversely.

(1.1) <u>Theorem.</u> <u>There exists a $C^2$ Kupka-Smale diffeomorphism $f: D^2 \to D^2$ which has neither periodic sources nor periodic sinks.</u>

From the construction it will be clear that it is possible to glue together two such examples to obtain a $C^2$ Kupka-Smale diffeomorphism of $S^2$ with no sources or sinks. The proof of (1) will be given in a sequence of lemmas.

Let D denote the unit disk in $R^2$. Inside D we consider four disks $D_i$, $1 \le i \le 4$, each of radius $r = 0.251$ (= $1/4 + \varepsilon$) situated symmetrically, one in each quadrant, as shown in figure 1.

<u>Lemma 1.</u> There exists a $C^\infty$ diffeomorphism $f:D \to D$ such that

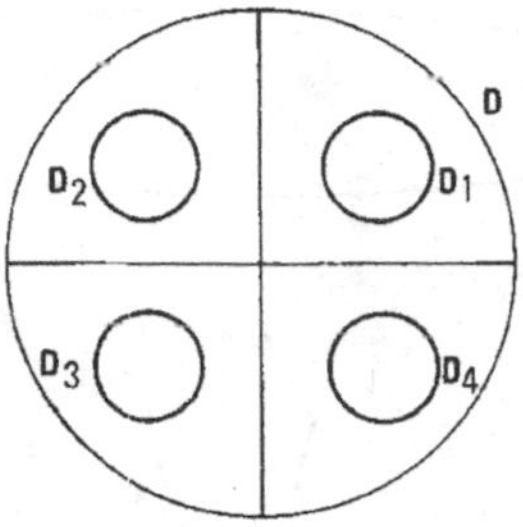


Figure 1.

(1) $f(D_i) = D_{i+1 \pmod 4}$.

(2) On a closed neighbourhood U of the boundary of D, $\partial D$, f is rotation by an angle $\alpha$, where $\alpha/\pi$ is irrational.

(3) On a closed neighbourhood $U_i$ of $D_i$, f is rotation by $\alpha/4$ plus a translation.

(4) Outside of U and $\cup U_i$, f has a hyperbolic fixed point of saddle type and a hyperbolic orbit of period two of saddle type, and no other points which are $\omega$-limit points.

Proof. We first construct a diffeomorphism $g: P \to P$ where P is D with four open disks removed such that

(1) $g|\partial D$ is rotation by $\pi$;

(2) g permutes the boundaries of the holes $\partial P_i$, $1 \le i \le 4$;

(3) the limit set L of g consists of the boundary of P, $\partial P = \partial D \cup (\bigcup_i \partial P_i)$, together with a hyperbolic saddle fixed point and a hyperbolic saddle orbit of period 2;

(4) g is Kupka-Smale.

This is accomplished by first taking the time one map of a gradient flow on the four-legged pants P (see figure 2). This has fixed points on the boundary and 3 hyperbolic fixed points, p, $q_1, q_2$. If this map is composed with a $180^\circ$ rotation about a central vertical axis then the map $g_0$ obtained has p fixed and $\{q_1, q_2\}$ an orbit of period 2. This map has all the properties we want for g except (2), the four pant's legs are not permuted. We remedy this by composing $g_0$ with a $180^\circ$ twist below the annulus A (see figure 2) about a veritcal axis through $q_2$. In A the twist is progressively less until no twist occurs above A. This has the effect of switching $\partial P_2$ and $\partial P_4$. Of course, no twisting occurs in the leg containing $\partial P_1$ and $\partial P_3$. The resulting map g cyclically permutes the $\partial P_i$ and has the other properties mentioned above.

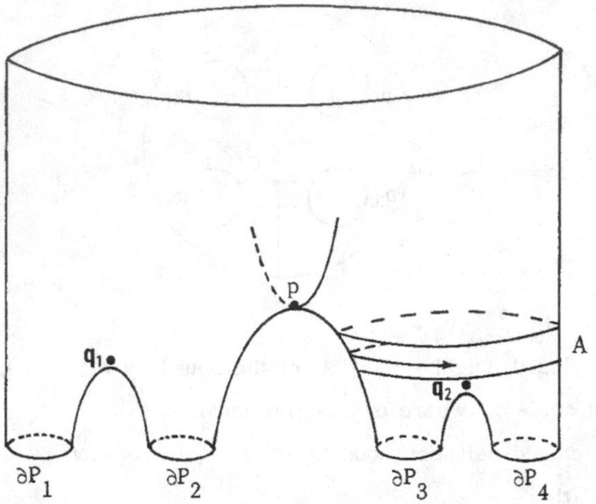


Figure 2.

Now to get f we first choose a diffeomorphism $h:P \to P'$ where $P'$ is the disk inside D of radius .999 with the interiors of disks $U_i$ concentric with $D_i$ and of radius .252 removed. We choose it so that $h(\partial P_i) = \partial U_i$. The map $f_0 = h \circ g \circ h^{-1} : P' \to P'$ has most of the properties we want. We alter it on a small neighbourhood of $\partial P'$ by twists so that the new map f satisfies

$$f|U_i = \text{rotation by } \alpha/4 \text{ plus a translation}$$

and on the outer boundary component f is rotation by $\alpha$. The map $f:P' \to P'$ is almost Kupka-Smale since all limit points except $h^{-1}(p)$, $h^{-1}(q_1)$, and $h^{-1}(q_2)$ are in $\partial P'$ where the map has no periodic points. By the Kupka-Smale theorem we can perturb f slightly away from $\partial P'$ so that the stable and unstable manifolds of the points p, $q_1, q_2$ all intersect transversely. Finally, we extend f to $f:D \to D$ by requiring

$$f|U_i = \text{rotation by } \alpha/4 \text{ plus a translation}$$

and requiring that f be rotation by $\alpha$ on the annulus between the outer boundary of $P'$ and $\partial D$.

Q.E.D.

Lemma 2. There is a $C^\infty$ isotopy $f_t: M \times I \to M$, $I = [0,1]$ such that

(1) $f_1 = f$ and $f_0 = id: D \to D$.

(2) On a neighbourhood of $\partial D$, $f_t$ is rotation by $t\alpha$, and

(3) On a neighbourhood of $D_i$, $1 \le i \le 4$, $f_t$ is rotation by $t\alpha/4$ plus a translation.

Proof. We again let U be the disk of radius .999 concentric with D, and recall that $U_i$ is the disk of radius .252 concentric with $D_i$. Let $\varphi_t^i: U \times I \to U$ be an isotopy which switches $U_i$ and $U_{(i+1)\mathrm{mod}\ 4}$ in a counterclockwise fashion, leaves the other two disks $U_j$ and $U_{j'}$ fixed and satisfies

$$\varphi_t^i | U_j : U_j \to \varphi_t^i(U_j) \text{ is an isometry for all } t \in [0,1] \text{ and } 1 \le j \le 4 ,$$

(see figure 3).

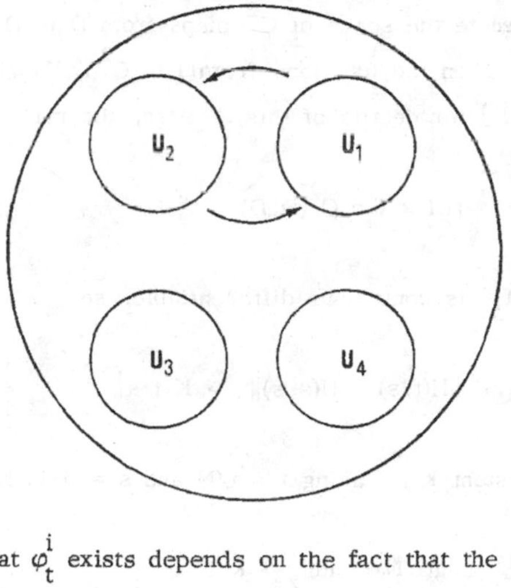


Figure 3.

The fact that $\varphi_t^i$ exists depends on the fact that the radius of $U_j$ is $.252 = 1/4 + 2\varepsilon$ (it would not be possible if the radius was 1/3).

Now any diffeomorphism $g: U \to U$ such that $g(\cup U_i) = \cup U_i$ is isotopic by an isotopy preserving $U_j$, $1 \le j \le 4$, to a product of the maps $\{\varphi_1^i\}$ and their inverses (see [B]). In particular, there is an isotopy from f to such a product which at each stage preserves $U_j$, $1 \le j \le 4$. Putting this together with the isotopy to id obtained from $\varphi_t^i$, we have the isotopy $h_t: U \quad I \to U$ satisfying $h_1 = f$, $h_0 = id$ and for each $t \in [0,1]$ and each $1 \le i \le 4$, $h_t(U_i)$ is a disk of radius .252. If we now adjust $h_t$ on

D - U and int $U_i$, $1 \leq i \leq 4$, to obtain $f_t$ satisfying $f_t$ = rotation by $t\alpha/4$ plus translation on a neighbourhood of $D_i$ and $f_t$ = rotation by $t\alpha$ on a neighbourhood of $\partial D$, we have the desired isotopy.

Q.E.D.

Given an integer $N > 0$ and the isotopy $f_t$ from Lemma 2, define a diffeomorphism $g(i,N): D \to D$ by $g(i,N) = f_{i/N} \circ f^{-1}_{(i-1)/N}$ so that

$$f = g(N,N) \circ g(N-1,N) \circ \ldots \circ g(2,N) \circ g(1,N) .$$

Lemma 3. There is a constant $K > 0$ such that for all $N > 0$ and all $i \leq N$ in the $C^2$ norm,

$$\|g(i,N) - id\|_2 \leq K/N .$$

Proof. Let $C^2(D,D)$ denote the space of $C^2$ maps from D to D with the topology given by the $C^2$ norm $\| \ \|_2$. Then the functions from I to $C^2(D,D)$ given by $t \to f_t$ and $t \to f_t^{-1}$ are smooth ($C^\infty$) (see [F] for details of this). Also, the map

$$H: I \times I \to C^2(D,D)$$

given by $H(t,s) = f_t \circ f_s^{-1}$ is continuous differentiable, so

$$\|H(t,s) - H(s,s)\|_2 \leq K|t-s|$$

for some Lipschitz constant K. Taking $t = i/N$ and $s = (i-1)/N$, we get

$$\|g(i,N) - id\|_2 \leq K/N .$$

Q.E.D.

We can now complete the proof of our theorem. We define a sequence of diffeomorphisms $\{f_n\}$ which is Cauchy in the $C^2$ norm and whose limit is the desired diffeomorphism. Let $f_1$ be the diffeomorphism f of lemma 1.

Let $p_i$ denote the centre of $D_i$ and recall that $r = .251 = (1/4 + \varepsilon)$ is the radius of $D_i$. We define

$$f_2(x) = \begin{cases} f_1(x) & \text{if } x \notin \cup D_i \\ rg(i,4)((x-p_i)/r) + p_{(i+1)\bmod 4} & \text{if } x \in D_i \,. \end{cases}$$

Thus $f_2$ cyclically permutes the four disks; however, $f_2^4|D_i$ is not a rotation but a scaled down version of $f_1$, i.e., $f_2^4: D_1 \to D_1$ is given by

$$\begin{aligned}(f_2)^4(x) &= rg(4,4) \circ g(3,4) \circ g(2,4) \circ g(1,4)(x-p_1)/r \\ &= rf(x-p_1)/r) + p_1 \,.\end{aligned}$$

Inside each $D_i$ are 4 disks of radius $r^2$ which are cyclically permuted by $(f_2)^4$. We name them by letting $D_1^2 = \beta(D_1)$ where $\beta(x) = r(x-p_1) + p_1$, and defining $D_i^2 = (f_2)^{i-1}(D_1^2)$, $1 \leq i \leq 16$ (see figure 4). Let $p_i^2$ denote the centre of $D_i^2$.

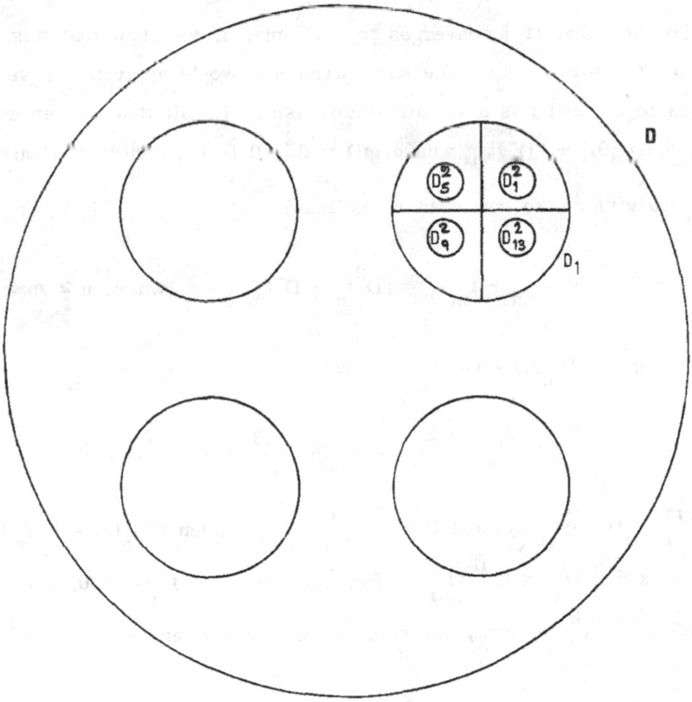


Figure 4.

The sequence of diffeomorphisms is obtained by iterating this construction. Thus suppose inductively we have defined $f_{n-1}: D \to D$ and a family of disks $D_i^{n-1}$, $1 \le i \le 4^{n-1}$ with radius $r^{n-1}$ and centres $p_i^{n-1}$. We then can define $f_n$ by

$$f_n(x) = \begin{cases} f_{n-1}(x) & \text{if } x \notin \bigcup_i D_i^{n-1} \\ r^{n-1} g(i,N)((x-p_i^{n-1})/r^{n-1}) + p_{(i+1)\bmod N}^{n-1} & \text{if } x \in D_i^{n-1}, \text{ where } N = 4^{n-1} . \end{cases}$$

Next define $\{D_i^n\}$ by $D_1^n = \beta(D_1^{n-1})$ where $\beta(x) = r(x-p_1^{n-1}) + p_1^{n-1}$, and $D_i^n = f_n^{i-1}(D_1^n)$, $1 \le i \le 4^n$. Thus $f_n$ cyclically permutes the $4^n$ disks $\{D_i^n\}$. Each $f_n$ is Kupka-Smale except on the disks $\{D_i^n\}$ (notice that if $x \in D_i^k$ and $y \in D_j^{k'}$, $k' < k$, $x, y \in \cup D_i^n$, then x and y are separated by the invariant circles $\cup \partial D_j^k$, so $W^s(x) \cap W^u(y)$ and $W^u(x) \cap W^s(y)$ are both empty. If we show that $\{f_n\}$ converges to a diffeomorphism f, then f will be Kupka-Smale if $x \in \bigcup_i D_i^n$) then x is not periodic (since $y \in D_i^n$ implies $f^j(y) \in D_i^n$, $1 \le j \le 4^n$, so if y is periodic its period is $\ge 4^n$).

To show that $\{f_n\}$ converges to a $C^2$ map f, we show that this sequence is Cauchy in the $C^2$ norm $\| \ \|_2$. The same argument would apply to the sequence $\{f_n^{-1}\}$, so it converges to $f^{-1}$ and f is a $C^2$ diffeomorphism. Recall that we can define $\|g\|$ $\|g\|_2 = |g(0)| + |Dg(0)| + \|D^2 g\|_0$ where $g: D \to R^2$, 0 is the centre of D and $\|D^2 g\|_0 = \sup_{x \in D} \|D^2 g(x)\|$. We now want to estimate

$$\|f_n - f_m\|_2 = \|D^2 f_n - D^2 f_m\|_0 \qquad \text{where } n \ge m > 0$$

since $f_n(0) = f_m(0)$ and $Df_n(0) = Df_m(0)$. Now

$$\|D^2 f_n - D^2 f_m\|_0 = \sup_{x \in D} \|D^2 f_n(x) - D^2 f_m(x)\|$$

and if $x \in \cup D_i^m$, $f_n(x) = f_m(x)$, and if $x \in \cup D_j^n \subset \bigcup_i D_i^m$ then $D^2 f_n(x) = D^2 f_m(x) = 0$, so we can consider $x \notin \bigcup_j D_j^n$, $x \in \bigcup_i D_i^m$. For such an x, $D^2 f_m(x) = 0$, and $f_n(x) = r^k f(j,N)((x-p_j^k)/r^k) + \text{const}$, for some k, $m \le k \le n$ and with $N = 4^k$. Therefore,

$$D^2 f_n(x) = r^{-k} D^2 g(j,N)((x-p_j^k)/r^k) .$$

So

$$\|D^2 f_n(x) - D^2 f_m(x)\| = \|D^2 f_n(x)\| \le r^{-k}\|g(j,N)\|_2 .$$

Thus

$$\begin{aligned}\|f_n - f_m\|_2 &\le r^{-k}K/N = K(r^{-k}4^{-k}) \\ &\le K(1/4r)^k \le K(1/4r)^m\end{aligned}$$

since $k \ge m$ and $(1/4r) < 1$. Thus $\{f_n\}$ is Cauchy in $\| \ \|_2$ and a similar argument is valid for $\{f_n^{-1}\}$ so $f = \lim f_n$ is a $C^2$ diffeomorphism.

Q.E.D.

<u>Remark.</u> In fact one can show that f is $C^{2+\varepsilon}$ for some $\varepsilon > 0$. The estimates are similar to what has already been done. We want an estimate like

$$\|D^2 f(x) - D^2 f(g)\| \le k|x-y|^{\varepsilon} .$$

There are three cases : (1) If $x, y \in \bigcap_{m=1}^{\infty}(\bigcup_i D_i^m)$, i.e., in the limit set, then $D^2 f(x) = D^2 f(y) = 0$; (2) If $x, y \in D_i^m - (\bigcup_j D_j^{m+2})$ use the mean value theorem (f is $C^3$ in this region and Lemma 3 holds for the $C^3$ norm; (3) In any other case, use the fact that $|x-y| \ge Cr^m$ where C is the minimum distance between the $D_i^1$'s and the boundary of D, and m is the smallest integer such that y is in $\bigcup_i D_i^m$ but x is not or vice versa.

## References.

[B] J. Birman, Braids, Links and Mapping Class Groups. Annals of Math. Studies 82 Princeton University Press, 1975.

[B-F] R. Bowen & J. Franks, The Periodic Points of Maps of the Disk and the Interval, Topology 15 (1976), 337-342.

[F] J. Franks, Manifolds of $C^r$ Mappings and Applications to Differentiable Dynamical Systems, Studies in Analysis, Advances in Math Studies 14 (1979), 271-290.

[S] S. Smale, Dynamical Systems and the topological conjugacy problem for diffeomorphisms, Proc. Int. Congr. of Math. (1962), 490-495.

J. Franks, Department of Mathematics, Northwestern University, Evanston, Illinois, USA.
L-S. Young, Mathematics Institute, University of Warwick, Coventry, England.

# On a Codimension Two Bifurcation.*

John Guckenheimer

## Introduction.

This paper studies the multiple bifurcations of certain systems of differential equations whose linearisation at an equilibrium has a double degeneracy. There is a rapidly growing literature about such systems, and they have been studied from different points of view. The particular equations studied here are perturbations of ones for which the linearisation at an equilibrium solution has a simple pair of pure imaginary eigenvalues and a simple zero eigenvalue, with no other eigenvalues on the imaginary axis. This problem has been studied recently by Langford [42] using different techniques, and his formulation of the problem is different from ours in ways which are explained below. It is our feeling that the approach here is conceptually clearer. Moreover, our results extend his in that one obtains substantially more information about the complex dynamics found in some perturbations.

Systems of the kind studied here arise in varied applications. Several are analysed by Langford, and we complete his analysis of the one example for which his results are not definitive. We also analyse an example drawn from the study of reaction-diffusion equations in considerable detail. This example is considerably more complicated than those considered by Langford, and there is an independent reason for its consideration. We are interested in the occurence of aperiodic (chaotic) motion in systems with an infinite number of degrees of freedom. Beginning with the speculations of Ruelle and Takens [43] and the subsequent experimental work of Gollub and Swinney [39], our views of how turbulence and aperiodic motion arises in fluid systems have changed. An outstanding question in this area is the extent to which the complicated dynamical phenomena seen in finite dimensional dynamical systems provide a good mathematical model for the aperiodic motions seen in turbulent physical systems. One aspect of this question involves finding solutions to partial differential equations which have solutions of the desired aperiodic (homoclinic) type. This we do here for the reaction-diffusion system known as the Brusselator as an application of our theory. While this example has been chosen for the completeness of its mathematical treatment, the approach which we use should be applicable to a wide range of problems involving

---

* Research partially supported by the National Science Foundation and the Volkswagen Foundation.

aperiodic motion in continuum and fluid mechanics. We note that Holmes and Marsden [45] have completed an analysis of this kind for a forced buckled beam, and that Holmes and Moon [46] have conducted highly suggestive experiments with a magneto-elastic system to test these ideas. The noteworthy feature of our reaction-diffusion example is that it provides a model system in which algebraic computations can be completed which indicate where aperiodic solutions of the system should be found. Numerical computations of Kuramoto [41] provide corroborating evidence that aperiodic solutions do exist for this system.

The remainder of this introduction is an outline of our approach to these bifurcation problems and a comparison with the formal techniques of Langford, Keener, et al. The starting point is a system of differential equations

$$\frac{dx}{dt} = f(x;\mu,\varepsilon) \qquad (*)$$

where $x \in \mathbb{R}^n$ $(n \geq 3)$ and $(\mu,\varepsilon)$ is a pair of real parameters. When $(\mu,\varepsilon) = (0,0)$, we assume that (*) has an equilibrium solution $x_0$ at which the matrix $Df(x_0)$ of the linearised equations has a simple zero eigenvalue $\lambda_1$ and a simple pair of pure imaginary eigenvalues $\lambda_2,\bar{\lambda}_2$. This is the multiple bifurcation point whose perturbations we want to study. The second assumption is that the parameters $(\mu,\varepsilon)$ behave in a non-degenerate way with respect to linearised equations. Here there are two possible choices, and we differ from Langford in making it. In his setting, one assumes that there is a family of equilibrium solutions $x_0(\mu,\varepsilon)$ which depends smoothly on $(\mu,\varepsilon)$. Then the non-degeneracy assumption $(\mu,\varepsilon)$ is that the real parts of the eigenvalues $\lambda_1(\mu,\varepsilon)$, $\mathrm{Re}\,\lambda_2(\mu,\varepsilon)$ of $Df(x_0)$ depend in a non-singular way on $(\mu,\varepsilon)$ :

$$\det\begin{pmatrix} \dfrac{\partial\lambda_1}{\partial\mu} & \dfrac{\partial\lambda_1}{\partial\varepsilon} \\ \dfrac{\partial \mathrm{Re}\lambda_2}{\partial\mu} & \dfrac{\partial \mathrm{Re}\lambda_2}{\partial\varepsilon} \end{pmatrix} \neq 0 \ .$$

In our setting, for some parameter values there will be no equilibrium solution near $x_0(0,0)$ and for other parameter values there will be two. Our assumption begins with the statement that the map $F:\mathbb{R}^n \times \mathbb{R}^2 \to \mathbb{R}^n$ has $(x_0,0,0)$ as a regular point (i.e. the matrix $Df(x_0,0,0)$ has rank n). This implies that the equilibria of (*) near $(x_0,0,0)$ form a smooth two dimensional surface (whereas in Langford's case they form two transversally intersecting surfaces). We then assume that $(\lambda_1, \text{Re}\lambda_2)$ vary in a non-singular way on this surface. In Section §4, we show how to derive bifurcation diagrams for the case of "trivial solutions" from this one. Whether or not a particular example has trivial solutions depends upon the way (*) depends upon the parameters. For the reaction-diffusion example of Section 5, one can have either case depending upon the boundary conditions of the problem.

The third and final assumption involves the nonlinear terms of the Taylor expansion of $f(x;0,0)$ at $x_0$. Several inequalities must hold which are specified in the course of our analysis. These inequalities arise in our attempts to "simplify" the equations (*) as much as possible by making appropriate changes of coordinates. Linear equations are the ideal we seek (but cannot achieve), and we try to remove as many nonlinear terms from the Taylor expansion as possible. The first step in this procedure invokes the Centre Manifold Theorem [19] to reduce our system (*) to a three dimensional system in which all of the bifurcation behaviour takes place. A theorem of Palis and Takens [25] guarantees that the dynamical behaviour of the system in complementary directions does not change with the parameters. Having reduced the system to a three dimensional one, the next step is to employ the method of <u>normal forms</u> (Section §1). This procedure determines which nonlinear terms in the equation can be removed by coordinate changes and which cannot. The result of this analysis is a set of model equations, the <u>normal forms,</u> whose dynamics can be investigated in more detail. One has reduced the study of our general system to these special normal forms without changing the original problem in any essential way.

The normal forms still contain extra parameters which are coefficients of nonlinear terms, but there are many fewer of these than in the original problem. We expect that near the bifurcation, the lowest order terms will determine the dynamical behaviour of solutions. If this is true, then it is worthwhile to invest the effort to study the dynamics of these particular equations. If these behave in a robust or "structural stable" way, then our expectations will be met. If the dynamics of the normal form families are not robust within the appropriate context, then higher order terms in the Taylor expansions or yet further information about the equations must be retained. Considerations of this kind do play an important role in some of the cases which we study and indicate some of the limitations of formal analysis.

The dynamics of three dimensional systems is difficult to determine because the possibility of complicated aperiodic solutions is present. Two dimensional systems are much easier to handle. For the equations of interest here, the normal form computations give us a welcome means of using a two dimensional phase plane analysis for the three dimensional systems. Introduce coordinates near the bifurcation point $x_0$, so that the solution of the linearised equations are rotations about an axis. In cylindrical coordinates $(r, \theta, z)$ based upon this geometry, one outcome of the normal form analysis is that all $\theta$ dependence of any finite order can be removed from the right hand side of the differential equations (*). In other words, to all orders in a Taylor expansion, the system of equations near bifurcation appears to have a circular symmetry and the equations for $\dot{r}$ and $\dot{z}$ decouple from the equation for $\theta$. Therefore, we can study the dynamics of the equations for $(r, z)$ first, and then infer the full three dimensional dynamics from this analysis.

Most of the two dimensional analysis is straightforward. The only difficulty comes when the $(r, z)$ equations allow periodic solutions. (These solutions represent the tori which are the subject of Langford's paper.) The existence of these periodic solutions can be easily deduced from the Hopf Bifurcation Theorem, but the stability and how the full set of these periodic solutions sits relative to the parameter space is harder to determine and was not accomplished by Langford. We follow Takens [36] in our strategy here. A parameter dependent rescaling of the $(r, z)$ equations yields an integrable system which has a family of periodic solutions which correspond to the full family in the original equations. A variational technique of Andronov et al. [38] determines (to first order in the rescaling parameters) which parameter values $(\mu, \varepsilon)$ yield a particular periodic solution.

An additional interesting feature of this analysis is that one can see the kind of bifurcations which yield these periodic solutions of the $(r, z)$ equations. In one case, there is an "infinite period" bifurcation in which the family of periodic solutions terminates in a closed curve consisting of two equilibrium saddle points and two of their separatrices (see Figure 6 F/A for a phase portrait). The corresponding three dimensional flow is structurally unstable if we leave the class of systems which have an exact circular symmetry. Asymmetry in the three dimensional equations leads to transverse homoclinic "chaotic" solutions as the invariant tori in these systems split apart. Because the formal expansion of the equations at bifurcation does have an exact symmetry, this feature that periodic and steady-state mode interactions lead to chaotic solutions cannot be detected by formal methods.

Our analysis extends Langford's results by giving a much more complete description of the stability of tori and their location relative to parameter values as well as finding completely new dynamical phenomena in problems involving the interaction of periodic and steady-state modes. It is perhaps worthwhile to compare also our approaches to point out the advantages and limitations of each. Langford begins as we do with a linear change of coordinates so that the solutions of the three dimensional linearised system are given by rotations around one of the coordinate axes. He then represents solutions of the nonlinear system as perturbations of the solutions of the linearised equations and proceeds with a formal asymptotic analysis. In this asymptotic analysis, he specifically searches for periodic solutions by solving a two point boundary value problem with periodic boundary conditions. In terms of the parameters used in the asymptotic analysis, the first order terms yield a system of algebraic equations for locating the periodic solution. The coefficients here are trigonometric integrals and the reduction seems equivalent to that obtained by applying the <u>method of averaging</u> to the original equations viewed as a perturbation of the linearised system. The corresponding step in our analysis is the decoupling of $(\dot{r}, \dot{z})$ from $\theta$ which occurs in the normal form analysis. The everaging process yields approximate equations for the Taylor series expansion at equilibrium while the normal form procedure involves no approximations. The solutions of Langford's auxiliary bifurcation equations correspond to the equilibria of our $(r,z)$ equations.

Langford detects secondary bifurcation leading to tori by examining the solutions of the auxiliary bifurcation equations further. With his formal procedures, only the algebraic solutions of these bifurcation equations have meaning for the dynamics of (*). In our analysis, the $(r,z)$ equations are differential equations whose dynamics are immediately relevant to those of (*).

An additional difference between the two pieces of work is that our normal forms include cubic terms. If only quadratic terms are retained, the Hopf bifurcation of the $(r,z)$ equations is highly degenerate and one obtains integrable systems with a whole family of periodic solutions of particular parameter values. Langford's third example falls into this category of degenerate families and has the integral

$$H(u_1, u_2, u_3) = \frac{\lambda}{2} u_1 (u_2^2 + u_3^2) - \frac{u_1^2}{2}(u_2^2 + u_3^2) - \frac{1}{4}(u_2^2 + u_3^2)$$

when $2\sigma = 3\lambda$. It appears to be much easier to retain cubic terms with our approach (and automatically obtain global information about periodic solutions) since a whole new order of terms must be examined in the formal scheme to determine the type of secondary Hopf bifurcation which occurs. The penalty for our method is that one must compute the cubic term in the normal form. This computation is of the same level of difficulty as that involved in the Hopf bifurcation theorem. In particular examples, it can be lengthy and the formal procedures may be easier to implement. Finally, as we noted above, the existence of aperiodic homoclinic solutions is undetectable by techniques which rely solely upon formal analysis.

Despite these differences, the computations required to analyse the bifurcations in any particular problem are similar in the two approaches. As we saw above, the auxiliary bifurcation equations of the formal analysis are closely related to the (r, z) equations in the normal forms. In any application, the principal computational effort involves determining the coefficients of one of these systems, and the work involved will be comparable from both points of view. For half the cases, tori do not occur and the two approaches yield the same results. It is only with regard to some of the more delicate dynamical aspects of these problems that there is a substantial difference in the methods.

The principal content of our results is contained in the figures and bifurcation diagrams. These give a comprehensive description (modulo details of the quasiperiodic and homoclinic solutions) of all the dynamical behaviour which one expects to find in perturbations of systems of equations in which the doubly degeneracy described above occurs in the simplest possible way. After allowing for scale changes and time reversals, there are four qualitatively different cases which depend upon the signs of certain coefficients in the normal forms of these equations. One of these cases further splits into two subcases. In two of the four cases, two dimensional invariant tori are present in the flows of some perturbations. We do not examine the expected flow on these invariant tori, though we note here that the results of Arnold [44] and Herman [40] apply. These show that both periodic and quasiperiodic motion will be present for sets of parameter values of positive measure. In one of the four cases, the invariant tori give way to systems in which there are transversal homoclinic solutions. Here the results of Smale [31] imply the existence of periodic solutions with arbitrarily long periods as well as solutions with aperiodic asymptotic behaviour. The determination of which case corresponds to a particular example requires the calculation of the normal form equations for the example. In Section §5, these calculations are carried through in detail for one version of the Brusselator reaction-diffusion equation.

We record here our results in a formal manner.

Theorem 1 : Let $X_{\mu,\varepsilon}$ be a 2 parameter family of $\mathfrak{X}^k$, the space of $C^k$ vector fields on $\mathbb{R}^3$ (k≥3) for which the following properties hold :

(1) $X_{0,0}$ has an equilibrium p at which $DX_{(0,0)}(p)$ is the matrix

$$\begin{pmatrix} 0 & -\omega & 0 \\ \omega & 0 & 0 \\ 0 & 0 & 0 \end{pmatrix}$$

(2) The map $(x,\mu,\varepsilon) \to X_{\mu,\varepsilon}(x)$ is transversal to the variety in $\mathbb{R}^3 \times \mathfrak{X}^k(\mathbb{R}^3)$ defined by the equations

$$X_{\mu,\varepsilon}(x) = 0 \quad \text{and} \quad \mathrm{Tr}DX_{\mu,\varepsilon}(x) = \det DX_{\mu,\varepsilon}(x) = 0 \ .$$

Then there is a smooth change of coordinates in $\mathbb{R}^3 \times \mathbb{R}^2$ commuting with the projection map $\mathbb{R}^3 \times \mathbb{R}^2 \to \mathbb{R}^2$ so that the normal form expression of $X_{\mu,\varepsilon}$ in cylindrical coordinates is given by

$$\begin{aligned} r\dot{\theta} &= \omega r + 0(1) \\ \dot{r} &= r(\varepsilon + a_2 z + b_1 r^2 + b_2 z^2) + 0(3) \\ \dot{z} &= \mu + a_3 r^2 + a_4 z^2 + b_3 r^2 z + b_4 z^3 + 0(3). \end{aligned}$$

Theorem 2 : Let $X_{\mu,\varepsilon}$ be a two parameter family of vector fields satisfying the hypotheses of Theorem 1. If each of the coefficients $a_i \neq 0$, then the dependence of equilibria and periodic orbits with their stability is described by the bifurcation diagrams in Figures 3-6. When $a_2 a_3 < 0$, there is a region in the $(\mu,\varepsilon)$ plane for which more complicated asymptotic behaviour occurs through secondary Hopf bifurcation.

Theorem 3 : Let $X_{\mu,\varepsilon}$ be a two parameter family of vector fields in $\mathbb{R}^3$ which satisfies the hypotheses of Theorem 1 and is equivariant with respect to rotations around the z-axis. If $a_2 a_3 < 0$ in the normal form of $X_{\mu,\varepsilon}$, then there is a family of invariant two dimensional tori occuring in the flows $X_{\mu,\varepsilon}$.

These tori are perturbations of the level surfaces $\Sigma$ of $H(\theta, r, z, \mu) = r^{\beta}(\frac{\mu}{\beta} + \frac{a_3}{\beta+2} r^2 + \frac{a_4}{\beta} z^2)$ with $\beta = {}^{-2a_4}/a_2$. The parameters for which a given torus exists are approximately determined by the equations

$$\int_D r^{\beta-1}(\beta\varepsilon + ((\beta+2)b_1 + b_3)r^2 + (\beta b_2 + 3b_4)z^2) = 0 \qquad \text{(V)}$$

with D the interior of the intersection of $\Sigma$ with a half plane $\theta$ = constant. For fixed $\mu$, equation (V) defines a function $\varepsilon(c)$, c the value of H. The stability of the invariant torus is determined by $\varepsilon'(c)$. If $\varepsilon(c)$ has a nondegenerate critical point, then there is a corresponding curve in parameter space which separates parameter regions in which the number of invariant tori differs by two.

Theorem 4 : Let $X_{\mu,\varepsilon}$ be a two parameter family of vector fields on $\mathbb{R}^3$

(1) satisfying the hypotheses of Theorem 1,
(2) equivariant with respect to rotations around the z-axis,
(3) having non-zero coefficients in its normal form, and
(4) such that the function $\varepsilon(c)$ is a Morse function.

Then the family $X_{\mu,\varepsilon}$ is structurally stable at (p, 0, 0) with respect to the following equivalence relation in the space of vector fields equivariant with respect to rotations around the z-axis : two families $X_{\mu,\varepsilon}$ and $Y_{\mu,\varepsilon}$ are equivalent at $(p_0, \mu_0, \varepsilon_0)$ if there are neighbourhoods U of $p_0$ in $\mathbb{R}^3$ and V of $(\mu_0, \varepsilon_0)$ in $\mathbb{R}^2$, and a homeomorphism $h: V \to \mathbb{R}^2$ so that $X_{\mu,\varepsilon}$ and $Y_{h(\mu,\varepsilon)}$ are topologically equivalent in U.

Remark : The following features will not be persistent with respect to perturbations away from equivariant vector fields :

(1) invariant tori which are close to neutrally stable invariant tori or invariant sets composed of stable and unstable manifolds of hyperbolic equilibria, and
(2) periodic flows on invariant tori.

Invariant tori with quasiperiodic motion will persist in the general non-equivarant perturbation for sets of parameter values having positive measure. These issues are more fully discussed for the more complicated situation of bifurcations with two pairs of pure imaginary eigenvalues in a subsequent paper. Here we focus upon the existence of transversal homoclinic orbits, obtaining the following theorem as a corollary of the work of Silnokov.

Theorem 5 : Let $\mathfrak{X}$ be the class of two parameter families of vector fields on $\mathbb{R}^3$ with normal form

$$\begin{aligned} \dot{\theta} &= \omega + \ldots \\ \dot{r} &= r(\varepsilon + az + \ldots) \\ \dot{z} &= \mu - r^2 - z^2 + \ldots \end{aligned}$$

and $a < 2$. Then there is an open, dense subset of $\mathfrak{X}$ containing vector fields having transversal homoclinic orbits.

Theorem 6 : There is a system of autonomous nonlinear partial differential equations (the "Brusselator") with an equilibrium whose spectrum has a zero eigenvalue, a pair of pure imaginary eigenvalues, and the remainder of its spectrum in the left half plane. There are parameters such that, on the centre manifolds of this equilibrium, the normal form of this equation satisfies the hypothesis of Theorem 5.

Remark : Unless there are unsuspected symmetries in this equation which prevent the assumptions of Silnikov's theorem from being satisfied, then the Brusselator equation has transversal homoclinic solutions.

The development of this work has been influenced by a number of people. Phil Holmes has been vigourous in employing this approach to searching for aperiodic behaviour in mechanical systems, Jerry Marsden patiently explained to me how the centre manifold theory works, Jim Keener made me aware that there are quasiperiodic dynamical phenomena associated with the bifurcation we study, and Marty Golubitsky explained his work with Dave Schaeffer on the "Brusselator" to me. I am also grateful to the Courant Institute for Mathematical Sciences and the Institut des Hautes Etudes Scientifiques for their hospitality while this work was done and to the Volkswagen Foundation and the National Science Foundation for financial support. An earlier version of this paper was distributed which included an incorrect treatment of the relative magnitude of various terms in the rescaling arguments.

## §1. Normal Forms : Derivation of the nonlinear terms.

We shall use Cartesian coordinates $(x, y, z)$ and cylindrical coordinates $(r, \theta, z)$ throughout this paper. Following a linear change of coordinates, we study degenerate vector fields with linear part given by $L = \omega(-y\frac{\partial}{\partial x} + x\frac{\partial}{\partial y}) = \omega\frac{\partial}{\partial\theta}$. In the space of linear vector fields, L has codimension 2 with respect to the relation of topological equivalence. The perturbations $L_{\mu,\varepsilon} = L + \varepsilon(x\frac{\partial}{\partial x} + y\frac{\partial}{\partial y}) + \mu\frac{\partial}{\partial z}$ give an unfolding of L within the space of affine vector fields : any affine vector field $\tilde{L}$ close to L is topologically equivalent to an $L_{\mu,\varepsilon}$.

In this section we shall use the computation of "normal forms" to deduce which nonlinear terms should be added to "stabilise" the singular point of a vector field $X_0$ whose linearisation is given by L above. This computation also shows which terms in the Taylor expansion of $X_0$ at the singular point can be eliminated by a local change of coordinates. The procedure we use is well known [34]. The lowest order nonlinear terms which cannot be removed by a coordinate change are the ones which will play a significant role in our later analysis.

We recall briefly the procedure for computing normal forms. Denote by $H^k$ the space of homogeneous vector fields of degree k in $\mathbb{R}^n$. These are vector fields of the form $\Sigma x_1^{i_1} x_2^{i_2} \dots x_n^{i_n} \frac{\partial}{\partial x_j}$ with $i_1 + i_2 + \dots + i_n = k$. The (kth) Taylor expansion of a vector field X having a singularity at the origin is $X = X_1 + X_2 + \dots + X_k + R_{k+1}$ with $X_i \in H^i$ and $R_{k+1} = 0(|x|^k)$. If $X_i \in H^i$ and $X_j \in H^j$ are two homogeneous vector fields, then their Lie bracket $[X_i, X_j]$ is in $H^{i+j-1}$. Therefore a linear vector field L induces a map of each $H^i$ into itself by $X_i \to [L, X_i]$. Split $H^i = B^i + G^i$ where $B^i$ is the image of this map and $G^i$ is a complement. Then if $X = X_1 + \dots + X_k + R_{k+1}$ is the kth Taylor expansion of X having a singularity at the origin and $L = X_1$, there is a $C^k$ change of coordinates $\varphi: \mathbb{R}^n \to \mathbb{R}^n$ fixing the origin such that the vector field $Y = D\varphi X \varphi^{-1}$ has the Taylor decomposition $Y = Y_1 + \dots + Y_k + R_{k+1}$ with $Y_1 = L$ and $Y_i \in G^i$ for $2 \leq i \leq k$. With respect to the splitting of $H^i$ described above, Y is a <u>normal form</u> for X. Nonlinear terms of X lying in $G^i$ cannot be eliminated by local coordinate changes.

Let us now examine these computations for a vector field X on $\mathbb{R}^3$ with linear term $L = \omega(-y\frac{\partial}{\partial x} + x\frac{\partial}{\partial y})$. The space $H^k$ of homogeneous vector fields of degree k on $\mathbb{R}^3$ has dimension $\frac{3}{2}(k+1)(k+2)$ and is spanned by vector fields of the form $x^a y^b z^c \frac{\partial}{\partial x}$,

$x^a y^b z^c \frac{\partial}{\partial y}$, and $x^a y^b z^c \frac{\partial}{\partial z}$ with $a + b + c = k$. To compute the map induced by Lie brackets L on $H^k$, it is useful to introduce complex numbers. Write $\xi = x+iy$ and $\bar{\xi} = x-iy$. Then $\frac{\partial}{\partial \xi} = \frac{1}{2}(\frac{\partial}{\partial x} - i\frac{\partial}{\partial y})$, $\frac{\partial}{\partial \bar{\xi}} = \frac{1}{2}(\frac{\partial}{\partial x} + i\frac{\partial}{\partial y})$ and $L = i\omega(\bar{\xi}\frac{\partial}{\partial \bar{\xi}} - \xi\frac{\partial}{\partial \xi})$. One then has the following computations:

$$[L, \xi^j \bar{\xi}^\ell z^{k-(j+\ell)} \frac{\partial}{\partial \xi}] = i\omega(\ell-j+1)\xi^j \bar{\xi}^\ell z^{k-(j+\ell)} \frac{\partial}{\partial \xi}$$

$$[L, \xi^j \bar{\xi}^\ell z^{k-(j+\ell)} \frac{\partial}{\partial \bar{\xi}}] = i\omega(\ell-j-1)\xi^j \bar{\xi}^\ell z^{k-(j+\ell)} \frac{\partial}{\partial \bar{\xi}}$$

$$[L, \xi^j \bar{\xi}^\ell z^{k-(j+\ell)} \frac{\partial}{\partial z}] = i\omega(\ell-j)\xi^j \bar{\xi}^\ell z^{k-(j+\ell)} \frac{\partial}{\partial z} \quad .$$

These computations give the action of L (as a complex vector field) in terms of a basis of eigenvectors in the complex $H^k$. An inspection of these formulas shows that the kernel of the action is spanned by the vector fields $|\xi\bar{\xi}|^j z^{k-(2j+1)} \xi\frac{\partial}{\partial \xi}$, $|\xi\bar{\xi}|^j z^{k-(2j+1)} \bar{\xi}\frac{\partial}{\partial \bar{\xi}}$, and $|\xi\bar{\xi}|^j z^{k-2j} \frac{\partial}{\partial z}$. In real polar coordinates, the complement of the image is spanned by the fields $r^{2j} z^{k-(2j+1)} r\frac{\partial}{\partial r}$, $r^{2j} z^{k-(2j+1)} \frac{\partial}{\partial \theta}$, and $r^{2j} z^{k-2j} \frac{\partial}{\partial z}$.

For $k = 2$, the complement is spanned by $zr\frac{\partial}{\partial r}$, $z\frac{\partial}{\partial \theta}$, $r^2\frac{\partial}{\partial z}$, and $z^2\frac{\partial}{\partial z}$. The complement of the action on $H^3$ is spanned by multiples of these terms by $z(z^2 r\frac{\partial}{\partial r}$, $z^2\frac{\partial}{\partial \theta}$, $zr^2\frac{\partial}{\partial z}$, and $z^3\frac{\partial}{\partial z}$) together with the vector fields $r^3\frac{\partial}{\partial r}$ and $r^2\frac{\partial}{\partial \theta}$. Thus, the vector fields which we are led to study have Taylor expansions in polar coordinates which begin

$$\begin{aligned} \dot{\theta} &= \omega + \ldots \\ \dot{r} &= r(\varepsilon + a_1 z + \ldots) \qquad\qquad (*) \\ \dot{z} &= \mu + b_1 z^2 + b_2 r^2 + \ldots \, . \end{aligned}$$

Here $\varepsilon$ and $\mu$ represent two free parameters which we use to perturb the doubly degenerate singular point at $\varepsilon = \mu = 0$ and the dots indicate higher order terms. The equations (*) define a 2-parameter family of vector fields $X_{\mu,\varepsilon}$ in $\mathbb{R}^3$ for each choice of the constants $a_1, b_1, b_2$. Our first goal is the description of the phase portraits and bifurcations of $X_{\mu,\varepsilon}$ for almost all choices of the constants $a_1, b_1, b_2$. We then examine the stability of these families in the space of 2-parameter families of vector fields which are equivariant with respect to rotations around the z-axis.

Before we proceed to the analysis of the phase portraits of these equations, we want to examine the effect of scale changes to reduce the number of choices of the different coefficients which enter into the equations. We shall later make more sophisticated use of scaling properties when we study the phase portraits of these vector fields. Consider the linear transformations defined by $r = \lambda_1 \rho$ and $z = \lambda_2 \zeta$, with $\lambda_1 > 0$. The equations (*) can be re-expressed in terms of $\rho$, $\zeta$, and $\tau$ :

$$\frac{d\theta}{dt} = \omega$$

$$\lambda_1 \frac{d\rho}{dt} = \lambda_1 \rho(\varepsilon + a_1 \lambda_2 \zeta + \ldots)$$

$$\lambda_2 \frac{d\zeta}{d\tau} = \mu + b_1 \lambda_2^2 \zeta^2 + b_2 \lambda_1^2 \rho^2 + \ldots$$

If each of $b_1$ and $b_2$ is non-zero, then we take $\lambda_1^2 = |b_1 b_2|^{-1}$, $\lambda_2 = -b_1^{-1}$. With these choices we then have

$$\frac{d\theta}{d\tau} = \omega$$

$$\frac{d\rho}{d\tau} = \rho(\varepsilon - a_1/b_1 \zeta + \rho^2)$$

$$\frac{d\zeta}{d\tau} = \tilde{\mu} - \zeta^2 \pm \rho^2$$

with $\tilde{\mu} = -b_1 \mu$ . This computation shows that we may take $a_2 = +1$, $b_1 = -1$, and $b_2 = \pm 1$ in equations (*) by a simple rescaling of the variables and parameters. The parameter $a = -a_1/b_1$ in these rescaled variables is a "modulus" which plays an important part in later considerations.

## §2. The Phase Portraits of the Unfolding.

In the last section, we derived the following 2-parameter family of vector fields as a candidate for a codimension 2 singularity within the class of $S^1$ symmetric systems: $X^{a,b}_{\mu,\varepsilon} = \omega\frac{\partial}{\partial\theta} + r(\varepsilon + az)\frac{\partial}{\partial r} + (\mu - z^2 + br^2)\frac{\partial}{\partial z}$. This family depends upon the choices of a and b, with $b = \pm 1$. In this section we begin to study the phase portraits of these vector fields in $\mathbb{R}^3$ and determine how they depend upon the parameters $(\mu, \varepsilon)$. There are a number of different cases which depend upon the choices of a and b. A noteworthy aspect of this analysis is that the magnitude of a as well as its sign plays an important role in determining the phase portrait in two cases.

All of the vector fields $X^{a,b}_{\mu,\varepsilon}$ are equivariant with respect to rotations around the z-axis of $\mathbb{R}^3$. This implies that if $\gamma(t) = (r(t), \theta(t), z(t))$ is an integral curve of X, then $\gamma_\alpha(t) = (r(t), \theta(t) + \alpha, z(t))$ is also an integral curve. Moreover, the projection of $\gamma(t)$ into the $(r,z)$ half plane $r \geq 0$ is an integral curve of the vector field $Y = r(\varepsilon+az+r^2)\frac{\partial}{\partial r} + (\mu-z^2+br^2)\frac{\partial}{\partial z}$. Thus, from the phase portrait of Y we can easily reconstruct the phase portrait of X. Since vector fields in $\mathbb{R}^2$ have relatively simple limiting behaviour, this simplifies our task considerably. Note, that in the translation of results about Y to results about X, equilibria for Y on the z-axis r=0 correspond to equilibria for X, while equilibria for Y with $r > 0$ correspond to closed orbits for X. Closed orbits for Y correspond to invariant two-dimensional tori for X.

We begin our analysis by describing the equilibria of Y as a function of $(\mu, \varepsilon)$. These are found by solving the equation $Y = 0$ :

$$\left.\begin{aligned} r(\varepsilon + az) &= 0 & \quad \text{(E1)} \\ \mu - z^2 + br^2 &= 0 & \quad \text{(E2)} \end{aligned}\right\} \quad \text{(E)}$$

Here $b = \pm 1$. There are four cases to consider, depending upon the choices of signs for a in (E1) and b in (E2). The solutions of (E1) consist of the z-axis $r = 0$ together with a horizontal line. The topology of the set of zeros of (E2) depends upon the signs of $\mu$ and b If $b < 0$, then the zero set of (E2) is empty, a point, or a semi-ellipse depending upon $\mu$. If $b > 0$, then the zero set is a semi-hyperbola of one or two components or two rays with vertex at the origin The different cases are illustrated in Figure 1.

We are concerned with the stability of solutions of (E) in addition to their location. At non-degenerate equilibria this is determined by the Jacobian J of the map $(r,z) \mapsto (r(\varepsilon + az), \mu - z^2 + br^2)$ :

$$J(r,z) = \begin{pmatrix} \varepsilon + az & ar \\ 2br & -2z \end{pmatrix}$$

The condition that an equilibrium point p with real eigenvalues be nondegenerate is that $\det(J(p)) \neq 0$. Degenerate equilibria also occur if J has pure imaginary eigenvalues.

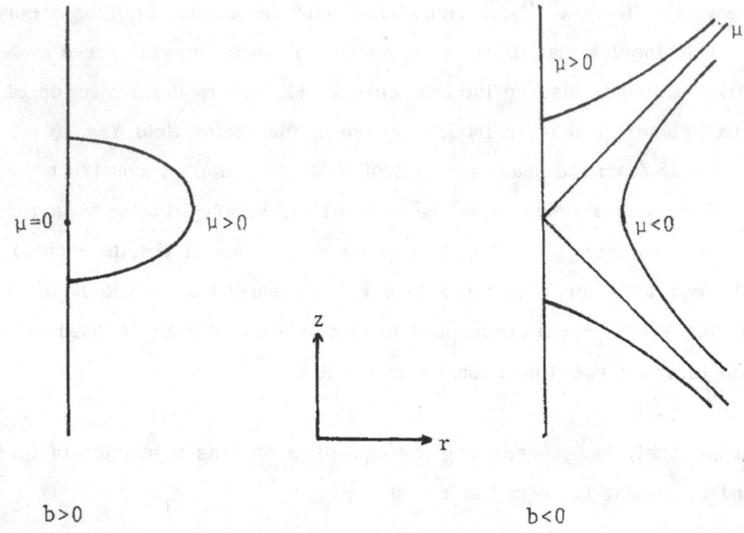


Figure 1

Zero sets of equation (E2)

This is the case of the Hopf bifurcation leading to periodic orbits in the flow of J and is discussed later. Now $\det J(r,z) = -2z(\varepsilon + az) - 2abr^2$. An equilibrium with $r = 0$ is degenerate if $z(\varepsilon + az) = 0$. If $z = 0$, then $\mu = 0$ giving the $\varepsilon$-axis as a curve of parameter values for which there are degenerate equilibria. If $z = -\varepsilon/a$, then the equation (E2) implies that $a^2\mu = \varepsilon^2$. This defines a second curve of parameter values for which there are degenerate equilibria. At an equilibrium with $r > 0$, equation (E1) implies that $\det J(r,z) = -2abr^2$ since $\varepsilon + az = 0$. Thus $\det J(r,z) \neq 0$ and only Hopf bifurcations occur when $r > 0$

We can now proceed to a description of the equilibria of $Y_{\mu,\varepsilon}$ in each of the regions of the parameter plane bounded by the curves described above. Before doing so, we pause to assess how far this information brings us toward a complete description of the phase portrait of $Y_{\mu,\varepsilon}$. As a plane vector field, the solutions of $Y_{\mu,\varepsilon}$ asymptotically approach either (1) an equilibrium, (2) a periodic orbit, or (3) a closed curve which is the union of saddle-point equilibria and their separatrices. The computations we have been doing determine the equilibria and their stability. Finding periodic orbits and saddle loops (closed curves composed of saddle separatrices and saddle equilibria) is more difficult, and we shall expend more effort in the process. In most cases we shall rely on the results of numerical computations in these arguments involving the disposition of periodic orbits.

Return now to a consideration of the equilibria of $Y^{a,b}_{\mu,\varepsilon}$. For the various cases which present themselves, we want to discuss the location and stability of the equilibria near the origin as a function of $(\mu, \varepsilon)$. In all cases, we have determined that the degenerate equilibria with real eigenvalues occur on one of the two curves in the $(\mu, \varepsilon)$ plane defined by

$$\mu = 0 \qquad \text{(B1)}$$

$$\mu = \varepsilon^2/a^2 \qquad \text{(B2)}$$

All of the points of (B1) and (B2) are bifurcation points of equilibria near the origin.

Thus there are four regions which we need to consider : the half plane, the two regions between the curves (B1) and (B2), and the region to the right of (B2).

The bifurcations represented by (B1) are the coallesence and birth/destruction of equilibria on the z-axis. For $\mu > 0$, there are two equilibria at $(r,z) = (0, \pm\sqrt{\mu})$ in all cases. For $\mu < 0$, there are no equilibria on the z-axis. The bifurcations represented by (B2) are ones which involve the coallesence of an equilibrium on the z-axis with one lying in the interior of the $(r, z)$ half plane. (Note that if we allow negative values of r, then the symmetry of Y with respect to reflection in the z-axis means that there is a corresponding equilibrium in the left half plane which also coallesces with the other two.) Which side of the curve $\mu = \varepsilon^2/4a^2$ represents parameter values for which there is an equilibrium of Y off the z-axis depends upon the sign of b. If we write $\tilde{z} = z - \varepsilon/a$ and $\tilde{\mu} = \mu - \varepsilon^2/a^2$, then (E1) and (E2) become $r(a\tilde{z}) = 0$ and $\tilde{\mu} - 2\varepsilon\tilde{z}/a - \tilde{z}^2 + br^2 = 0$. If $\tilde{z} = 0$ then we have $\tilde{\mu} + br^2 = 0$. For this to have a solution with r and $\varepsilon$ small, b and $\tilde{\mu}$ must have opposite signs. Thus if $b > 0$, the equilibria with $r > 0$ but small occur for parameter values to the left of the curve (B2).

If $b < 0$, these equilibria occur for parameter values to the right of the curve (B2).

Let us now examine the stability of the various equilibria of Y near the curves (B1) and (B2). On these curves, we have $\det J = 0$ at the bifurcating equilibrium. The simplest thing to do now is to give individually a discussion of the four cases I : $a > 0$, $b = 1$; II : $a > 0$, $b = -1$; III : $a < 0$, $b = 1$; IV : $a < 0$, $b = -1$.

Case I : ($a > 0$, $b = 1$) We know here that there is one equilibrium off the z-axis to the left of the curve $\mu = \varepsilon^2/a^2$. For this equilibrium, we have $\det J(r, z) = -2ar^2$. Here $\det J < 0$ so that the equilibrium is a saddle. For the equilibria on the z-axis, we have $\det J = -2z(\varepsilon + az)$ and $\operatorname{tr} J = \varepsilon + (a-2)z$. Therefore, the stability of the equilibrium is determined by the signs of $z$, $\varepsilon + (a-2)z$, and $\varepsilon + az$. The two possible values of z are $\pm\sqrt{\mu}$. In the region $\mu > \varepsilon^2/a^2$, $z$ and $\varepsilon + az$ have the same sign at both equilibria. Therefore $\det J < 0$ and both equilibria are saddles. In the regions with $0 < \mu < \varepsilon^2/a^2$, one of the equilibria has $\det J > 0$ and the other has $\det J < 0$. When $\varepsilon > 0$, the equilibrium with $z = -\sqrt{\mu}$ is a source while the equilibrium with $z = \sqrt{\mu}$ is a sink when $\varepsilon < 0$. This is all summarised in Figure 2 where we draw the bifurcation diagram of the $(\mu, \varepsilon)$ plane and the local phase portraits of Y for all parameter values. In this case, our description of the flow of Y must be complete since there are no equilibria off the z-axis which have positive index. If either saddle loops or closed orbits occur, then inside must be an equilibrium of positive index.

Case II : ($a > 0, b = -1$) In this case, the equilibria of Y near the origin with $r > 0$ occur for parameter values with $\mu > \varepsilon^2/a^2$. In the left half $\mu < 0$ of the parameter plane, the flow of Y is trivial with $\dot{z} < 0$ along all solutions. The equilibria on the z-axis are analysed as in Case I. We still have $\det J = -2z(\varepsilon + az)$ and $\operatorname{tr} J = \varepsilon + (a-2)z$. The possible values of z are again $\pm\sqrt{\mu}$. In the region defined by $0 < \mu < \varepsilon^2/a^2$, $\varepsilon > 0$, we have a saddle point at $(0, \sqrt{\mu})$ and a source at $(0, -\sqrt{\mu})$. In the region defined by $0 < \mu < \varepsilon^2/a^2$, $\varepsilon < 0$, there is a sink at $(0, \sqrt{\mu})$ and a saddle at $(0, -\sqrt{\mu})$. When $\varepsilon^2/a^2 < \mu$, both of these equilibria are saddles. There is the additional equilibrium with $r > 0$ when $\varepsilon^2/a^2 < \mu$. At this equilibrium, $\det J = 2ar^2 > 0$.

Now $\operatorname{tr} J = -2z$ at this equilibrium. There are parameter values for which $\operatorname{tr} J > 0$ and those for which $\operatorname{tr} J < 0$. The stability of the equilibrium changes from being a sink to a source as z changes sign. This behaviour together with the location

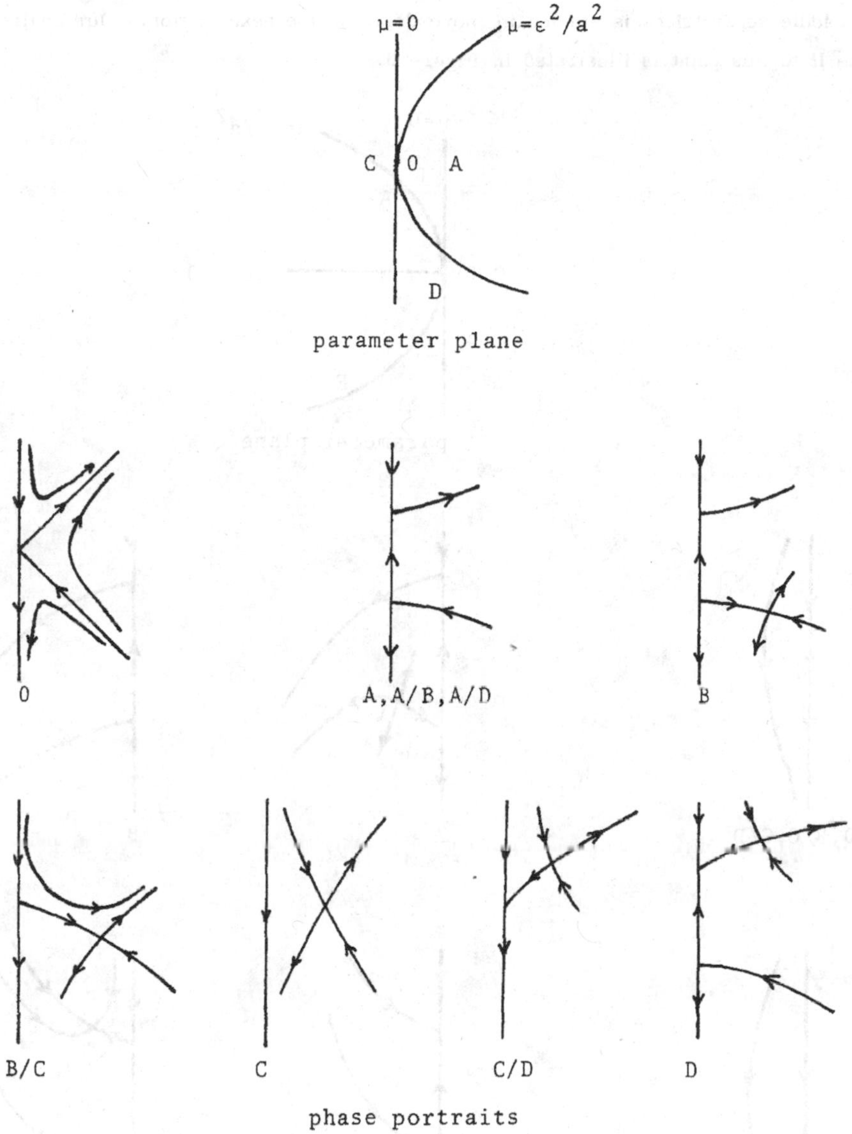


Figure 2
Birfurcation diagram of Case I and phase portraits

$$\dot{r} = r(\varepsilon + az + r^2), \quad a>0$$
$$\dot{z} = \mu - z^2 + r^2$$

of the saddle separatrices is studied in more detail in the next section. Our analysis of Case II to this point is illustrated in Figure 3.

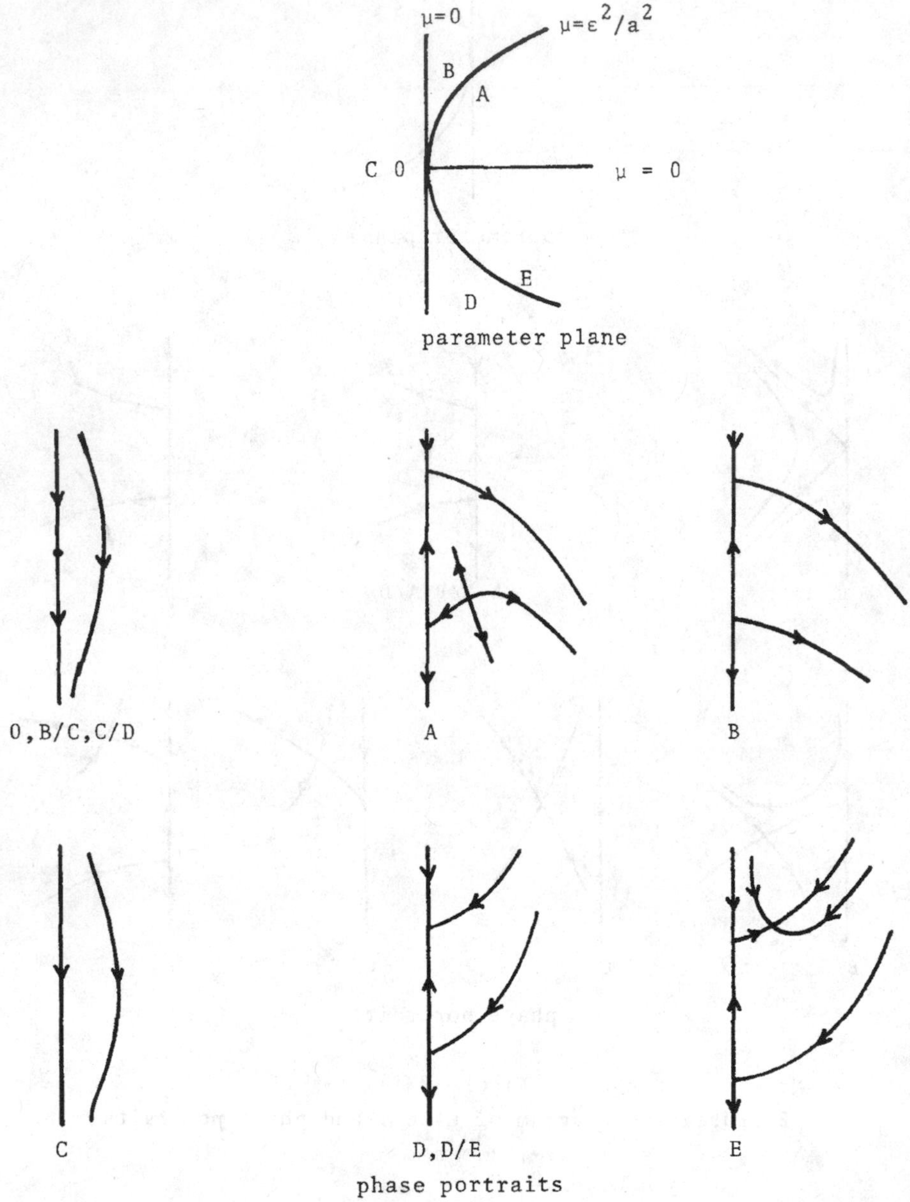


Figure 3

Partial phase portrait of Case II

$$\dot{r} = r(\varepsilon + az + r^2), \quad a > 0$$

$$\dot{z} = \mu - z^2 - r^2$$

Case III : (b = 1, a < 0) Since b > 0, the equilibria off the z-axis occur in the region $\mu < \varepsilon^2/a^2$. The stability analysis of the equilibria on the z-axis is slightly different from the preceding cases. We still have $z = \pm\sqrt{\mu}$, $\det J = -2z(\varepsilon + az)$, and $\operatorname{tr} J = \varepsilon + (a-2)z$. When $\varepsilon^2/a^2 < \mu$, $\det J > 0$ at both $(0, \pm\sqrt{\mu})$ with tr J having opposite signs at these two points. There is a sink at $(0, \sqrt{\mu})$ and a source at $(0, -\sqrt{\mu})$. In the region $0 < \mu < \varepsilon^2/a^2$, $\varepsilon > 0$, $(0, -\sqrt{\mu})$ is still a source, but $(0, \sqrt{\mu})$ is a saddle. In the region, $0 < \mu < \varepsilon^2/a^2, \varepsilon < 0$, $(0, -\sqrt{\mu})$ is a saddle and $(0, \sqrt{\mu})$ is a sink. It is clear from index considerations (or direct computation) that the equilibrium off the z-axis is not a saddle. We have $\det J = -2ar^2 > 0$ and $\operatorname{tr} J = -2z$ which has the opposite sign to z. Near $\mu = \varepsilon^2/a^2$, the equilibrium is a sink or a source depending upon whether $\varepsilon > 0$ or $\varepsilon < 0$. Once again, additional bifurcations must occur, and we consider the non-equilibrium behaviour of this case in more detail below. The phase portraits are shown in Figure 4.

Case IV : (a < 0, b = -1) This case has the simplicity of the first. For an equilibrium near the origin not on the z-axis, $\det J = 2ar^2 < 0$ and the equilibrium is a saddle. These occur for parameter values $\mu > \varepsilon^2/a^2$. The equilibria on the z-axis satisfy $z = \pm\sqrt{\mu}$, $\det J = -2z(\varepsilon + az)$ and $\operatorname{tr} J = \varepsilon + (a-2)z$. When $\mu > \varepsilon^2/a^2$ these are a sink and a source. When $0 < \mu < \varepsilon^2/a^2$ and $\varepsilon > 0$, $(0, -\sqrt{\mu})$ is a source and $(0, \sqrt{\mu})$ is a saddle. When $0 < \mu > \varepsilon^2/a^2$ and $\varepsilon < 0$, $(0, \sqrt{\mu})$ is a saddle and $(0, \sqrt{\mu})$ is a sink. The phase portraits are shown in Figure 5. No further analysis of this case is required for the same reasons as those which apply to Case I.

In Cases II and III, there is non-equilibrium asymptotic behaviour of some solutions. This remains to be analysed. Our technique for doing so corresponds to Taken's "Hamiltonian bifurcations" [35]. Let us describe this approach intuitively before examining the particular computations involved in the two examples we wish to study. The bifurcations of equilibria with r > 0 occur when $\varepsilon = 0$. We shall show that the corresponding system Y are integrable; i.e. there is a function constant along its trajectories. This behaviour is structurally unstable since these systems contain whole families of periodic orbits. Adding cubic terms to the vector fields $Y_{\mu,\varepsilon}$ usually destroys this structural instability and yields "robust" families. For $(\mu, \varepsilon)$ small, the cubic terms will be small compared to the quadratic ones. A variational argument describes how each periodic solution behaves (to first order in the parameter $\mu^2 + \varepsilon^2$).

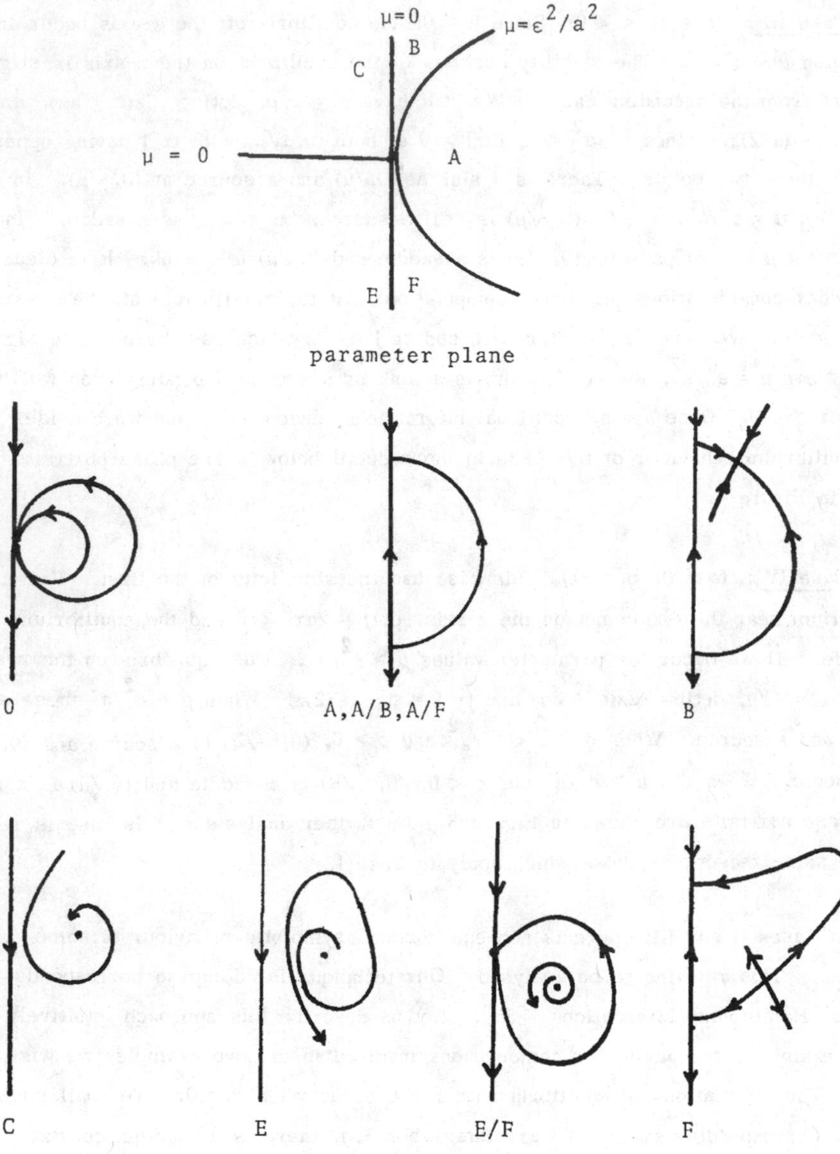


Figure 4

Partial phase portrait of Case III

$$\dot{r} = r(\varepsilon + az + r^2), \quad a<0$$

$$\dot{z} = \mu - z^2 + r^2$$

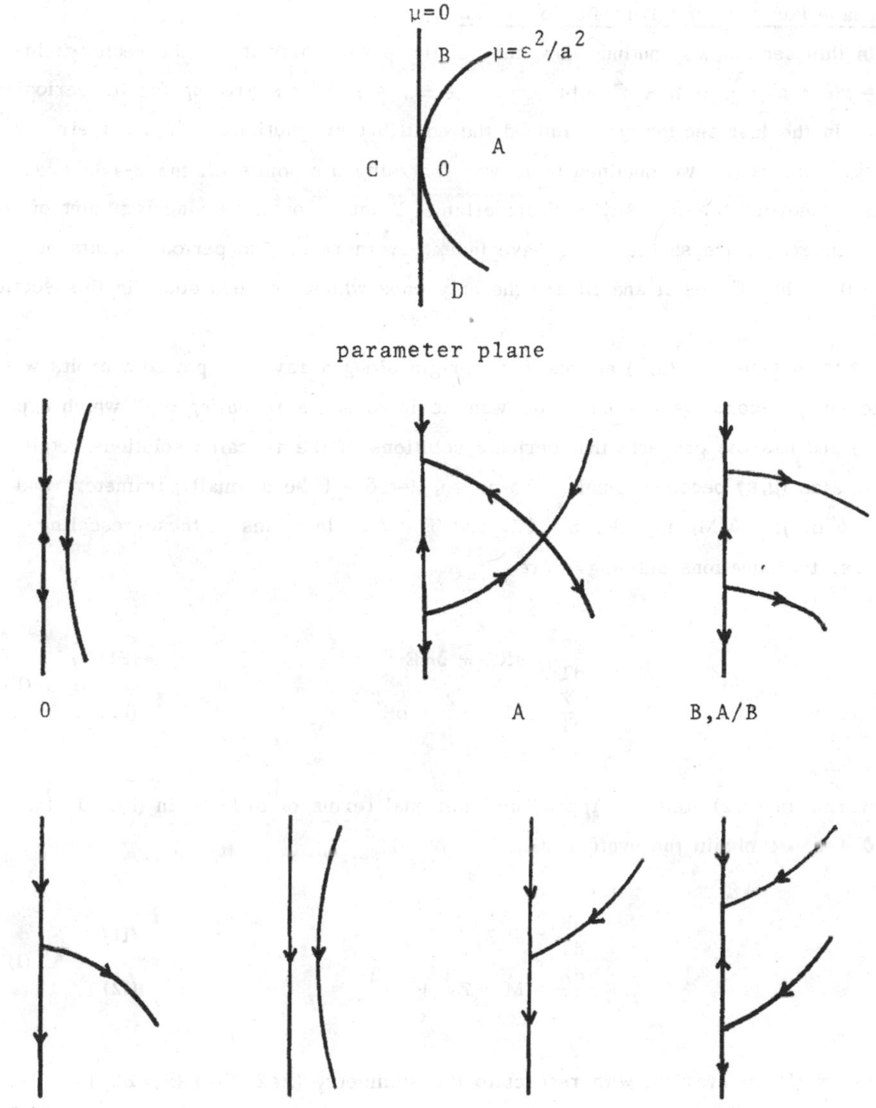


Figure 5

Phase portraits of Case IV

$$\dot{r} = r(\varepsilon + az + r^2), \quad a<0$$
$$\dot{z} = \mu - z^2 - r^2$$

## §3. Phase Portraits II : Periodic Solutions.

In this section we continue our study of the phase portrait of the vector field $Y^{a,b}_{\mu,\varepsilon} = r(\varepsilon + az)\frac{\partial}{\partial r} + (\mu - z^2 + br^2)\frac{\partial}{\partial z}$, $b = \pm 1$, $a \neq 0$ by searching for its periodic orbits. In the last section we examined the equilibrium solutions of Y and their stability. One result we obtained there was that singular points off the z-axis are saddles whenever $ab > 0$. Any periodic orbit of Y must contain a singular point of positive index. Since saddle points have index -1, there are no periodic orbits of Y if $ab > 0$. Thus Cases II and III are the only ones which we need study in this section.

As the parameters $(\mu, \varepsilon)$ approach the origin along a ray, the periodic orbits we want to study become very small. We want to introduce a rescaling of Y which depends on $(\mu, \varepsilon)$ and has the property that periodic solutions of the rescaled solutions retain their size as $(\mu, \varepsilon)$ becomes small. To do so, let $\delta > 0$ be a small parameter, and set $\varepsilon = \delta^2 E$, $\mu = \delta^2 M$, $r = \delta R$, $z = \delta Z$, and $T = \delta t$. In terms of these rescaling variables, the equations defining Y are

$$\left.\begin{aligned} \frac{dR}{dT} &= aRZ + \delta RE && \text{(P1)} \\ \frac{dZ}{dT} &= M - Z^2 + bR^2 && \text{(P2)} \end{aligned}\right\} \quad \text{(P)}$$

Cubic terms in $(r, z)$ added to Y produce additional terms of order $\delta$ in (P). In the limit $\delta \to 0$, we obtain the system

$$\left.\begin{aligned} \frac{dR}{dT} &= aRZ && \text{(I1)} \\ \frac{dZ}{dT} &= M - Z^2 + bR^2 && \text{(I2)} \end{aligned}\right\} \quad \text{(I)}$$

The system (I) is invariant with respect to the symmetry $(R, Z, T) \to (R, -Z, -T)$. This suggests that it is integrable because integral curves which intersect the R-axis are symmetric with respect to reflection in this axis.

We seek an integral of (I) by trying to find a new vector field parallel to Y which is Hamiltonian. Recall that the vector field $f\frac{\partial}{\partial x} + g\frac{\partial}{\partial y}$ is Hamiltonian $\frac{\partial f}{\partial x} + \frac{\partial g}{\partial y}$ is identically zero. In this case, we can find a function $H(x, y)$ such that $f = -\frac{\partial H}{\partial y}$ and $g = \frac{\partial H}{\partial x}$. For the equation (I), there is an integrating factor of the form $R^{\alpha}$ for the correct choice of $\alpha$. Thus the equations

$$\frac{dR}{dT} = R^{\alpha}(aRZ)$$
$$\frac{dZ}{dT} - R^{\alpha}(M - Z^2 + bR^2) \qquad (H)$$

are Hamiltonian if $a(1 + \alpha) R^{\alpha}Z - 2R^{\alpha}Z = 0$ or $\alpha = 2/a - 1$. The Hamiltonian $H(R,Z)$ of (H) is $\frac{a}{2}(MR^{2/a} - R^{2/a}Z^2 + \frac{b}{1+a}R^{2(1+a)/a}$, provided that $a \neq -1$. If $a = -1$, $H(R,Z) = -\frac{1}{2}(MR^{-2} - R^{-2}Z^2 - 2b \log R)$. Note that equations (I) have an equilibrium solution at $(R,Z) = (\sqrt{-M/b}, 0)$ if b and M have opposite signs and at the points $(0, \pm\sqrt{M})$ if $M > 0$. The singular point on the R-axis has pure imaginary eigenvalues if $ab < 0$. The two cases of interest in searching for periodic solutions of Y are II : ($a > 0$, $b = -1$, $M > 0$) and III : ($a < 0$, $b = 1$, $M < 0$).

To find periodic solutions of (P) with cubic terms added, it is convenient to replace these equations with ones which have the same integral parametrised differently and approach (H) as $\delta \to 0$ :

$$\frac{dR}{dT} = aR^{2/a}Z + \delta R^{2/a}E$$
$$\frac{dZ}{dT} = MR^{2/a-1} - Z^2R^{2/a-1} + bR^{2/a+1} \qquad (P')$$

Studying the periodic solutions of (P') as $\delta \to 0$ is now a classical problem in perturbation theory [1]. The solutions of (P') remain close to solutions of (H) for periods of time which are $0(1/\delta)$. Infinitesmally in $\delta$, we can calculate whether a particular solution of (H) remains periodic near $\delta = 0$.

Let us recall these basic results of perturbation theory. If $W = f\frac{\partial}{\partial x} + g\frac{\partial}{\partial y}$ is a vector field on $\mathbb{R}^2$, then the divergence of W is the function $\frac{\partial f}{\partial x} + \frac{\partial g}{\partial y}$. Denoting by $\omega$ the one form $-g\,dx + f\,dy$, we have $\omega(W)$ identically zero and $d\omega = \text{div}(W)dx \wedge dy$. Therefore, Stokes theorem implies that if $\Gamma$ is a periodic orbit of X, then $0 = \int_\Gamma \omega(W) = \int_D \text{div}(W)\,dxdy$ where D is the disk with boundary $\Gamma$. Consider now a periodic solution $\Gamma$ of (H) and consider $(P'_\delta)$ with cubic terms as a deformation of (H) with deformation parameter $\delta$. If there is a continuous family $\Gamma_\delta$ of periodic solutions of these vector fields $W_\delta$ and $D_\delta$ is the interior of $\Gamma_\delta$, then $\int_{D_\delta} \text{div}(W_\delta)dRdZ$ vanishes identically. Now, if $W_\delta$ is the vector field of (P') then $\text{div}(W_\delta) = \frac{2\delta}{a}R^{2/a}(E+c_1R^2+c_2Z^2)$, so that $\int_{D_\delta} R^{2/a}(E+c_1R^2+c_2Z^2)dRdZ$ is zero for all $\delta$. In particular, this equation is satisfied for $\delta = 0$. Thus a necessary criterion for $\Gamma$ to lie in a family of periodic orbits $\Gamma_\delta$ of $W_\delta$ is that the function

$$\pi(\Gamma) = \int_{\text{int}\,\Gamma} R^{2/a}(E + c_1 R^2 + c_2 Z^2)\,dR\,dZ$$

defined on periodic solutions of (P') should vanish at $\Gamma$. Conversely, if we parametrise the periodic solutions of (P') smoothly (say, by a transverse coordinate or by the area contained inside an integral curve), then the conditions $\pi(\Gamma) = 0$ and $\frac{d\pi(\Gamma)}{d\Gamma} \neq 0$ imply that there will be a one parameter family $\Gamma_\delta$ of periodic orbits of $W_\delta$ with $\Gamma_0 = \Gamma$.

The behaviour of periodic solutions (of period $0(1/\delta)$) of (P') is therefore determined by the behaviour of the function $\pi$.

We have computed the relevant integrals numerically for a sampling of values of a when $c_2 = 0$ and $c_1 = 1$. The following remarks are therefore observations based upon numerical evidence rather than proved facts. There are three intervals on the a-line which yield different qualitative behaviour. These are $(0, \infty)$, $(-1, 0)$, and $(-\infty, -1)$. The first of these intervals gives Case II examples and the second and third intervals give Case III examples. The value $a = -1$ appears as exceptional. For a in each of these intervals, the solutions E(c) of L appear to vary monotonically with c or $R_0$. Figures 6, and 7 indicate the completion of the bifurcation diagrams for these cases with $c_2 = 0$.

To find the value of E for which a given level curve $\Gamma$ of H is the limit of a family of periodic orbits, one solves the equation $\pi(\Gamma) = 0$ for E. The results of this calculation depend upon the values of $c_1$ and $c_2$. For most values of $c_1$ and $c_2$, one expects the equation $\pi(\Gamma)$ to implicitly define E as a Morse function of $c = H(\Gamma)$. When E(c) is a Morse function, then the qualitative behaviour of the periodic orbits with respect to parameters will remain the same with perturbations of $W_\delta$.

The simplest behaviour for E(c) to have is that E is a monotone function of c. This situation corresponds to the existence of at most one limit cycle for each vector field in the family $W_\delta$. Since there are two independent coefficients $c_1, c_2$ on which $\pi(\Gamma)$ depends linearly, it is apparent that there are values of $(c_1, c_2)$ for which E(c) is not monotone. There are different choices of cubic terms in our normal forms

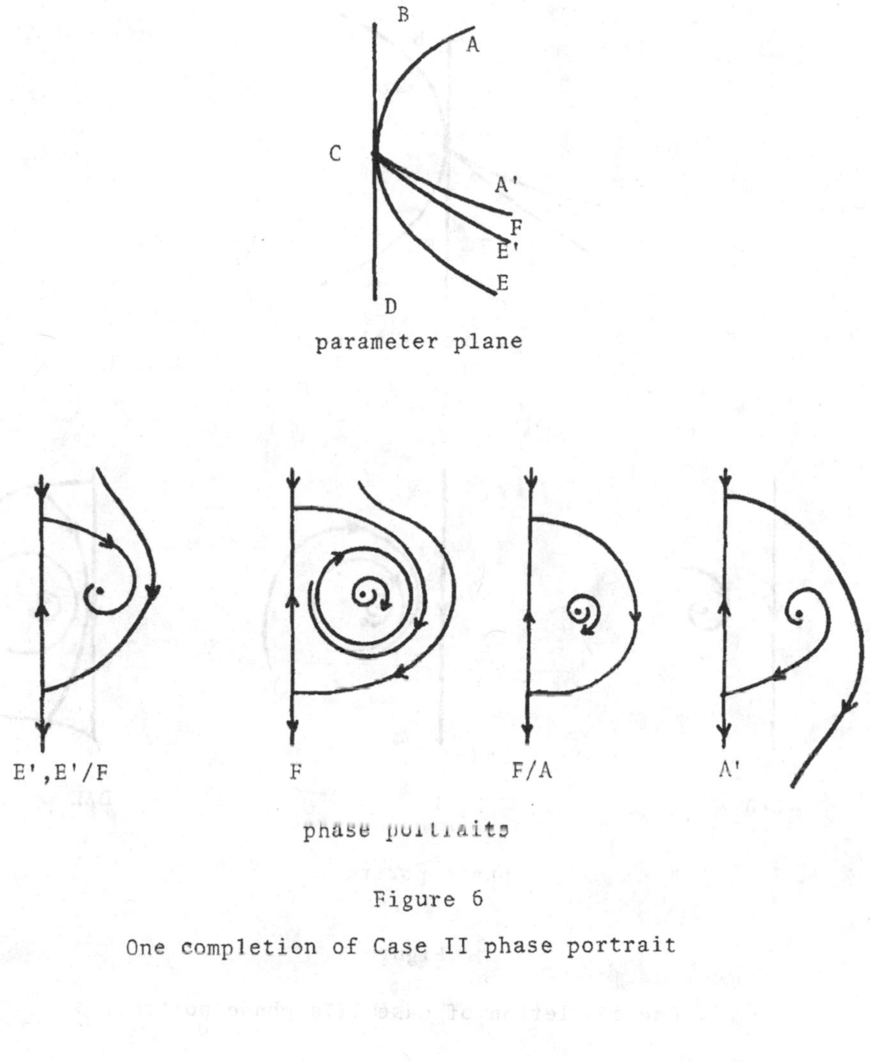


Figure 6

One completion of Case II phase portrait

which produce subtly different bifurcations diagrams, but we have not undertaken a systematic study of these.

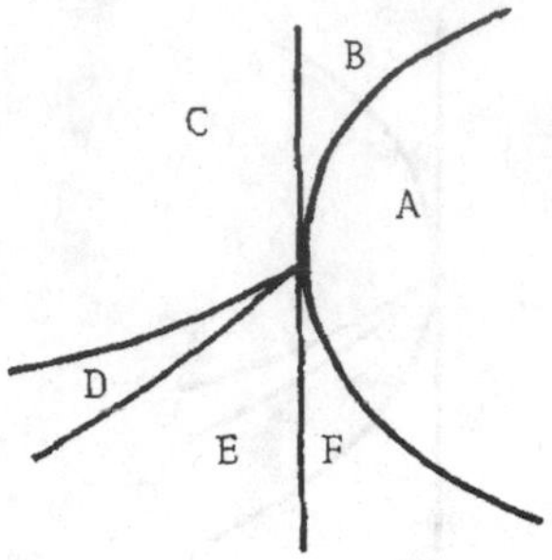


parameter plane

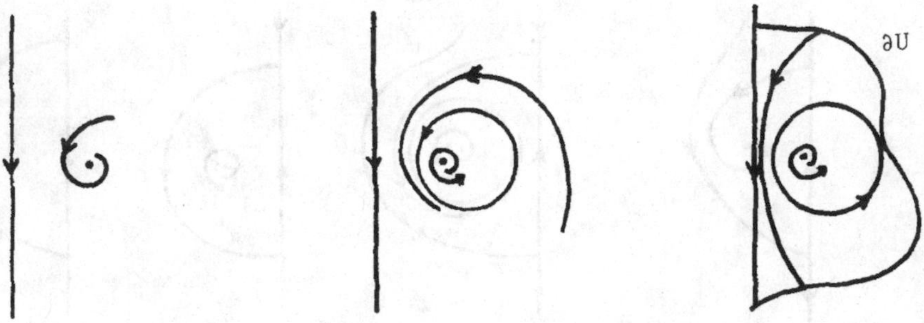


C/D D D/E

phase portraits

Figure 7

One completion of Case IIIa phase portrait

## §4. Remarks and Aperiodic Behaviour.

In the two previous sections we have studied the orbit structure of the two dimensional vector fields $Y_{\mu,\varepsilon}$. In this section, we shall make a few remarks about variants of this problem and the relationship with the bifurcations of the three dimensional vector field X which was the beginning of our discussion. Leaving the class of $S^1$ symmetric vector fields, we shall find interesting aperiodic solutions.

There are at least three variants of the problem we study which seem to have substantial interest. These come from restricting our attention to classes of vector fields which satisfy some additional constraints beyond those we have considered. In the first of these, one examines vector fields X which are equivariant with respect to the group $S^1 \times \mathbb{Z}/2\mathbb{Z}$. Here $\mathbb{Z}/2\mathbb{Z}$ acts by reflection along the z-axis of $\mathbb{R}^3$ and $S^1$ acts by rotation around the z-axis. These fields satisfy the equation $X(-x) = -X(x)$. The effect of this restriction on our analysis is that the two dimensional system for $(r, z)$ must have the same sort of symmetry in z that it does for r. The expression of $\dot{r}$ is odd in r and even in z, while the expression of $\dot{z}$ is even in r and odd in z. The unfolding of this problem is discussed in [48]. Holmes gives information about some perturbations of the degenerate vector field.

The second variant of the bifurcation problem considers volume preserving vector fields of $\mathbb{R}^3$. These are important as models for the velocity field of an incompressible fluid. Volume preserving flows do not have stable limit cycles, so one expects that the bifurcation structure of these examples will be quite different. If the vector fields also preserve an $S^1$ symmetry, then the reduced vector fields on the plane will also be area preserving with respect to the density $r dr \wedge dz$. This places stringent conditions on the lower order terms of the two dimensional vector fields one studies. Broer has some results on this case [47].

The third variant of the problem involves assuming that there is a "trivial" equilibrium solution for all values of the parameter. This has the effect of inducing one to look at the two dimensional family $W_{\alpha,\beta}$ defined by

$$\begin{aligned} \dot{r} &= r(\alpha + az) \\ \dot{z} &= \beta z - z^2 + br^2 \end{aligned} \qquad \text{(T)}$$

If we make the simple coordinate change $\bar{z} = z - \beta/2$ then these equations become

$$\begin{aligned} \dot{r} &= r(\alpha - a\beta/2 + a\bar{z}) \\ \dot{\bar{z}} &= \beta^2/4 - \bar{z}^2 + br^2 \end{aligned}$$

Thus each vector field $W^{a,b}_{\alpha,\beta}$ is equivalent by translation to the vector field $Y^{a,b}_{\mu,\varepsilon}$ with $\mu = \beta^2/4$ and $\varepsilon = \alpha - \alpha\beta/2$. This allows us to easily deduce the bifurcation diagram of W from that of Y. For all cases, we obtain only values of $\mu \geq 0$. The curves (B1) and (B2) given by $\mu = 0$ and $\mu = \frac{\varepsilon^2}{a^2}$ become $\beta = 0$ and $\alpha^2 - \alpha\beta/a = 0$. These describe 3 lines in the $\alpha, \beta$ plane : $\alpha = 0$, $\beta = 0$, and $a\alpha = \beta$. Only in Case II are periodic solutions possible. These occur in two cusp shaped sectors tangent to the line $\alpha = a\beta/2$. When $a = \sqrt{2}$, then we have the degenerate situation that these sectors are tangent to the bifurcation curve $a\alpha = \beta$.

We next make a few remarks about the unfoldings of vector fields X in $\mathbb{R}^3$ which are not equivariant with respect to an $S^1$ symmetry. The normal form theorem [34] can be used to introduce coordinates on $\mathbb{R}^3$ (which will depend smoothly on $(\mu, \varepsilon)$) so that the Taylor polynomial of X of degree k in cylindrical coordinates has no dependence on $\theta$. This is true for any k if X is $C^\infty$. Consequently, when $(\mu, \varepsilon)$ is small, the vector field $X_{\mu,\varepsilon}$ differs from one which is symmetric by an amount which can be expressed as a flat function of $(\mu^2 + \varepsilon^2)$. This means that even after rescaling X so that its non-equilibrium behaviour occurs on a fixed scale as $(\mu, \varepsilon)$ becomes small, X is still very close to being symmetric.

Nonetheless, X will generally not be symmetric with respect to an $S^1$ action, and this will have an important qualitative effect on its phase portrait in some instances. The most dramatic of these occurs when Y belongs to Case II and the parameter values of Y are chosen so that its flow has a saddle loop. The symmetric vector field X has a flow whose phase portrait can be deduced from the phase portrait of Y. The flow of X has an invariant surface $\Sigma$ homeomorphic to the two dimensional sphere. The intersections of this sphere with the z-axis are two equilibria p and q. The top one p is a saddle with two dimensional unstable manifold $\Sigma - \{q\}$ and one dimensional stable manifold in the z-axis. The bottom equilibrium q has two dimensional stable manifold $\Sigma - \{p\}$ and one dimensional unstable manifold in the z-axis. Trajectories starting inside $\Sigma$ and not on the z-axis have trajectories which asymptotically spiral down $\Sigma$ and then return to the top of the sphere near the z-axis. Each time a trajectory makes this circuit it gets closer to $\Sigma$. Trajectories outside $\Sigma$ spiral down past it on their way from $z = +\infty$ to $x = -\infty$. The flow is illustrated in Figure 8.

The intersections of stable and unstable manifolds of the saddle points in the flow described above can be drastically changed by non-symmetric perturbations of the normal form equations. We want to explore conditions in which a lack of symmetry is associated with the occurrence of trajectories having aperiodic asymptotic behaviour. Our results here rely upon a theorem of Silnokov in which he proves that homoclinic

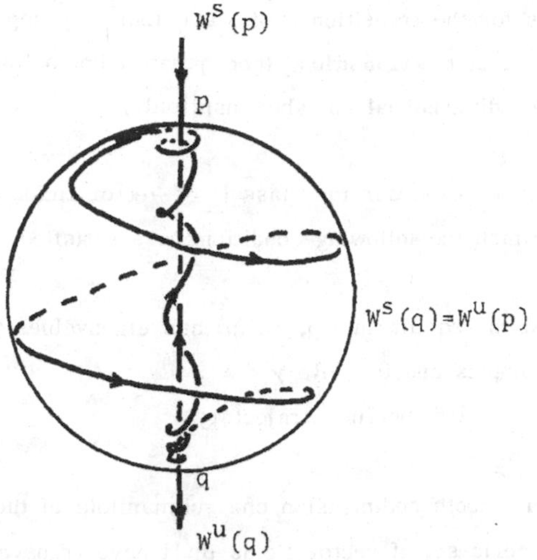


Figure 8

The symmetric homoclinic flow of X
One orbit inside the sphere is shown .

trajectories for certain saddle points in a three dimensional flow are the limits of sets of transversal homoclinic orbits.

We first establish that the requisite kinds of homoclinic orbits occur in generic perturbations of the symmetric flows. The vector field X has p and q with dim $W^s(p)$ = dim $W^u(q) = 1$ and $W^s(p) \cap W^u(q)$ a segment of the z-axis. In a generic one parameter family of three dimensional vector fields, two one dimensional stable and unstable manifolds will not intersect. Thus generic nonsymmetric perturbations of X will have $W^s(p) \cap W^u(q) = \emptyset$. Inside the family $X_{\mu,\varepsilon}$, there is a transition between vector fields for which trajectories which pass near q with z increasing remain bounded and vector fields for which these trajectories tend to $z = -\infty$ after passing near q. Perturbations of $X_{\mu,\varepsilon}$ still have trajectories which make this transition. The two possibilities for the transition orbits are that they approach p or q as $t \to \infty$. If $W^s(p) \cap W^u(q) = \varphi$ at the transition, then q has a homoclinic orbit (which is the upper branch of its one dimensional unstable manifold.)

Theorem (Silnikov) Consider the class H of vector fields X on $\mathbb{R}^3$ which have an equilibrium at which the following conditions are satisfied.

(1) At the equilibrium p, DX(p) has eigenvalues $\gamma$, $\bar{\gamma}$ and $\lambda$ with $\gamma$ complex and $0 < -\mathrm{Re}\,\gamma < \lambda$ .

(2) p has a homoclinic trajectory.

Then H forms a smooth codimension one submanifold of the space of $C^\infty$ vector fields on $\mathbb{R}^3$ and a generic set of vector fields in H have transverse homoclinic trajectories.

We give a heuristic description of the proof of this theorem which indicates the conditions which imply the existence of homoclinic trajectories. The argument relies upon an examination of the return map T around the homoclinic trajectory. Introduce two cross sections $\Sigma_0$ and $\Sigma_1$ to the homoclinic trajectory which lie near p and decompose T into $T = T_1 \circ T_0$ with $T_0 : \Sigma_0 \to \Sigma_1$ and $T_1 : \Sigma_1 \to \Sigma_0$. If the trajectories flowing from $\Sigma_0$ to $\Sigma_1$ pass near p, then $T_0$ describes the flow behaviour near p while $T_1$ describes the flow around the "global" portion of the homoclinic trajectory. Figure 9 shows the location of $\Sigma_0$ and $\Sigma_1$.

We describe the properties of $T_0$ by doing calculations which assume that the vector field is linear in the neighbourhood U of p which contains trajectory segments from $\Sigma_0$ and $\Sigma_1$. Coordinates are chosen so that the vector field is in (real) Jordan normal form, and $\Sigma_0$ and $\Sigma_1$ are chosen to be parallel to coordinate hyperplanes in cylindrical coordinates. This special case gives the correct qualitative behaviour of $T_0$ for the general case. In cylindrical coordinates on U, the flow of X is defined explicitly by

$$\Phi(r, \theta, z; t) = (\exp(t\mathrm{Re}\gamma)r, \qquad \theta + t(\mathrm{Im}\gamma), \qquad \exp(t\lambda)z) .$$

The sections $\Sigma_0$ and $\Sigma_1$ are defined by $r = c_0$, $z = c_1$ respectively. The map $T_0$ sends the point $(c_0, \theta, z)$ with $z > 0$ to the point $(\bar{r}, \bar{\theta}, \bar{c}_1)$ where

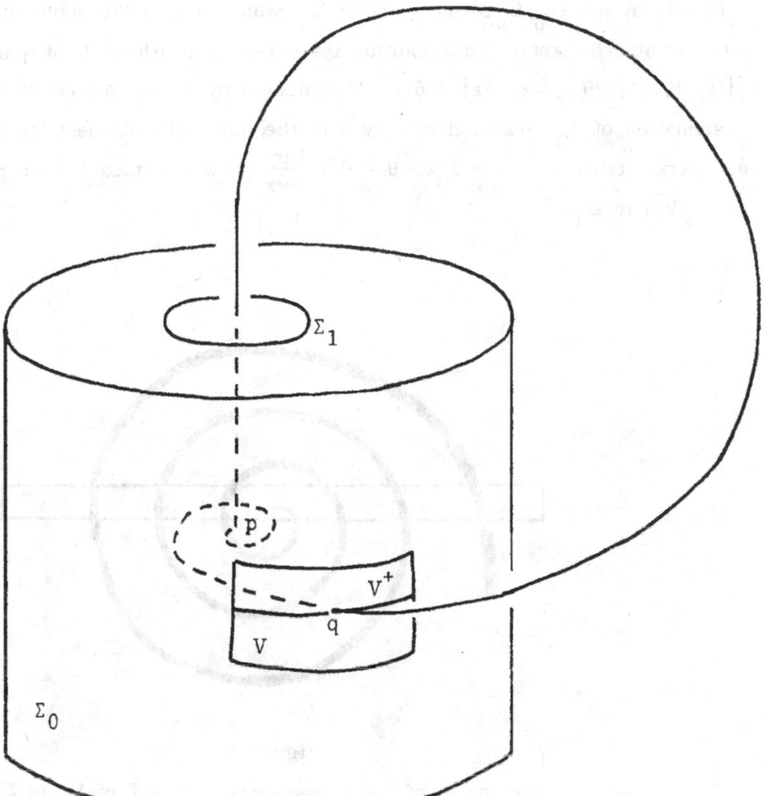


Figure 9

The cross sections to the homoclinic trajectory of Silnikov's theorem

$$\bar{r} = c_0\left(\frac{z}{c_1}\right)^{\frac{-\mathrm{Re}\gamma}{\lambda}}$$

$$\bar{\theta} = \theta + \frac{\mathrm{Im}\gamma}{\lambda}\log\left(\frac{c_1}{z}\right)$$

Since $0 < \frac{-\mathrm{Re}\gamma}{\lambda} < 1$, $\frac{\partial \bar{r}}{\partial z} \to \infty$ as $z \to 0$. This property is essential to the described properties of T.

Denote by $q = (c_0, \theta_0, 0)$ the point of $\Sigma_0$ which lies in the homoclinic orbit. We want to examine the return behaviour of a small neighbourhood V of q in $\Sigma_0$ defined by $V = \{(c_0, \theta, z) \mid |\theta-\theta_0| < \varepsilon, |z| < \delta\}$. Also denote by $V^+$ the subset of V with $z > 0$. The calculation of $T_0$ implies that $T_0(V^+)$ is the union of congruent logarithmic spirals defined parametrically by $\bar{r} = c_0 w$, $\bar{\theta} = \theta + \frac{\mathrm{Im}\gamma}{\mathrm{Re}\gamma} \log w$ for fixed $\theta$ near $\theta_0$. Figure 10 shows $T_0(V^+)$ in $\Sigma_1$.

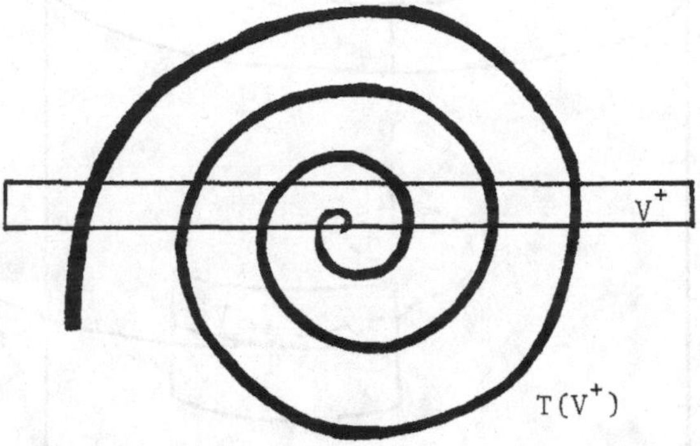


Figure 10.

The image of the cross-section map T of $V^+$ in $\Sigma_0$.

The map $T_1$ is nonsingular and sends a neighbourhood of $(0, \theta, c_1)$ in $\Sigma_1$ to a neighbourhood of q in $\Sigma_0$. We want to examine $V^+ \cap T_1(T_0 V^+)$. Since $T(V^+)$ looks like a union of spirals converging to q, $V^+ \cap T(V^+)$ will contain an infinite number of components. The image of each vertical segment in $V^+$ runs through each component of $V^+ \cap T(V^+)$.

To finish our sketch of the proof, we estimate the derivative of T in an effort to establish its hyperbolicity. We have $T = T_1 \circ T_0$ so $DT = DT_1 \cdot DT_0$. Since $T_1$ is nonsingular, our initial estimate for $DT_1$ is that it be a constant, nonsingular matrix $D = (d_{ij})$. As for $DT_0$, we have

$$T_0\binom{\theta}{z} = \binom{\bar{r} \cos \bar{\theta}}{\bar{r} \sin \bar{\theta}}$$

in a Cartesian coordinate system for $\Sigma_1$. Therefore

$$DT = D\begin{pmatrix} -\bar{r}\sin\bar{\theta} & \bar{r}_z\cos\bar{\theta}-\bar{r}\sin\bar{\theta}\,\bar{\theta}_z \\ \bar{r}\cos\bar{\theta} & \bar{r}_z\sin\bar{\theta}+\bar{r}\cos\bar{\theta}\,\bar{\theta}_z \end{pmatrix}$$

$$= \begin{pmatrix} \bar{r}(-d_{11}\sin\bar{\theta}+d_{12}\cos\bar{\theta}) & \bar{r}_z(d_{11}\cos\bar{\theta}-d_{12}\sin\bar{\theta}) + \bar{r}\bar{\theta}_z(d_{11}\sin\bar{\theta}+d_{12}\cos\bar{\theta}) \\ \bar{r}(-d_{21}\sin\bar{\theta}+d_{22}\cos\bar{\theta}) & \bar{r}_z(d_{21}\cos\bar{\theta}-d_{22}\sin\bar{\theta}) + \bar{r}\bar{\theta}_z(d_{21}\sin\bar{\theta}+d_{22}\cos\bar{\theta}) \end{pmatrix}$$

using $\bar{r}_\theta = 0$ and $\bar{\theta}_\theta = 1$. Compute

$$\bar{r}_z = \frac{-\mathrm{Re}\gamma}{\lambda}\, c_0 c_1^{\frac{\mathrm{Re}\gamma}{\lambda}}\, z^{-\frac{\mathrm{Re}\gamma}{\lambda}-1}$$

$$\bar{\theta}_z = -\frac{\mathrm{Im}\gamma}{\lambda c_1}\, z^{-1} \quad .$$

Therefore, the expected magnitude of the terms in the second column of DT is $z^{\frac{-\mathrm{Re}\gamma}{\lambda}-1}$ which becomes large as $z \to 0$. The magnitude of the terms in the first column in DT is $z^{-\frac{\mathrm{Re}\gamma}{\lambda}}$ which becomes small as $z\to 0$. Provided that the coefficient of $z^{-\frac{\mathrm{Re}\gamma}{\lambda}-1}$ in the lower right corner of DT is not close to 0, DT has a small eigenvalue whose eigenvector is almost horizontal and a large eigenvalue whose eigenvector is far from horizontal. If we choose V with $\delta/\varepsilon$ small, then the image of $V^+$ intersects itself in many components which cut across $V^+$ in directions far from horizontal. The map T has a hyperbolic structure for this intersection which contracts in a direction that is approximately horizontal and stretches in the direction of the eigenvectors of DT. Figure 10 shows $T(V^+) \cap V^+$.

## §5. An example :

In this section we consider a model problem to which the theory developed in the earlier sections can be applied. The example is a reaction-diffusion equation with highly idealised kinetics. We are able to demonstrate the existence of aperiodic and homoclinic solutions in this system of reaction-diffusion equations by locating parameter values for which the evolution of the system reduces to the Case II studied above. Diffusion plays an essential role here insofar as spatially homogeneous aperiodic solutions of these equations cannot exist. There must be variations of the spatial structure of the chemical concentrations so that the qualitative appearance of the "dissipative structure" which is formed will change with time. The reader might be interested to compare this analysis with other recent experimental literature involving aperiodic behaviour in chemical systems [7, 23, 27, 28, 31]

Reaction-diffusion equations have the form

$$\frac{\partial w}{\partial t} = D\nabla^2 w + f(w) \tag{RD}$$

where $w: \mathbb{R}^+ \times \mathbb{R}^m \to \mathbb{R}^n$ with coordinates $(t, \xi)$ in the domain, $\nabla^2$ is the Laplacian in the $\xi$-coordinates, $D$ is an $n \times n$ diagonal matrix of "diffusion constants", and $f: \mathbb{R}^n \to \mathbb{R}^n$ defines the vector field of "chemical kinetics" in concentration space. Typically $w$ will be defined on some bounded region $\Omega \subset \mathbb{R}^n$ with suitable boundary conditions imposed for (RD). We shall work with a particular set of reaction-diffusion equations whose bifurcations have already been studied at considerable length by various people [2, 3, 4, 11, 12, 14, 16, 17, 26]. The equation is commonly known as the "Brusselator" and was created to illustrate the "pattern formation" possibilities of a reaction-diffusion equation.

In this example $m = 1$ and the domain $\Omega$ is the interval $[0, \pi]$. There are two chemical species (n=2), and the equation (RD) is

$$\begin{aligned} \frac{\partial X}{\partial t} &= D_1 \frac{\partial^2 X}{\partial \xi^2} + X^2 Y - (B+1)X + A \\ \frac{\partial Y}{\partial t} &= D_2 \frac{\partial^2 Y}{\partial \xi^2} - X^2 Y + BX \end{aligned} \tag{BR}$$

with boundary conditions $X(0) = X(\pi) = A$, $Y(0) = Y(\pi) = B/A$. The kinetic terms of (BR) correspond to a hypothetical, autocatalytic reaction scheme. Our strategy in locating aperiodic solutions in this example will proceed as follows. Solutions of (BR) can be represented as Fourier series $\sum_{n=0}^{\infty} W_n(t) \sin n\xi$, $W_n(t) = (X_n(t), Y_n(t))$. Using this representation, we find values of the various parameters for which the linearisation of (BR) at the equilibrium solution $w \equiv (A, B/A)$ has pure imaginary (or zero) eigenvalues. These locate bifurcation values of the parameters. Using the centre manifold theory as outlined in [13 or 26], we obtain a system of differential equations on a finite dimensional manifold with degenerate singular point. We can compute the coefficients of the non-linear terms in this equation. Our task is accomplished by showing that, with the appropriate choice of coefficients, we obtain a Case II example of the theory described in earlier sections and that variations of the parameters give an unfolding of the sort we have described.

Employing the Fourier series decomposition of w, we now examine the spectrum of the linearised operator

$$L = \begin{pmatrix} D_1 & 0 \\ 0 & D_2 \end{pmatrix} \frac{\partial}{\partial \xi^2} + \begin{pmatrix} B-1 & A^2 \\ -B & -A^2 \end{pmatrix}$$

at the equilibrium $w = (A, B/A)$. Its eigenvalues are those of the matrices.

$$E_\ell = \begin{pmatrix} B-1-\ell^2 D_1 & A^2 \\ -B & -A^2-\ell^2 D_2 \end{pmatrix}$$

as $\ell$ ranges over the positive integers. We seek parameter values with no eigenvalues having positive real parts (and some having zero real parts.) Observe that $\operatorname{tr} E_\ell = B-1-A^2-\ell^2(D_1+D_2)$ is a decreasing function of $\ell$ because $D_1$ and $D_2$ are positive. If there is an $\ell > 1$ for which $E_\ell$ has pure imaginary eigenvalues, then $\operatorname{tr} E_1 > 0$ and $E_1$ has an eigenvalue with positive real part. Thus we seek pure imaginary eigenvalues only for $E_1$. These occur when $\operatorname{tr} E_1 = B-1-A^2-(D_1+D_2) = 0$ and $\det E_1 = A^2 + A^2 D_1 + D_2 + D_1 D_2 - BD_2 > 0$.

For $\ell > 1$, we have $\det E_\ell = A^2 + \ell^2(A^2 D_1 + D_2 - BD_2) + \ell^4 D_1 D_2$. Thus $\det E_\ell = 0$ if $B = 1 + A^2 D_1/D_2 + D_1 \ell^2 + A^2/D_2 \ell^2$. As a function of $\ell^2$, $\det E_\ell$ is convex and it assumes its minimum value on the integers for one or two values of $\ell$, depending upon $(A, B, D_1, D_2)$. There are two minimum values of 0 for $\det E_\ell$ at k and (k+1) if $A^2 = D_1 D_2 k^2 (k+1)^2$ in addition to B satisfying the equation above. Thus the most severe degeneracy for (BR) occurs when we have a pair of pure imaginary eigenvalues for $E_1$ and 0 eigenvalues at $\ell = k, k+1$. The parameter values for which this degeneracy occurs are given by

$$D_2 = \frac{D_1^2 k^2 (k+1)^2 + 2D_1 k(k+1)}{1 + D_1 k^2 (k+1)^2}$$

$$A^2 = D_1 D_2 k^2 (k+1)^2$$

$$B = 1 + A^2 + D_1 + D_2$$

These computations give us parameter values for which (BR) has more degeneracy than we seek. However, there certainly will be values of $(A, B, D_1, D_2)$ near these for which $E_\ell$ has a zero eigenvalue at $\ell = k$, pure imaginary eigenvalues at $\ell = 1$, and all other eigenvalues have negative real parts. These solutions satisfy the equations

$$A^2 = D_2 k^2 \left( \frac{D_1 + D_2 - D_1 k^2}{1 + D_1 k^2 - D_2 k^2} \right) \qquad \text{(D)}$$

$$B = 1 + A^2 + D_1 + D_2$$

and the inequalities

$$D_2 k^2 \left( \frac{D_1 + D_2 - D_1 k^2}{1 + k^2 (D_1 - D_2)} \right) (1 + (k \pm 1)^2 (D_1 - D_2)) - (k \pm 1)^2 D_2 (D_1 + D_2) + (k \pm 1)^4 D_1 D_2 > 0 .$$

Having determined the parameter values corresponding to the spectrum we seek, we next want to "reduce" the equations (BR) to a finite dimensional invariant subspace in which all of the non-equilibrium asymptotic behaviour takes place. The centre manifold theory [19] guarantees the existence of such a subspace in this problem. On this subspace, the equations (BR) define a flow. For the appropriate parameter choice, this flow has a degenerate equilibrium whose linear part has a zero eigenvalue and a pair of pure imaginary eigenvalues. We want to calculate the coefficients of the higher order terms of this equation in the normal form coordinates of Section §2.

Introduce now coordinates $w = (u, v)$ in concentration space with $u = X - A$ and $v = Y - B/A$. Then the equilibrium of (BR) in these new coordinates is $u = v = 0$ and (BR) can be rewritten as

$$w_t = L(w) + N(w)$$

with $$L(w) = (D_1 \frac{\partial^2 u}{\partial \xi^2} + (B-1)u + A^2 v,\ \ D_2 \frac{\partial^2 u}{\partial \xi^2} - Bu - A^2 v)$$

and $$N(w) = (\frac{Bu^2}{A} + 2Auv + u^2 v)\ \ (1,-1)\ .$$

We choose parameter values satisfying (D), so that there is a pair of pure imaginary eigenvalues and a zero eigenvalue of L, a three dimensional space E spanned by the eigenfunctions of these eigenvalues and $P:C_0^2([0,\pi],\mathbb{R}^2) \to E$. We want to express in E the equations $(Pw)_t = P(Lw + Nw)$ or $w_t = Lw + PNw$ for $w \in E$. These are the "truncated" equations of (BR) which give an approximate description of the flow on the centre manifold. Following the argument in [19, Section 4], the qualitative structure of the truncated and centre manifolds are the same to the extent determined by the fact that the low order terms in their Taylor expansions are the same.

Let us determine E. This space will be spanned by three functions $f_1(\xi) = (\sin \xi)v_1$, $f_2(\xi) = (\sin \xi)v_2$ and $f_3 = (\sin k\xi)v_3$ where $v_3$ is a zero eigenvector of $E_k$ and $\{v_1, v_2\}$ is a basis of $\mathbb{R}^2$. For $v_3$, we make the explicit choice $v_3 = (A^2 + k^2 D_2, -B)$, and for $v_1$ and $v_2$ we take $v_1 = (1,-1)$ and $v_2 = (\det E_1)^{-1/2}(D_2, -(1+D_1))$. This gives $L(f_3) = 0$, $L(f_1) = \lambda f_2$, and $L(f_2) = -\lambda f_1$ with $\lambda^2 = \det E_1$. If $w = \Sigma\alpha_i f_i$, we can calculate Nw and PNw by direct substitution, leading to lengthy formulae.

We write $w = (u, v)$ with

$$u = \beta_1 \sin \xi + \beta_2 \sin k\xi$$
$$v = \beta_3 \sin \xi + \beta_4 \sin k\xi\ .$$

Here

$$\beta_1 = \alpha_1 + (\det E_1)^{-1/2} D_2 \alpha_3$$
$$\beta_2 = (A^2 + k^2 D_2)\alpha_3$$
$$\beta_3 = \alpha_1 + (\det E_1)^{-1/2}(1+D_1)\alpha_3$$
$$\beta_4 = -B\alpha_3\ .$$

Now $N(w) = n(w)\ (1,-1)$ with

$$n(w) = (\frac{B}{A}\beta_1^2+2A\beta_1\beta_3)\sin^2\xi + (\frac{2B}{A}\beta_1\beta_2+2A\beta_1\beta_4+2A\beta_2\beta_3)\sin\xi\sin k\xi$$
$$+ (\frac{B}{A}\beta_2^2+2A\beta_2\beta_4)\sin^2 k\xi + \beta_1^2\beta_3\sin^3\xi + (2\beta_1\beta_2\beta_3+\beta_1^2\beta_4)\sin^2\xi\sin k\xi$$
$$+ (\beta^2\beta_3+2\beta_1\beta_2\beta_4)\sin\xi\sin^2 k\xi + \beta_2^2\beta_4\sin^3 k\xi \quad .$$

From these formulae we want to compute the relevant terms in the normal form of the truncated equations for w. This computation involves taking the projection of N(w) onto the space spanned by the eigenfunctions $f_1, f_2$, and $f_3$ and then extracting the combinations of terms which yield the coefficients of the normal form developed in Section §2. Recalling that X was of the form $\frac{\partial}{\partial\theta} + r(r^2+a_1 z)\frac{\partial}{\partial r} + (a_2 z^2+a_3 r^2)\frac{\partial}{\partial z}$ in cylindrical coordinates, there are three coefficients to compute. We proceed to describe the calculation of these. There are three parts to these calculations : (1) we identify those combinations of terms of a vector field of $\mathbb{R}^3$ with the correct linear part which yield each coefficient in the normal form, (2) we pick out those terms in the expression of n(w) which contribute to each of the combinations identified in (1) and compute the projection of these onto the two dimensional spaces of functions of the form $(\sin\xi)v$ and $(\sin k\xi)v$, and (3) we compute the coordinates of each of these two dimensional spaces with respect to the bases $\{v_1, v_2\}$ and the eigenvectors of $E_k$, respectively. This yields the coefficients of the normal form of the reduced equation.

Let $X(x,y,x) = (X^1, X^2, X^3)$ be a vector field on $\mathbb{R}^3$ with $X(0) = 0$ and $DX(0) = \begin{pmatrix} 0 & \lambda & 0 \\ -\lambda & 0 & 0 \\ 0 & 0 & 0 \end{pmatrix}$. We want explicit expressions for the coefficients of the normal form of X at 0. There are 3 coefficients to compute which are quadratic. The coefficient of $z^2\frac{\partial}{\partial z}$ in the normal form is simply given by $\frac{1}{2}\frac{\partial^2 Z^3}{\partial z^2}$. To find the coefficient of $r^2\frac{\partial}{\partial z}$, consider the space **P** of homogeneous polynomial vector fields of degree 2 of the form $P(x,y)\frac{\partial}{\partial z}$ A basis of **P** is given by $(x^2+y^2)\frac{\partial}{\partial z}$, $(x^2-y^2)\frac{\partial}{\partial z}$, and $2xy\frac{\partial}{\partial z}$. Now $(x^2-y^2)\frac{\partial}{\partial z}$ and $2xy\frac{\partial}{\partial z}$ span the image of DX(0) acting on **P**, and $(x^2+y^2)\frac{\partial}{\partial z}$ is a basis for the complement. Thus it determines the coefficient of $r^2\frac{\partial}{\partial z}$ in the normal form and the expression for the coefficient is $\frac{1}{4}(\frac{\partial^2 X^3}{\partial x^2} + \frac{\partial^2 X^3}{\partial y^2})$. The final coefficient in the normal form is that of $rz\frac{\partial}{\partial r}$. This comes from the space of vector fields of the form zA where A is a linear vector field on $\mathbb{R}^2$. The appropriate basis for this space of vector fields is $\{z(x\frac{\partial}{\partial x} + y\frac{\partial}{\partial y}), z(-y\frac{\partial}{\partial x} + x\frac{\partial}{\partial y}), z(x\frac{\partial}{\partial x} - y\frac{\partial}{\partial y}), z(y\frac{\partial}{\partial x} + x\frac{\partial}{\partial y})\}$. The last two span the image and the first two span the kernel of the action induced by Lie derivatives of DX(0). The vector field $z(-y\frac{\partial}{\partial x} + x\frac{\partial}{\partial y})$ is $z\frac{\partial}{\partial \theta}$ in cyclindrical coordinates, and $z(x\frac{\partial}{\partial x} + y\frac{\partial}{\partial y})$ is the term $rz\frac{\partial}{\partial r}$ we want. Thus the coefficient of $rz\frac{\partial}{\partial r}$ is $\frac{1}{2}(\frac{\partial^2 X^1}{\partial x \partial z} + \frac{\partial^2 X^2}{\partial y \partial z})$.

Next we use the expressions of $\alpha_i, \beta_i$, and N(w) to compute these various coefficients for the Brusselator example. Only certain terms in the formula for n(w) will contribute to each coefficient. We record these.

For coefficient of $z^2\frac{\partial}{\partial z}$ : $(\frac{B}{A}\beta_2^2 + 2A\beta_2\beta_4)\int_0^\pi \sin^3 k\xi d\xi$

For coefficient of $r^2\frac{\partial}{\partial z}$ : $(\frac{B}{A}\beta_1^2 + 2A\beta_1\beta_3)\int_0^\pi \sin^2\xi \sin k\xi d\xi$

For coefficient of $rz\frac{\partial}{\partial r}$ : $(\frac{\partial B}{A}\beta_1\beta_2 + 2A\beta_1\beta_4 + 2A\beta_2\beta_3)\int_0^\pi \sin^2\xi \sin k\xi d\xi$ .

At this point, note that the trigonometric integrals in these formulae are heavily dependent upon the parity of k. If k is even, then the three coefficients are zero because there is the symmetry $\sin k(\pi - \xi) = -\sin k\xi$. Thus to obtain the example we seek, we must take k odd. If k is even then the bifurcation problem has an approximate $S^1 \times \mathbb{Z}/2\mathbb{Z}$ symmetry, and this variant of the theory must be used to analyse the bifurcation structure.

The next step in the computation is the insertion of the expressions of the $\beta_i$ in terms of the $\alpha_i$, so that we obtain the coefficients of the proper part of N(w).

For $z^2\frac{\partial}{\partial z}$ we have

$$(\frac{B}{A}(A^2+k^2D_2)^2 - 2AB(A^2+k^2D_2))\alpha_3^2\int_0^\pi \sin^3 k\xi d\xi$$

and the coefficient is $\frac{1}{2}(\frac{B}{A}k^4D_2^2 - A^3B)\int_0^\pi \sin^3 k\xi d\xi$ .

Denote $d = (\det E_1)^{-1}$. For $r^2\frac{\partial}{\partial z}$, we have

$$\{\frac{B}{A}(\alpha_1+dD_2\alpha_2)^2 - 2A(\alpha_1+dD_2\alpha_2)(\alpha_1+d(1+D_1)\alpha_2\}\int_0^\pi \sin^2\xi \sin k\xi d\xi$$

and the coefficient is $\frac{1}{4}[\frac{B}{A}(1+d^2D_2^2) - 2A(1+d^2D_2)]\int_0^\pi \sin^2\xi \sin k\xi d\xi$ .

For $rz\frac{\partial}{\partial r}$, we have

$$\{\frac{\partial B}{A}(\alpha_1+dD_2\alpha_2)(A^2+k^2D_2)\alpha_3 - 2AB(\alpha_1+dD_2\alpha_2)\alpha_3 - 2A(\alpha_1+d(1+D_1)\alpha_2)(A^2+k^2D_2)\}\int_0^\pi \sin^2 k \sin k\xi d\xi$$

and the coefficient of the $\alpha_1\alpha_3$ term is

$$\frac{1}{2}[\frac{2B}{A}k^2D^2 - 2A(A^2+k^2D_2)] \int_0^\pi \sin^2\xi \sin k\xi d\xi \quad .$$

In computing the coefficient for $rz\frac{\partial}{\partial r}$ we have taken note that the projection of N(w) into a plane spanned by $f_1$ and $f_2$ is parallel to $f_1$. Therefore the vector field PN(w) has no second component relative to the basis $\{f_1, f_2, f_3\}$. The final computation which remains is the expression of the vector (1, -1) in terms of the basis for $\mathbb{R}^2$ $v_3 = (A^2+k^2D_2, -B)$, $v_4 = (A^2, -A^2-k^2D_2)$ consisting of eigenvectors of $E_k$. If we write $(-1,1) = \gamma v_3 + \delta v_4$, then $\gamma = ((A^2 + k^2D_2)^2 - A^2B)^{-1}(A^2+k^2D_2-B)$. This factor $\gamma$ must be multiplied by the formulae above to obtain the final expressions for the coefficients of $z^2\frac{\partial}{\partial z}$ and $r^2\frac{\partial}{\partial r}$ in the normal form of the reduced equation. The results of the computations in this section are summarised in the following table

Table 5.1 : Computation of normal form of reduced Brusselator equations

$$A^2 = D_2k^2(\frac{D_1 + D_2 - D_1k^2}{1 + D_1k^2 - D_2k^2})$$

$$B = 1 + A^2 + D_1 + D_2$$

$$\det E_\ell = A^2 + (A^2D_1 + (1-B)D_2)\ell^2 + D_1D_2\ell^4$$

$$\det E_{k\pm 1} > 0$$

$$d^2 = (A^2 + A^2D_1 + D_2 + D_1D_2 - BD_2)^{-1}$$

$$\gamma = ((A^2 + k^2D_2)^2 - A^2B)^{-1}(A^2 + k^2D_2 - B)$$

coefficients

$$z^2\frac{\partial}{\partial z} \ : \ \frac{\gamma B}{2A}(k^4D_2^2 - A^4)\int_0^{\pi}\sin^3 k\xi d\xi$$

$$r^2\frac{\partial}{\partial z} \ : \ \frac{\gamma}{4A}[B(1+d^2D_2^2) - 2A^2(1+d^2D_2+d^2D_1D_2)]\int_0^{\pi}\sin^2\xi \sin k\xi d\xi$$

$$rz\frac{\partial}{\partial r} \ : \ \frac{1}{A}[Bk^2D_2 - A^2(A^2+k^2D_2)]\int_0^{\pi}\sin^2\xi \sin k\xi d\xi$$

$$r^3\frac{\partial}{\partial r} \ : \ \frac{3\pi d}{4}\{[-1+d^2(d_2^2+D_2(1+D_1))]\int_0^{\pi}\sin^4\xi d\xi - \frac{2d^2}{A^2}[(BD_2-A^2(1+D_1D_2)$$

$$(B-2A^2+d^2(BD_2^2-2A^2D_2(1+D_1)](\int_0^{\pi}\sin^3\xi d\xi)^2\} .$$

Finding parameter values $(D_1, D_2, k)$ for which we obtain a Case II example requires some effort. We merely point out that they do exist with $(D_1, D_2, k) = (.02, .09, 5)$ being one such choice. Here we obtain $A^2 = 1.17$, $B = 2.28$, and the approximate values $d^2 = .926$ and $\gamma = .126$. We have approximate values .16 and .19 for det $E_4$ and det $E_6$. Approximate values of the coefficients in the normal form of the reduced equation are .218 for $z^2\frac{\partial}{\partial z}$, .000268 for $r^2\frac{\partial}{\partial z}$, and -.0397 for $rz\frac{\partial}{\partial r}$. Upon rescaling this gives a Case II normal form with $a < 2$.

Finally we note that this version of the Brusselator equations always has a trivial solution. Thus the unfolding will be given by the variant described in Section §4. To obtain a "regular" Case II example, we can modify the kinetic equations, as in [26], by replacing the constant A by the spatially dependent function $A\dfrac{\cos h\sqrt{\varepsilon}\,(\xi-\frac{\pi}{2})}{\cos h\sqrt{\varepsilon}\,\frac{\pi}{2}}$. For small choices of the parameter $\varepsilon$, there will be perturbations of the above example which yield two parameter families depending upon A and B which give Case II examples of our theory.

References.


[1] V.I. Arnold, Lectures on Bifurcations in Versal Families, Russian Math. Surveys 27 (1972) 54-123.

[2] J.F.G. Auchmuty, Bifurcating waves, Annals NY Acad. Sci. 316 (1979) 263-278.

[3] J.F.G. Auchmuty & G. Nicolis, Bifurcation analysis of non-linear reaction-diffusion equations I, Bull. Math. Biology 37 (1975), 323-365.

[4] J.F.G. Auchmuty & G. Nicolis, Bifurcation analysis of reaction-diffusion equations III, Chemical Oscillations, Bull. Math. Biology 38, (1976), 325-249.

[5] R.I. Bogdanov, Orbital equivalence of singular points of vector fields on the plane, Funct. Anal. Appl. 10, 316-317, 1976.

[6] R.I. Bogdanov, Versal deformations of a singular point of a vector field on a plane in the case of zero eigenvalues, Proceedings of the I.G. Petrovskii Seminar, 2, 37-65, 1976.

[7] H. Degn, L. Olsen & J. Perram, Bistability, Oscillation, and Chaos in an enzyme reaction, Annals N.Y. Acad. Sci. 316, (1979) 623-637.

[8] F. Dumortier, Singularities of vector fields in the plane, Journal of Differential Equations, 23, 53-106, 1977.

[9] N.K. Gavrilov & L.P. Silnikov, On three dimensional dynamical systems close to systems with a structurally unstable homoclinic curve, Math. USSR, Sb. 17, 467-485, 1972, and 19, 139-156, 1973.

[10] M. Golubitsky & D. Schaeffer, A theory for imperfect bifurcation via singularity theory, Comm. Pure Appl. Math. 32 (1979) 21-98.

[11] M. Herschkowitz-Kaufman, Bifurcation analysis of reaction-diffusion equations II, Bull. Math. Biology 37 (1975) 589-636.

[12] M. Herschkowitz-Kaufman & T. Erneux, The bifurcation diagram of model chemical reactions, Annals N.Y. Acad. Sci. 316 (1979) 296-313.

[13] P. Holmes & J. Marsden, Bifurcation to divergence and flutter in flow-induced oscillations : an infinite dimensional analysis, Automatica 14 (1978) 367-384.

[14] J.P. Keener, Secondary bifurcation in non-linear diffusion reaction equations II, Studies in Appl. Math. 55 (1976), 187-211.

[15] N. Kopell & L. Howard, Target patterns and horseshoes from a perturbed central force problem : some temporally periodic solutions to reaction diffusion equations. Preprint 1979.

[16] R. Lefever & I. Prigogine, Symmetry breaking instabilities in dissipative systems II, J. Chem. Phys. 48 (1968), 1695-1700.

[17] T. Mahar & B. Matkowsky, A model chemical reaction exhibiting secondary bifurcation, SIAM J. Appl. Math. 32 (1977), 394-404.

[18] J. Marsden, Qualitative methods in bifurcation theory, Bull. Am. Math. Soc. 84 (1978) 1125-1148.

[19] J. Marsden & M. McCracken, The Hopf Bifurcation Theorem and Its Applications, Springer Verlag, 1976.

[21] S. Newhouse, Diffeomorphisms with Infinitely Many Sinks, Topology, 12, 1974, 9-18.

[22] S. Newhouse, The abundance of wild hyperbolic sets and non-smooth stable sets for diffeomorphisms, mimeographed, IHES, 1977.

[23] S. Newhouse, J. Palis, & F. Takens, Stable Arcs of Diffeomorphisms, Bull. Am. Math. Soc. 82, 499-502, 1976, and to appear.

[24] L.F. Olsen & H. Degn, Chaos in an enzyme reaction, Nature, 267, 1977, 177-178.

[25] J. Palis & F. Takens, Topological equivalence of normally hyperbolic dynamical systems, Topology, 16, 335-345, 1977.

[26] D. Ruelle, Sensitive dependence on initial condition and turbulent behaviour of dynamical systems, conference on Bifurcation Theory and its Applications, N.Y. Academy of Sciences, 1977.

[27] D. Schaeffer & M. Golubitsky, Bifurcation analysis near a double eigenvalue of a model chemical reaction, MRC Technical Report, #1859, 1978.

[28] R.A. Schmitz, K.R. Graziani & J.L. Hudson, Experimental evidence of chaotic states in the Belousov-Zhabotinskii reaction, J. Chem. Phys. 67, 1977, 3040-3044.

[29] R. Schmitz, G. Renola & P. Garrigan, Observations of complex dynamic behaviour in the $H_2$-$O_2$ reaction on nickel. Annals N.Y. Acad, Sci. 316 (1979), 638-651.

[30] L.P. Sil'nikov, A contribution to the problem of the structure of an extended neighbourhood of a structurally stable equilibrium of saddle-focus type, Math. USSR Sb. 10, 91-102, 1970.

[31] S. Smale, Diffeomorphisms with many periodic points, in Differential and Combinatorial Topology, Princeton, 1965, 63-80.

[32] P. Grace Sørensen, Experimental investigations of behaviour and stability properties of attractors corresponding to burst phenomena in the open Belousov reaction, Annals N.Y. Acad. Sci. 316 (1979) 667-675.

[33] J. Sotomayor, Bifurcations of vector fields on two dimensional manifolds, Publ. IHES, #43, 1-46, 1973.

[34] J. Sotomayor, Generic bifurcations of dynamical systems, Dynamical Systems, ed. M. Peixoto, Academic Press, 1973, 561-582.

[35] F. Takens, Singularities of Vector Fields, Publ., IHES, #43, 47-100, 1973.

[36] F. Takens, Forced Oscillations and Bifurcations, Applications of Global Analysis, Communications of Maths. Institute, Rijksuniversiteit, Utrecht, 3, 1974, 1-59.

[37] R. Thom, Structural Stability and Morphogenesis, W.A. Benjamin Inc., Reading, Mass. 1975.

[38] A. Andronov, E. Leontovich, I. Gordon & A. Maier, The Theory of Bifurcation of Plane Dynamical Systems, 1971.

[39] J. Gollub & H. Swinney, Onset of Turbulence in a Rotating Fluid, Physical Review Letters, 35, 1975, 927-930.

[40] M. Herman, Sur la Conjugasion Differentiable des Diffeomorphismes du Cercle a des Rotations, Publ. IHES, 49, 1979, 5-234.

[41] Y. Kuramoto, Diffusion Induced Chaos in Reacting Systems, Suppl. Prog. Theo. Phys. 64, 1978, 346-367.

[42] W. Langford, Periodic and Steady-State Mode Interactions Lead to Tori, SIAM J. Appl. Math., 37, 1979, 22-48.

[43] D. Ruelle & F. Takens, On the Nature of Turbulence, Comm. Math. Phys. 20, 1971, 167-192.

[44] V.I. Arnold, Small denominators I, Mappings of the circle onto itself, Izv. Akad. Nauk, SSSR Ser. Mat. 25, 1961, 21-86.

[45] P. Holmes & J. Marsden, A Partial Differential Equation with infinitely many periodic orbits : chaotic oscillations of a forced beam, preprint.

[46] P. Holmes & F. Moon, A Magnetoelastic Strange Attractor, J. Sound and Vibration 65, 1979, 275-296.

[47] H.W. Broer, Bifurcations of singularities in volume preserving vector fields, thesis, Rijksuniversiteit te Groningen, 1979.

[48] W.F. Langford & G. Iooss, Interactions of Hopf and pitchford bifurcations, Workshop on Bifurcation Problems, Birkhauser Lecture Notes, 1980.

[49] P. Holmes, A strange family of three dimensional vector fields near a degenerate singularity, J. Diff. Eq. 37, 382-403, 1980.

J. Guckenheimer : Department of Mathematics, University of California, Santa Cruz, California, USA.

# Stability and bifurcation in a parabolic equation.

J.K. Hale. *

Abstract.

Recent results on the stability of equilibrium solutions of a parabolic equation are given with indications of the proofs. Particular attention is devoted to dependence of the stability properties on the shape of the domain and the manner in which nonhomogeneous stable equilibria can occur through a bifurcation induced by varying the domain.

In this paper, we present a few recent results on the asymptotic behaviour of the solutions of a parabolic equation of the form

$$\begin{aligned} u_t &= \Delta u + f(u) \quad \text{in } \Omega \\ \frac{\partial u}{\partial n} &= 0 \quad \text{on } \partial\Omega \end{aligned} \tag{1}$$

where $\Omega$ is a bounded open set in $\mathbb{R}^n$ with smooth boundary. We also discuss how the qualitative behaviour of the stable equilibria depend upon the shape of $\Omega$ and the nonlinear function f. The function f is supposed to satisfy conditions which ensure that Eq. (1) defines a strongly continuous semigroup $T_f(t)$ on $H^1(\Omega)$ .

We remark that the boundary conditions in (1) are not important as far as the spirit of the problems to be discussed. Of course, the details will depend generally in a very significant manner upon the boundary conditions.

Let $E_f(\Omega)$ be the set of equilibrium solutions of (1); that is, the set of solutions of the equation

$$\begin{aligned} \Delta u + f(u) &= 0 \quad \text{in } \Omega \\ \frac{\partial u}{\partial n} &= 0 \quad \text{on } \Omega \ . \end{aligned} \tag{2}$$

If $\Omega = (0,\lambda) \subseteq \mathbb{R}$, then the set $E_f(0,\lambda)$ is given by

*This research was supported in part by the National Science Foundation under MCS-79-0-774, in part by the United States Army under AROD DAAG 27-79-C-0161, and in part by the United States Air Force under AF-AFOSR 76-3092C.

$$E_f(0,\lambda) = \{\text{ periodic solutions of period } 2\lambda \text{ of the equation } u_{xx} + f(u) = 0\} \tag{3}$$

In the following, we let $W^u(\varphi)$, $W^s(\varphi)$ denote, respectively, the unstable and stable manifolds for an equilibrium point $\varphi$ of (1). The following result for $n = 1$ is due to a number of authors. The references are in the proof.

Theorem 1. ($n=1$) If $\Omega = (0,\lambda) \subset \mathbb{R}$, then

(i) the $\omega$-limit set of any bounded solution of (1) is in $E_f(0,\lambda)$.

(ii) the $\omega$-limit set of any bounded solution of (1) is a single point in $E_f(0,\lambda)$.

(iii) the only stable equilibrium points of (1) are constants.

(iv) If $\int_0^u f(s)ds \to -\infty$ as $u \to \pm\infty$, then every solution of (1) is bounded. If, in addition, $E_f(0,\lambda)$ is a bounded set, then there is a maximal compact invariant set $A_f(0,\lambda)$ of (1),

$$A_f(0,\lambda) = \cup_{\varphi \in E_f(0,\lambda)} W^u(\varphi),$$

$A_f(0,\lambda)$ is uniformly asymptotically stable and, for any bounded set $B$ in $H^1(\Omega)$, $\text{dist}(T_f(t)B, A_f(0,\lambda)) \to 0$ as $t \to \infty$ .

(v) If, in addition to the hypothesis in (iv), all $\varphi \in E_f(0,\lambda)$ are hyperbolic, then, for any bounded set $B$ in $H^1(\Omega)$, the set

$$(\cup_{\varphi \text{ stable}} W^s(\varphi)) \cap B$$

is open and dense in $B$.

Proof. (i) This is due to Chafee [4] and is independent of the boundary conditions. The idea is very simple. If

$$V(u) = \int_0^\lambda (u_x^2 - \int_0^u f(s)ds)dx$$

then the derivative of $V$ along the solutions of (1) satisfies

$$\dot{V}(u) = -\int_0^\lambda u_t^2 dx \le 0 .$$

Since every bounded orbit is precompact, a simple application of the invariance principle implies the result.

(ii) This result is due to Matano [13] and is independent of the boundary conditions He used a rather sophisticated application of the invariance principle. We sketch a proof based on the theory of dynamical systems. The details will appear in Hale and Massatt [6]. The idea is very simple and can be traced to Malkin [12], Hale and Stokes [7] and perhaps even further. If $\varphi$ is an element of an $\omega$-limit of an orbit which is not a single point, then $\varphi$ belongs to a continuum in $E_f(0,\lambda)$ and $\varphi$ is not hyperbolic. The linear variational equation about $\varphi$ has the simple eigenvalue zero with all other eigenvalues being nonzero. Thus, $\varphi$ belongs to a smooth one dimensional submanifold M of $E_f$. In modern terminology $\varphi$ is normally hyperbolic. One now can show that any solution of (1) which remains in a sufficiently small neighbourhood of M for t sufficiently large must be on $W^s(M)$, the stable manifold of M. Finally, one shows that each orbit in $W^s(M)$ approaches a single point.

(iii) This result is due to Chafee [4]. The following proof is taken from a preprint of Bardos, Matano and Smoller [1]. Suppose $\varphi$ is a nonconstant equilibrium solution of (1). Then $v = d\varphi/dx \not\equiv 0$, $v = 0$ at $x = 0$, $x = \lambda$, and v satisfies the equation

$$v_{xx} + f'(\varphi)v = 0 .$$

Let $\sigma_N, \sigma_D$ be the spectrum of this differential operator with, respectively, homogeneous Neumann and Dirichlet boundary conditions. Since $\inf \sigma_N < \inf \sigma_D$ and $0 \in \sigma_D$, the result is proved.

(iv) Using the function V(u) in part (i), one easily shows that every solution is bounded. Thus, the $\omega$-limit set of every solution of (1) belongs to $E_f(0,\lambda)$. Since $E_f(0,\lambda)$ is bounded, there is a bounded set B such that every solution of (1) eventually enters B; that is, Eq. (1) is point dissipative. Since the semigroup $T_f(t)$ is compact for $t > 0$, the results follow from the general theory of dissipative processes (see, for example, Hale [9,10]).

(v) This result is due to Henry [11]. The idea of the proof is to observe first that the solution operator for the linear variational equation of (1) about any point is one-to-one. This can then be used to show that, for any $\varphi \in E_f(0,\lambda)$ which is not stable, the set $W^s(\varphi) \cap B$ is nowhere dense in B for any bounded set B. As remarked by Mañé, the one-to-oneness of this solution operator implies that the stable manifold $W^s(\varphi)$ can actually be given globally a manifold structure. This also gives a proof of the assertion in (v).

For $\Omega = (0, \lambda)$, we have remarked that the set $E_f(0,\lambda)$ coincides with the set of $2\lambda$-periodic solutions of $u_{xx} + f(u) = 0$. For any $a \in \mathbb{R}$, let $u(x,a)$ be the solution of this equation with $u(0,a) = a, u_x(0,a) = 0$. If a is such that $u(x,a)$ is periodic in x, let $2\lambda_f(a)$ be the period. The function $u(.,a) \in E_f(0,\lambda_f(a))$. For f(u) an arbitrary cubic polynomial in u, Smoller and Wasserman [15] have shown that the function $\lambda_f(a)$ has a finite number ($\leq 2$) of maxima and minima and the second derivative of $\lambda$ at these points is different from zero; that is, $\lambda_f(a)$ is a Morse function.

The above result has important implications for the applications. In fact, for f any cubic polynomial in u, and for $\lambda$ fixed and different from the maxima and minima of the function $\lambda_f(a)$, the set $E_f(0,\lambda)$ consists only of hyperbolic points. For $\lambda$ equal to one of the extreme values of $\lambda_f(a)$, there is a bifurcation of the saddle-node type.

The following qualitative result of Brunovsky and Chow [2] has recently been proved.

**Theorem 2.** There is a residual set $\mathcal{F} \in C^2(\mathbb{R})$ with the Whitney topology such that, for any $f \in \mathcal{F}$, the function $\lambda_f(a)$ above is a Morse function.

The proof is not trivial because the function f depends only on u and not on (x, u). The proof is based on a detailed analysis of an analytic expression of $\lambda_f(a)$ as a function of f, a. It is not a trivial exercise because there is no simple way to determine the qualitative properties of the derivatives of this function in a from the derivatives of f. In fact, there are nonlinear functions f for which $\lambda_f(a)$ is constant (see, for example, Urabe [16]).

Theorem 2 can be appropriately generalised to other boundary conditions (see Brunovsky and Chow [2]). Smoller and Wasserman [15] have also considered other boundary conditions.

Our next objective is to discuss the extent to which the above results are valid for a bounded set $\Omega$ in $\mathbb{R}^n$.

**Theorem 3.** If $\Omega$ is a bounded set in $\mathbb{R}^n$ with smooth boundary, then

(i) the $\omega$-limit set of a bounded orbit is in $E_f(\Omega)$.

(ii) the $\omega$-limit set of a bounded orbit is a single point if the following condition is satisfied :

If $\varphi \in E_f(\Omega)$ is not hyperbolic and k is the dimension of the null space of the operator $\Delta + f'(\varphi)$ in $\Omega$ with homogeneous Neumann conditions, then $\varphi$ belongs to a smooth submanifold of dimension k.

(iii) $\Omega$ convex implies the only stable equilibrium points are constants.

(iv) Same statement as (iv) in Theorem 1 holds.

(v) Same statement as (v) in Theorem 1 holds.

Proof. The proof of (i), (ii), (iv) and (v) are essentially the same as the corresponding assertions in Theorem 1.

(iii) This result was independently discovered by Casten and Holland [3], Matano [14]. The proof exploits special properties of the Laplacian on convex regions to prove that the linear variational equation has a negative eigenvalue for any nonconstant equilibrium.

An analogue of Theorem 2, as far as hyperbolicity of equilibrium and saddle node bifurcations, is not known for $\Omega$ in $\mathbb{R}^n$ and seems to be rather difficult.

We remark that part (iii) of Theorem 3 is also valid for $\Omega$ convex and the equations

$$\begin{aligned} u_t &= \Delta u + f(u), \quad \text{in } \Omega \\ v_t &= -g(u, v) \\ \frac{\partial u}{\partial n} &= 0 \quad \text{on } \partial\Omega \end{aligned}$$

provided that the spectrum of the operator $\partial g(\xi,\eta)/\partial v$ belongs to the set $\{z: \mathrm{Re}\, z > 0\}$ uniformly in $\xi, \eta$ (see Bardos, Matano and Smoller [1]).

The remainder of the discussion centres around the case when $\Omega$ is not convex and the objective is to understand more about the set of stable equilibrium. Before doing this, we make the important remark that, when $\Omega$ is convex, the qualitative structure of the stable equilibria is independent of the nonlinearity f. When $\Omega$ is not convex, this will no longer be the case.

The following result is due to Matano [14].

Theorem 4. There is a nonlinear function f and $\Omega \subseteq \mathbb{R}^n$, $n \geq 2$, such that (1) has a stable nonconstant equilibrium.

Idea of proof. Suppose f has zeros only at $a < 0 < b$, they are simple and $\int_0^u f(s)ds \to -\infty$ as $u \to \pm\infty$. Suppose $\Omega$ has the shape shown in Figure 1 and let $\lambda_2$ be the minima of the

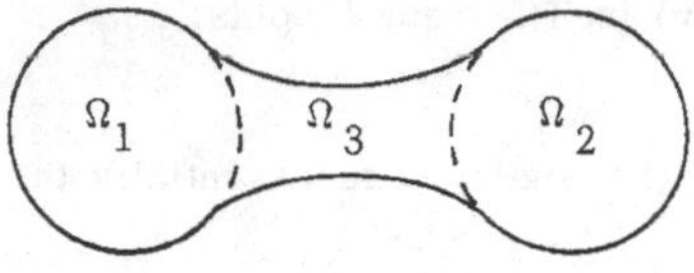


Figure 1.

second eigenvalues of the Laplacian on $\Omega_1$ and $\Omega_2$. Matano [14] gives a specific continuous function $G:\mathbb{R}^2 \to \mathbb{R}$ such that the set where $G < 0$ is nonempty with the property that, if $\lambda_2, \Omega_3$ are such that $G(\lambda_2, \text{meas } \Omega_3) < 0$, then there is a nonconstant stable equilibrium of (1). For fixed $\Omega_1, \Omega_2$, he shows there is an $\Omega_3$ such that the above inequality is satisfied.

The idea of the proof is the following. Let Y be the subset of functions u in $H^1(\Omega)$ such that $\int_{\Omega_1} u > 0$, $\int_{\Omega_2} u < 0$. If meas $\Omega_3$ is small enough, it is then shown that the set Y has a certain invariance property with respect to $T_f(t)$. A careful application of the maximum principle gives a minimal equilibrium solution $v_m$ in Y stable from below and a maximal equilibrium solution $v_M$ in Y stable from above. If it were known that there are only a finite number of equilibrium solutions, then the proof would be complete. Since this is not known, another argument must be used. Matano first proves that any solution unstable from above must be strongly unstable from above in the sense that it can be isolated from equilibrium solutions from above uniformly in Y. He then puts an ordering on the solutions from above, uses Zorn's lemma and the above property of solutions unstable from above.

In the proof of Theorem 4, the nonlinear function f has three simple zeros at $a < 0 < b$. The equilibrium points a, b are stable and zero is a saddle point. A stable, nonconstant equilibrium solution was shown to exist. The argument of Matano can be used to show there must be another nonconstant equilibrium solution with $\int_{\Omega_1} u < 0$, $\int_{\Omega_2} u > 0$. Thus, there are at least four stable equilibrium solutions and one unstable equilibrium solution. This is impossible dynamically and there must be some other equilibrium solutions which are unstable. In fact, an index argument implies there must be at least three unstable equilibria. Using more of the detailed information from the paper of Matano, one can show there must be at least five unstable equilibrium solutions. Consequently, there are at least nine equilibrium solutions for this nonlinear function f and region $\Omega$.

The basic problem is to understand in more detail how variations in the shape of the domain $\Omega$ causes these additional solutions to appear. We now summarise some work of Hale and Vegas [8] which give a possible explanation.

Let us begin with an intuitive discussion of how the stable nonconstant equilibrium solutions could appear as secondary bifurcations. Suppose $\mu \in [0,\infty)$, $\Omega_\mu$ is a bounded set in $\mathbb{R}^2$ with smooth boundary with the property that $\Omega_0$ is convex and the second eigenvalue $\lambda_2(\mu)$ of $-\Delta$ on $\Omega_\mu$ is a monotone decreasing function of $\mu$, approaching zero as $\mu \to \infty$. Also, suppose the third eigenvalue $\lambda_3(\mu)$ of $-\Delta$ on $\Omega_\mu$ satisfied $\lambda_3(\mu) \geq \delta > 0$ for all $\mu$. Let $f(\nu, u) = \nu^2 u - u^3$, $\nu > 0$, and let $A_{\nu,\mu}$ be the maximal compact invariant set for the equation

$$u_t = \Delta u + f(\nu, u) \quad \text{in } \Omega_\mu$$

$$\frac{\partial u}{\partial n} = 0 \quad \text{in } \partial\Omega_\mu \quad .$$

Fix $\nu$ sufficiently small so that $\nu^2 < \delta$ and the only equilibrium solutions in $\Omega_0$ are the constant functions 0, $\pm\nu$. The set $A_{\nu,0}$ is then the constant functions $\pm\nu$ and the unstable manifold of 0, which is one dimensional. Let $\mu_0$ be such that $\lambda_2(\mu_0) = f'(0) = \nu^2$. At the point $\mu = \mu_0$, the zero solution bifurcates creating two new equilibrium solutions which are unstable. They are unstable because the unstable manifold of zero becomes two dimensional - the direction of bifurcation is independent of the direction of the original unstable manifold in $A_{\nu,\mu}$ for $\mu < \mu_0$. The set $A_{\nu,\mu}$ for $\mu > \mu_0$ but close to $\mu_0$ is then two dimensional with three unstable and two stable equilibria.

Now suppose that $\Omega_\mu$ has the shape shown in Figure 1 and that meas $\Omega_\mu \to 0$ as $\mu \to \infty$. Then we can find a $\mu_1$ such that the inequality in the proof of Theorem 4 is satisfied. Thus, there is a stable nonconstant equilibrium solution. It is conjectured that this occurs as a secondary bifurcation from the unstable nonconstant equilibria discussed above. In Figures 2 and 3 we have depicted, respectively, the set $A_{\nu,\mu}$ as a function of $\mu$ and the conjectured bifurcation diagram.

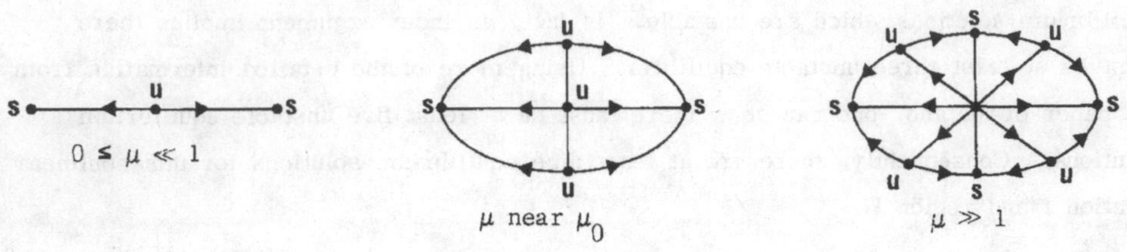


Figure 2.

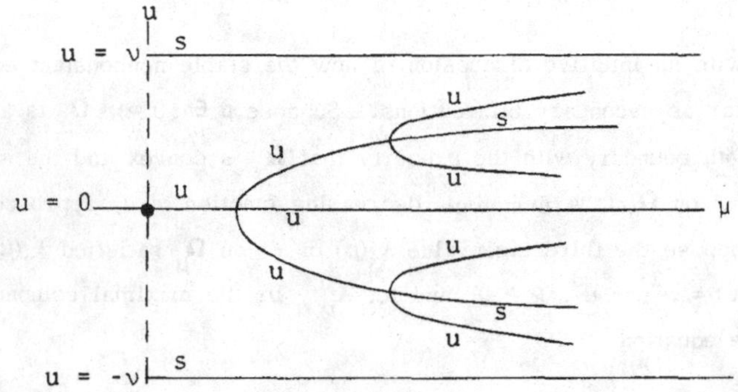


Figure 3

To do things more analytically, we have tried to discuss the neighbourhood of $\mu = \infty$ treating it as a bifurcation problem from a double eigenvalue zero. More specifically, suppose $\mu = \varepsilon^{-1}$ and the region $\Omega_\varepsilon$ is shown in Figure 4, two circles $\Omega_1$, $\Omega_2$ and a small channel between them.

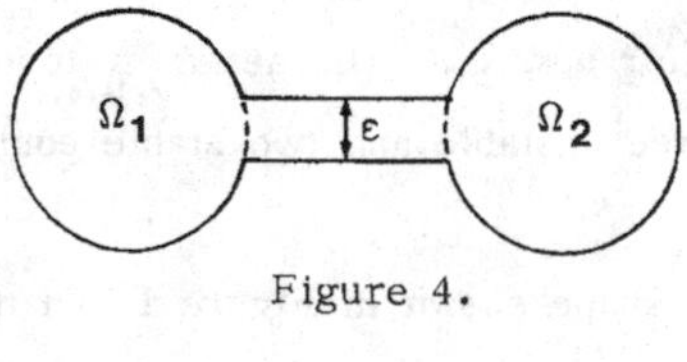


Figure 4.

For $\varepsilon = 0$, it is clear there are nine solutions consisting of all combinations of $0, \pm\nu$ on $\Omega_1$ and $\Omega_2$. Five of these are saddles and four are stable nodes. Let $\lambda_2(\varepsilon)$ be the second eigenvalue of $-\Delta$ on $\Omega_\varepsilon$ and let $w_\varepsilon$ be a unit eigenvector corresponding to $\lambda_\varepsilon$. Let $u_\varepsilon$ be the constant function $(\text{meas } \Omega_\varepsilon)^{-1}$; that is, $u_\varepsilon$ is a unit eigenvector for the eigenvalue 0 of the Laplacian on $\Omega_\varepsilon$.

One can now show that it is possible to apply the method of Liapunov-Schmidt for the solutions of (1) near $u = 0$ for $\varepsilon, \nu$ near zero. More specifically, for real $\alpha, \beta$ sufficiently small and $\varepsilon, \nu$ sufficiently small, there is a function $u^*(\alpha,\beta,\varepsilon,\nu)$ continuously differentiable in $\alpha,\beta,\nu$ and continuous in $\varepsilon$ such that $u^*(0,0,0,0) = 0$, $\partial u^*(0,0,0,0)/\partial(\alpha,\beta) = 0$ and

$$\Delta u^* + f(\nu, \alpha u_\varepsilon + \beta w_\varepsilon + u^*) - \pi f(\nu, \alpha u_\varepsilon + \beta w_\varepsilon + u^*) = 0$$

where $\pi u$ is the projection of $u$ onto the span of the constant function $u_\varepsilon$ and the function $w_\varepsilon$; that is,

$$\pi u = \int_{\Omega_\varepsilon} u + w_\varepsilon \int_{\Omega_\varepsilon} w_\varepsilon u.$$

If $u^*(\alpha,\beta,\varepsilon,\nu)$ satisfies the above, then

$$u = \alpha u_\varepsilon + \beta w_\varepsilon + u^*(\alpha,\beta,\varepsilon,\nu)$$

is a solution of (1) if and only if $(\alpha,\beta,\varepsilon,\nu)$ satisfy the bifurcation equations

$$\int_{\Omega_\varepsilon} f(\nu, \alpha u_\varepsilon + \beta w_\varepsilon + u^*(\alpha,\beta,\varepsilon,\nu))dx = 0$$

$$\int_{\Omega_\varepsilon} w_\varepsilon f(v, \alpha u_\varepsilon + \beta w_\varepsilon + u^*(\alpha,\beta,\varepsilon,\nu))dx = 0$$

are satisfied.

If we let $(\alpha,\beta) = \gamma$, then one can show that these equations have the form

$$c(\gamma) + \varepsilon L_1 \gamma + \nu L_2 \gamma + \text{h.o.t.} = 0$$

where h.o.t. denotes higher order terms in $\gamma, \varepsilon, \nu$, $L_1, L_2$ are constant two by two matrices and $c(\gamma)$ is homogeneous cubic two vector in the components of $\gamma$. One can adapt the method in, for example, Chow, Hale and Mallet-Paret [5] to obtain the complete bifurcation diagram for the solutions of (4). These are shown in Figure 5. Figure 5a shows the number of solutions for a fixed $(\varepsilon,\nu)$ and Figure 5b shows the way

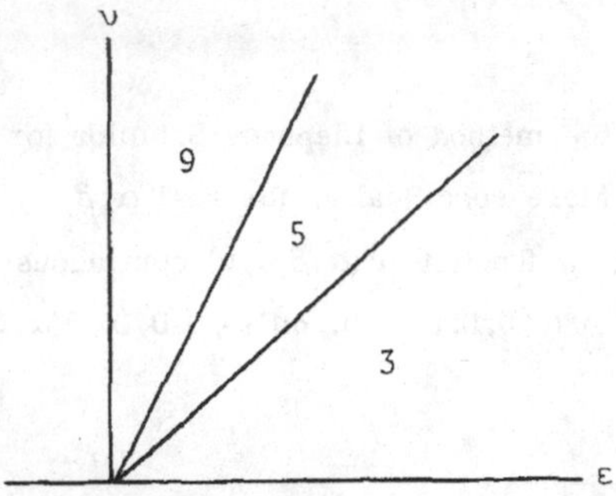


Figure 5a

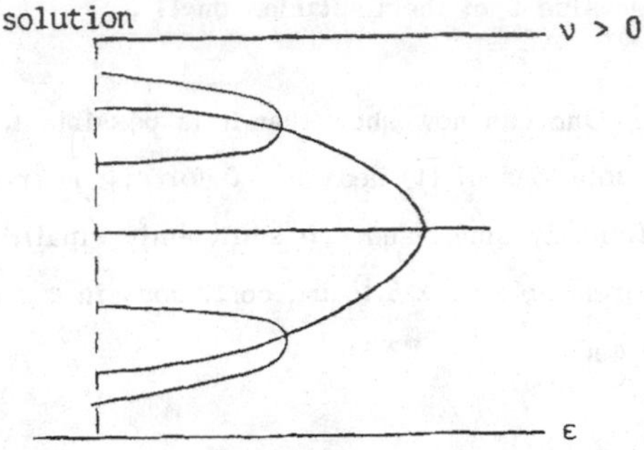


Figure 5b

the solutions bifurcate as a function of $\varepsilon$ for a fixed $\nu$.

In the verification of the previous results, it is crucial to show that the third eigenvalue of the Laplacian on $\Omega_\varepsilon$ is bounded away from zero. It would be very interesting to obtain general geometric conditions on a region $\Omega_\varepsilon$ to have this latter property satisfied.

## References.


1. C. Bardos, H. Matano & J. Smoller, Some results on the instability of the solutions of reaction diffusion equations. Preprint.

2. P. Brunovsky & S.N. Chow, Generic properties of stationary states of reaction diffusion equations. To be submitted.

3. R.G. Casten & C.J. Holland, Instability results for a reaction diffusion equation with Neumann boundary conditions. J. Differential Equations 27 (1978), 266-273.

4. N. Chafee, Asymptotic behaviour of a one-dimensional heat equation with homogeneous Neumann boundary conditions. J. Differential Equations 18 (1975), 111-134.

5. S.N. Chow, J.K. Hale & J. Mallet-Paret, Applications of generic bifurcation, II. Arch. Rat. Mech. Anal. 62 (1976), 209-236.

6. J.K. Hale & P. Massatt, Convergence of solutions in gradient-like systems. To be submitted.

7. J.K. Hale & A. Stokes, Behaviour of solutions near integral manifolds. Arch. Rat. Mech. Anal. 6 (1960), 133-170.

8. J.K. Hale & J. Vegas, Bifurcation with respect to domain in a parabolic equation. Submitted to Arch. Rat. Mech. Anal.

9. J.K. Hale, Functional Differential Equations. Appl. Math. Sci. Vol. 3, Springer-Verlag, 1977.

10. J.K. Hale, Some recent results on dissipative processes. Proc. Symp. on Functional Differential Equations and Dynamical Systems. São Carlos, Brazil, (1979). To appear in Lecture Notes in Math., Springer-Verlag.

11. D. Henry, Geometric Theory of Semilinear Parabolic Equations. To appear in Lecture Notes in Math., Springer-Verlag.

12. I.G. Malkin, Theory of Stability of Motion. Moscow 1952.

13. H. Matano, Convergence of solutions of one-dimensional semi-linear parabolic equations. J. Math. Kyoto University 18 (1978), 221-227.

14. H. Matano, Asymptotic behaviour and stability of solutions of semilinear diffusion equations. Res. Inst. Math. Sci., Kyoto 15 (1979), 401-454.

15. J. Smoller & A. Wasserman, Global bifurcation of steady-state solutions. J. Differential Equations. To appear, 12/1980.

16. M. Urabe, Nonlinear Autonomous Oscillations, Academic Press, 1967.

Jack K. Hale, LCDS - Box F, Brown University, Providence, R.I. 09212, U.S.A.

# Wandering Intervals.

J. Harrison

Let I be the closed interval of real numbers fron -1 to +1. A differentiable function $F:I \to I$ is said to be convex if it has just one critical point at the origin 0, say, and if it is monotone decreasing to the left of 0 and monotone increasing to the right.

In this paper we construct a $C^1$ convex function F which has a "wandering" interval in the sense of Denjoy. That is, there exists a closed interval $J \subset I$ such that the set of forward and inverse images of J under F are disjoint and the complement of the union of their interiors is a Cantor set. This Cantor set is an exceptional minimal set since it is closed and invariant under F, contains no such proper subsets, and is neither periodic nor the entire interval I. (Coven and Nitecki [1] have recently constructed a related example with two turning points by adapting the Denjoy diffeomorphism of the circle.)

It turns out that F is not topologically conjugate to any $C^2$ convex function of I. In fact, if G is $C^2$ and topologically conjugate to F then G has a inflection point in its nonwandering set. It is not known if such a G exists or if there are any $C^2$ maps of the interval with exceptional minimal sets.

I wish to thank H. Whitney for telling me about this problem which is stated as a question in logic by H. Friedman [3]. I also thank J. Milnor and W. Thurston for helpful conversations and finally the Institute for Advanced Study for its support.

## §1. Basic Facts about Kneading.

Apart from Denjoy analysis, the main techniques we use are based on the kneading invariant of Milnor and Thurston [4]. This is a topological invariant which is defined in terms of the behaviour of the critical point of a convex function and characterises much of the dynamical behaviour of continuous families of $C^1$ functions such as $f(x) = x^2 - a$.

If f is convex and $x \in I$ let $\varepsilon_i(x)$ be -1, 0 or +1 according to whether $f^i(x) > 0$, $=0$, or $<0$. The sequence $\varepsilon_i(x)$ is called the itinerary of x. Let $\theta_i(x) = \prod_{j=0}^{i} \varepsilon_j(x)$. Then the formal power series $\theta(x) = \sum_{j=0}^{\infty} \theta_j(x)\, t^j$ is called the invariant coordinate of x. The map $x \mapsto \theta(x)$ is monotone decreasing if we endow the ring $\mathbb{Z}[[t]]$ with the lexicographical ordering. Let $\Lambda$ denote the subset of $\mathbb{Z}[[t]]$ consisting of those formal power series whose coefficient lie in $\{-1, 0, +1\}$. Then it follows from the monotonicity of $\theta$ that

$$\theta(x^+) = \lim_{y \downarrow x} \theta(y) \quad \text{and} \quad \theta(x^-) = \lim_{y \uparrow x} \theta(y)$$

exist in the topology on $\Lambda$ induced by the metric

$$\rho\Big(\sum_{j=0}^{\infty} \theta_j t^j, \sum_{j=0}^{\infty} \theta'_j t^j\Big) = \sum_{j=0}^{\infty} |\theta_j - \theta'_j| 2^{-j}$$

<u>Definition.</u> The <u>kneading invariant</u> V of f, denoted by V(f) is the formal power series $\theta(0^+)$.

The n'th lap number is defined as follows. Let $\ell_n - 1$ be the number of local maxima and minima of $f^n$ within the interior of I. These points divide the interval into $\ell_n$ subintervals, each mapped homeomorphically by $f^n$. Milnor and Thurston proved that $\ell_n$ can be explicitly derived from the kneading invariant [4].

## §2. <u>Blowing up orbits.</u>

Consider any convex function $f: I \to I$ such that $f(1) = f(-1)$. Each point $x \in (f(0), f(1)]$ has two inverse images, denoted a(x) and b(x) where $a(x) < b(x)$. Extend a and b to f(0) by letting $a(f(0)) = b(f(0)) = 0$. For $x \in I$, define $G_x$ to be the semi-group consisting of all words of the form $\alpha f^n(x)$ where $\alpha$ is a word in the letters a and b, $n \geq 0$.

Call $G_x$ the <u>entire orbit</u> of x. The set of points $\{f^n(x), n \geq 0\}$ is called the <u>forward orbit</u> of x, and the set of points $\{\alpha(x), \alpha$ a word in a and b$\}$ is called the backward orbit of x. If $p \in G_x$ let $\|P\|$ be the number of symbols in the word $\alpha$ plus n.

A new function g can be made from f, roughly speaking, by replacing the entire orbit of x by a set of disjoint intervals $I_y$ and requiring $g(I_y) = I_{f(y)}$. The itinerary of any point remains unchanged. We call f the base function of g. The problem is to choose g to be as smooth as possible.

If the growth of the lap numbers is bounded by a polynomial then it is well known that the $\omega$-limit set of $x \in I$ is a periodic point of period $2^n$. In this case it is not too difficult to blow-up $G_x$ and obtain a $C^1$ function g.

We examine the special case when the lap numbers grow faster than any polynomial and slower than any exponential.

Theorem 1. Let $f: I \to I$ be a convex function such that the lap numbers $\ell_n$ satisfy $\ell_{n+1}/\ell_n \to 1$. Then there exists a $C^1$ convex function g possessing an invariant set of disjoint intervals $I_x$ with the critical point $c \in \text{Int } I_c$ such that $\nu f = \nu g$.

Proof. There are two possibilities corresponding to whether or not $G_c$ is dense in I. When it is not, $I - \bar{G}_c$ contains a maximal open interval U. Since U does not contain c it is mapped homeomorphically onto its image. Also $f(U) \cap U = \emptyset$, otherwise a point $p \in G_c$ would be in the interior of one or the other which implies $U \cap G_c \neq \emptyset$. Hence the maximal connected intervals in the complement of $\bar{G}_c$ are mapped homeomorphically onto one another. We simultaneously crush these down to points and blow-up the orbit of c. If $G_c$ is dense in I, our methods merely blow-up $G_c$.

Let $x \in G_c$. We construct disjoint intervals $I_x$ with length

$$a_x = k/\|x\|^3 \ell^2_{\|x\|}$$

where k is a constant such that

$$\text{(i)} \quad \sum_{x \in G_c} a_x = 1.$$

Such a k exists since

$$\sum_{x\in G_c} 1/\|x\|^3 \ell^2_{\|x\|} \le \sum_{n=1}^{\infty} (\ell_1 + \ell_2 + \dots + \ell_n)/n^3 \ell_n^2$$

$$< \sum_{n=1}^{\infty} n.\ell_n/n^3 \ell_n^2$$

$$< \sum_{n=1}^{\infty} 1/n^2 \quad .$$

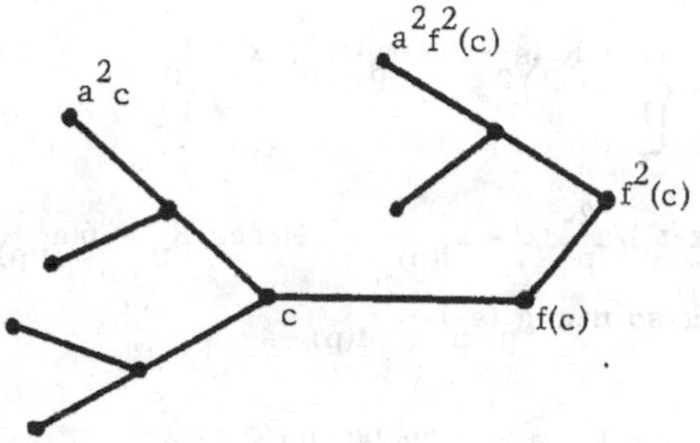


Then

(ii) $\lim_{\|x\|\to\infty} a_{f(x)}/a_x$

$$= \lim_{\|x\|\to\infty} (\|x\|\pm 1)^3 \ell^2_{\|x\|\pm 1} / \|x\|^3 \ell^2_{\|x\|}$$

$$= \lim_{n\to\infty} (n\pm 1)^3 \ell^2_{n\pm 1}/n^3 \ell_n^2$$

$$= \lim_{n\to\infty} \ell^2_{n\pm 1}/\ell_n^2$$

$$= 1 \; .$$

Both of these conditions are useful in establishing the differentiability of our final function g. The first must hold in order that I remains compact, but, less obviously, it is useful to ensure that the invariant set of intervals have full measure in I. This will make it easier to calculate the derivative of g at a point not in $\cup I_x$ since lengths can be expressed as sums (possibly infinite) of the $a_x$.

The second condition (ii) is necessary for a canonical smoothing of g since, for some base functions f, arbitrarily close to an endpoint of $I_x$ there will be both contracting and expanding intervals.

Define $I_x$ by

$$I_x = [\sum_{\substack{y<x \\ y\in G_c}} a_y,\ 1 - \sum_{\substack{y>x \\ y\in G_c}} a_y] = [s_x, t_x].$$

Then $g_p = g|I_p$ is obtained by integrating $g_p'$ which satisfies $\int_{I_x} g_p' = a_{f(x)}$ and $g_p'(x) = +1$ or $-1$ on the endpoints of $I_p$, $p \neq c$, depending on whether $p > c$ or $p < c$. For example if $p > c$, let

$$g_p'(x) = \begin{cases} 1 + R_p(s_p - x)(x - t_p)/a_p^2, & x \in I_p \\ 1 & , x \notin I_p \end{cases}$$

Then $\int_{s_p}^{t_p} [1 + R_p(s_p - x)(x - t_p)/a_p^2] dx = a_{f(p)}$. Hence $R_p = 6(a_{f(p)}/a_p - 1)$. Finally integrate $g_p'(x)$ and add a constant so that $g_p(s_p) = s_{f(p)}$.

Similarly construct $g_p$ over $I_p$, $p < c$ by letting

$$g_p'(x) = \begin{cases} -1 + R_p(s_p - x)(x - t_p)/a_p^2, & x \in I_p \\ -1 & , x \notin I_p \end{cases} .$$

For $x = c$ let $g_c : I_c \to I_{f(c)}$ be any smooth function satisfying $g_c'(s_c) = -1$ and $g_c'(t_c) = +1$.

Define $g: I \to I$ by

$$g(x) = \begin{cases} g_p(x), & x \in I_p,\ p \in G_c \\ \lim g_p(s_p) & x = \lim s_p \end{cases} .$$

It follows from conditions (i) and (ii) above that $g$ is $C^1$. For if $x \notin \cup I_p$ then

$$\begin{aligned} g'(x) &= \lim_{y\to x} \frac{|g(x)-g(y)|}{|x-y|} \\ &= \lim_{y\to x} \pm \sum_{x<p<y} a_{f(p)} \Big/ \sum_{x<p<y} a_p \quad \text{(the sign depending on whether } x \gtrless c.) \end{aligned}$$

since $\cup I_p$ has full measure in $I$ and $x \notin \cup I_p$. In general if $b_k$, $c_k \to 0$ are positive real numbers satisfying $b_k / c_k \to 1$ then

$$\lim \sum_{k=n}^{\infty} b_k \Big/ \sum_{k=n}^{\infty} c_k = 1 .$$

Hence, for $x \notin \cup I_p$, $g'(x) = \begin{cases} 1 & x > c \\ -1 & x < c \end{cases}$. Thus g is continuously differentiable.

Finally observe that $I_c$ is a wandering interval. □

## §3. The Base Function.

Consider any continuous family of quadratics such as $f(x) = x^2 - a$.

Theorem 2. There exists a value A such that $f_A(x) = x^2 - A$ satisfies $\ell_{n+1}/\ell_n \to 1$ and $f_A^{2^k}(0)$ alternates on either side of 0, drawing monotonically and arbitrarily closer.

Proof. Let A be $\sup\{a$ : critical point of $f_a$ converges to a periodic orbit of period $2^n\}$. Milnor and Thurston discuss this in [4, §9.6] and prove that $\ell_{n+1}/\ell_n \to 1$. Using their notation, we verify the rest of the theorem.

Writing f for $f_A$, observe that the reciprocal zeta function

$$\hat{\zeta}(f,t)^{-1} = \prod_0^\infty (1-t^{2^k}) = \Sigma(-1)^{\alpha(n)} t^n$$

where $\alpha(n)$ is the number of 1's in the diadic expansion of n. According to Milnor and Thurston, §9.6, in this case the kneading determinant

$$D(f,t) = \hat{\zeta}(f,t)^{-1} \quad .$$

Hence

$$\theta_n(0^+) = (-1)^{\alpha(n)}$$

$$\Rightarrow \theta_n(f^p(0)) = \theta_{n+p}(0^+)/\theta_{p-1}(0^+) = (-1)^{\alpha(n-1)+\alpha(n+p)} \quad .$$

Therefore if $k > 1$ $\theta_0(f^{2^{2k}}(0)) = (-1)^{\alpha(2^{2k}-1)+\alpha(2^{2k})} = (-1)^{2k+1} = -1$

and $\theta_0(f^{2^{2k+1}}(0)) = (-1)^{\alpha(2^{2k+1}-1)+\alpha(2^{2k+1})} = (-1)^{2k+2} = +1$ .

Since $\theta_0(p) = \varepsilon_0$, the "address" of p,

$$f^{2^{2k}}(0) < 0 < f^{2^{2k+1}}(0) .$$

We compare $\theta(f^{2^{2k}}(0))$ to $\theta(f^{2^{2(k+1)}}(0))$. Note that the first $2^{2k}$ terms are the same since for $n < 2^{2k}$,

$$\theta_n(f^{2^{2k}}(0)) = (-1)^{2k+2+\alpha(n)}$$

and

$$\theta^n(f^{2^{2(k+1)}}(0)) = (-1)^{2(k+1)+1+\alpha(n)} .$$

But if $n = 2^{2k}$, $\quad \theta_n(f^{2^{2k}}(0)) = (-1)^{2k+1} = -1$

and $\quad \theta_n(f^{2^{2(k+1)}}(0)) = (-1)^{2(k+1)+2} = +1 .$

Therefore $\quad f^{2^{2k}}(0) < f^{2^{2L}}(0)$, $L > k$. In a similar fashion, $f^{2^{2+1}}(0) > f^{2^{2L+1}}(0)$, $L > k$.

Finally, note that $\lim \theta(f^{2^{2k}}(0) = \theta(0^-)$. Hence $\theta(\lim(f^{2^{2k}}(0))) = \theta(0^-)$ since $\lim(f^{2^{2k}}(0)$ exists. In fact, $\lim(f^{2^{2k}}(0)) = 0$ as in this example 0 is the unique point p such that $\theta^-(p^{\mp}) = \theta(0^-)$. Similarly $\lim \theta(f^{2^{2(k+1)}}(0)) = \theta(0^+)$ so that $\lim f^{2^{2(k+1)}}(0) = 0$.□

## §4. Unsmoothability of g.

With the results of Denjoy on the circle in mind, one might naturally suspect that a $C^2$ map f of I with growth rate faster than polynomial has no subinterval J such that $f^k|J$ is a homeomorphism for all $k \geq 0$. (See also the work of Guckenheimer and Misieurwicz for discussion of the case when f has negative Schwarzian derivative.) We apply Theorem 1 to $f_A$, creating g, and show that there is no $C^2$ convex map conjugate to g.

If $x, y \in I$ we denote the open interval with endpoints x and y by (x, y) regardless of the order.

Lemma 3. Let $f(x) = x^2 - A$. There exists an increasing sequence of positive integers $n_i$ and points $c_{-1}, \ldots, c_{-n_i}$ such that $f(c_{-k}) = c_{-k+1}$, $c_0 = 0$, and the intervals $(c_k, c_{-n+k})$ are disjoint where $c_k = f^k(0)$.

Proof. Let $i \in z$. Consider all points x in both the forward and backward orbits of $c_0$ such that $\|x\| \leq 2^i$. Suppose $x_i$ is a member of this set with minimum distance to $c_0$. Denote it by $c_{n_i}$ if $f^{n_i}(c_0) = x_i$ and by $c_{-n_i}$ if $f^{n_i}(x_i) = c_0$.

In the former case, we choose $c_{-n_i}$ inductively. It helps to observe that if $c_1 \notin (x, y)$ and $f(x') = x$ is given, then there exists y' such that $f(y') = y$ and $f(x', y') \cong (x, y)$. Thus there exists $c_{-1}$ such that $f(c_{-1}, c_{n_i-1}) \cong (c_0, c_{n_i}) = I_+$ since $c_1 \notin I_+$. Furthermore $c_1 \notin (c_{-1}, c_{n_i-1})$ otherwise $c_2 \in I_+$. Hence we can choose $c_{-2}$ such that $f(c_{-2}, c_{n_i-2}) \cong (c_{-1}, c_{n_i-1})$ and inductively $c_{-p}$ such that $f(c_{-p}, c_0) \cong (c_{-p+1}, c_1)$ $p \leq n_i$ since $c_p \notin I_+$.

In the latter case f maps $I_- = (c_0, c_{-n_i})$ diffeomorphically onto $(c_1, c_{-n_i+1})$ since $c_0 \notin I_-$. Inductively, $c_0 \notin f^n(I_-)$ otherwise $c_{-1} \in f^{n-1}(I_-) \Rightarrow c_{-n} \in I_-$. Hence $f(c_p, c_{-n_i+p}) \cong (c_{p+1}, c_{-n_i+p+1})$.

It is easy to check that in both cases, these intervals are disjoint by considering all three possible intersections : If $x_i = c_{-n}$,

(i) $c_p \in (c_m, c_{m-n})$ and $c_{m-n} \in (c_p, c_{p-n}) \Rightarrow c_{p-m}$ or $c_{p-m}$ or $c_{m-p-n} \in I_-$

(ii) $c_{p-n} \in (c_m, c_{m-n})$ and $c_{m-n} \in (c_p, c_{m-n}) \Rightarrow c_{p-m-n}$ or $c_{m-p-n} \in I_-$

or

(iii) $c_p \in (c_m, c_{m-n})$ and $c_{p-n} \in (c_m, c_{m-n}) \Rightarrow c_{p-m}$ or $c_{p-m-n} \in I_-$ .

None of these are possible since $|p-m| < n$. A similar argument holds if $x_i = c_n$. Note that Theorem 2 implies the $n_i$ are unbounded so choose an increasing subsequence. □

We are now in a position to apply Denjoy analysis in which the following straightforward lemma is crucial.

Lemma 4. If f' has bounded variation and $|1/f'(x)|$ is bounded, then log f' has bounded variation.

If there is a critical point then the Denjoy analysis might not apply. However, the critical point of our blown-up function g is contained in the interior of a wandering interval $I_0$. This turns out to be strong enough to prove :

Theorem 5. The blown-up function g is not $C^0$ conjugate to any convex $C^2$ function G.

Proof. We construct a restricted function avoiding the critical point c. Let $U_0 \subset I_0$ be a small open interval containing c, and $J_0$ be a connected component of $I_0 - U_0$. Let J = I - (the entire orbit of $U_0$) and $F:J \to J$ be defined by F(x) = G(x). Let $J_n = f^n(J_0)$ and denote by $J_{-n}$ the n'th inverse image of $J_0$ under consideration. Let $a_n$ denote the length of $J_n$.

Consider the sequence $n_i \to \infty$ in Lemma 3. Then for each i there exists $x_0 \in J_0$ such that

$$\frac{a_{n_i}}{a_0} = (F^{n_i})'(x_0) = G'(x_{n_i-1})G'(x_{n_i-2}) \cdot G'(x_0).$$

There also exists $\hat{x}_{-n_i} \in J_{-n_i} \subset I_{-n_i}$ (where $I_{-n_i}$ is the blown-up interval corresponding to $c_{-n_i}$ which, in turn, is defined in Lemma 3) such that

$$\frac{a_0}{a_{-n_i}} = (F^{n_i})'(\hat{x}_{-n_i}) = G'(\hat{x}_{-1})G'(\hat{x}_{-2}) \dots G'(\hat{x}_{-n_i})$$

$$\text{Hence} \quad \log \frac{a_0^2}{a_{n_i} a_{-n_i}} = \sum_{k=1}^{n_i} |\log G'(\hat{x}_{-k}) - \log G'(x_{n_i-k})|$$

$$\leq \sum_{k=1}^{\infty} |\log G'(\hat{x}_{-k}) - \log G'(x_{n_i-k})|$$

$$< \text{constant} \quad .$$

The last inequality holds since $\hat{x}_{-k}$ and $x_{n_i-k}$, $k = 1,\dots,n_i$, are pairs of points in disjoint intervals (according to Lemma 3), all derivatives are bounded away from 0 and G' has bounded variation.

Therefore
$$\frac{a_0^2}{a_{n_i} a_{-n_i}} < e^k$$

$$\Rightarrow a_{n_i} . a_{-n_i} > \text{constant}$$

$$\Rightarrow \sum_{i=-\infty}^{+\infty} a_{n_i} = \infty . \qquad \square$$

## References.


1. E. Coven & Z. Nitecki, Nonwandering sets of the powers of maps of the interval, preprint.

2. A. Denjoy, Sur les courbes definies par les équations différentielles à la surface du tore, J. Math. Pures Appl. [9], 11, 333-375, 1932.

3. H. Friedman, 102 problems in Math'l logic, J. of Symbolic Logic, vol. 40, No. 2, 1975, p. 113.

4. J. Milnor & W. Thurston, On iterated maps of the interval I and II, preprint, Princeton 1977.


J. Harrison : Mathematical Institute, Oxford University, Oxford, England,
and Department of Mathematics, University of California, Berkeley, California, U.S.A.

# Space- and Time-Periodic Perturbations of the Sine-Gordon Equation.

Philip Holmes*

**Abstract.** We study the Sine-Gordon equation $\varphi_{tt} - \varphi_{zz} + \sin\varphi = 0$ subject to two classes of small perturbations : (1) spatially periodic perturbations on an infinite domain and (2) weak dissipation and temporally periodic perturbations on the boundary of a finite spatial domain. In the former case we prove the existence of a countable set of spatially periodic stationary solutions in addition to an uncountable set of non-periodic stationary solutions. In the latter case we prove that, if the excitation is sufficiently large compared with dissipation, a countable set of time-periodic motions of all periods as well as non periodic motions exist. All these stationary and periodic solutions are unstable (of saddle type) and are expected to coexist with stable solutions. However, in the global bifurcation leading to creation of the latter (time-dependent) solutions, infinite sets of asymptotically stable periodic orbits with arbitrarily long periods are expected to appear.

## 1. Introduction.

In this paper we employ certain methods of differentiable dynamics, and in particular of invariant manifolds, to study the global solution structure of a perturbed partial differential equation (PDE), the sine-Gordon equation :

$$\varphi_{tt} - \varphi_{zz} + \sin\varphi = 0, \ z \in \Omega \subset \mathbb{R} . \qquad (1.1)$$
$$\varphi_{zz} = 0, \ z \in \partial\Omega .$$

The two specific problems we address are

I. Spatially periodic perturbations of stationary solutions on a infinite domain :

$$\varphi_{tt} - \varphi_{zz} + \sin\varphi = \varepsilon f(z), \ z \in \mathbb{R} \qquad (1.2)$$
$$\varphi_z = 0 \ \text{ at } \ z = \pm\infty$$

II. Temporally periodic perturbations at the boundaries of a finite domain :

$$\varphi_{tt} - \varphi_{zz} + \sin\varphi = -\varepsilon\sigma\varphi_t \qquad (1.3)$$
$$\varphi_{z|z=0} = \varepsilon H, \ \varphi_{z|z=1} = \varepsilon(H+I(t)).$$

---

* Research partially supported by NSF grants ENG-78-02891 and ENG-79-19817.

The methods used are, however, applicable to a wide class of problems in which the unperturbed system is fully or partially integrable. While the Hamiltonian structure possessed by (1.1) is an important simplifying feature of our analysis, it does not appear to be essential to our methods. It seems natural to apply the method to other integrable problems, such as the Korteweg-deVries equation, in the presence of weak periodic fields. Problems such as that of the "$\varphi^4$" system, which are not completely integrable, can also be treated by these methods.

For a general review of the role of solitons in condensed matter physics, see Bishop, Krumhansl and Trullinger [1980]; for discussion of the sine-Gordon system in superconductivity and from a mathematical viewpoint, see Matisoo [1969] and Levi, Hoppensteadt and Miranker [1978].

The main ingredients of our analysis are invariant manifold theory (cf. Hirsch, Pugh and Shub [1977]) and the perturbation methods developed by Melnikov [1963] for the study of time-periodic perturbations of planar ordinary differential equations (ODE's). For an account and application of this method see Holmes [1979a,b; 1980]. We provide a brief introduction in section two, below. Recent work of Chow, Hale and Mallet-Paret [1980] provides an alternative, and in some respects simpler, derivation of the Melnikov function, but does not appear directly to offer the generalisation to infinite dimensional evolution equations necessary for problem II. This extension was carried out by Holmes and Marsden [1980] and is also summarised in section two, where we give the main results.

In section three we describe the stationary solutions of (1.2) as a phase plane problem and show that the study of spatially periodic perturbations f(x) becomes the study of a Poincaré map. Melnikov's method may then be applied directly to prove the existence of a countable set of spatially periodic solutions, under suitable assumptions on f(z) (f(z) can, in fact be replaced by a certain first order differential operator).

In section four we recast (1.3) as an evolution equation in a Banach space of functions after applying a simple transformation which yields homogeneous boundary conditions for $\varepsilon \neq 0$. We then apply the infinite dimensional extension of Melnikov's method to detect the existence of Smale horseshoes (Smale [1963, 1967]) in the Poincaré map associated with the flow of (1.3). This implies that (1.3) possesses a countable set of time-periodic motions of arbitrarily long period in addition to non-periodic, "chaotic" motions.

In section five we close with comments on the stability of the solutions found and on the bifurcations in which the time-periodic solution of problem II are created. We also discuss some of the difficulties of extending our analysis to time-periodic perturbations on an infinite spatial domain.

Our methods should be seen as complementary (in that they offer a geometric interpretation) to the perturbation methods developed by Kaup and Newell in connection with the inverse scattering transform (Kaup and Newell [1978a, b], Newell [1980]). The relationships between these two approaches still remain to be worked out.

## 2. Global perturbations in two and infinite dimensions.

In this section we introduce the perturbation methods to be applied later. We first give a general discussion of the two dimensional case, without giving detailed results, which may be found in Melnikov [1963] and Holmes [1980]. We then sketch the infinite dimensional extension of Holmes and Marsden [1980] and give the hypotheses and results in detail for that case.

### 2.1. Melnikov's method for Planar systems.

Melnikov [1963] considered the following problem:

$$\dot{x} = f_0(x) + \varepsilon f_1(x,t), \quad x \in \mathbb{R}^2, \quad f_1(\cdot,t) = f_1(\cdot,t+T), \quad 0 \leq \varepsilon \ll 1 \quad . \tag{2.1}$$

His main hypothesis was that, for $\varepsilon = 0$, (2.1) possesses a homoclinic orbit $\Gamma_0$ to a non-degenerate hyperbolic saddle point $p_0$, or a homoclinic cycle connecting several such saddle points (Figure 1). Non degeneracy implies that the eigenvalues of the linearised

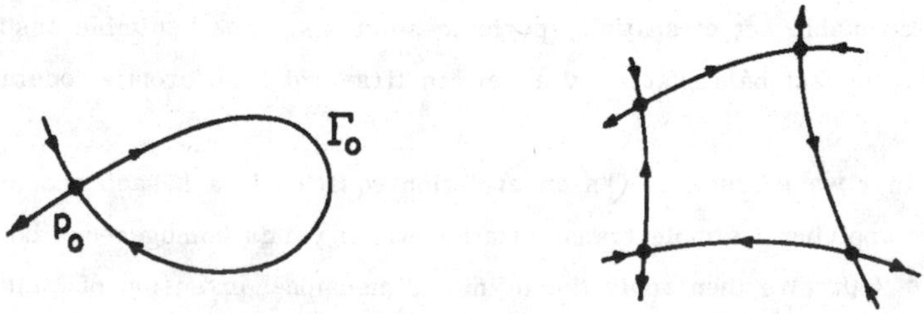


Figure 1. Homoclinic cycles.

operator (2×2 matrix) $Df_0(p_0)$ are real and of opposite sign and that there are linearly independent eigenvectors. The stable manifold theorem (Hartman [1964], Hirsch-Pugh-Shub [1977]) then implies that $p_0$ possesses local stable and unstable manifolds $W^s_0(p_0)$, $W^u_0(p_0)$, often called separatrices in the planar case.

Associated with (2.1) we have the autonomous system

$$\dot{x} = f_0(x) = \varepsilon f_1(x,\theta), \qquad \dot{\theta} = 1; \quad (x,\theta) \in \mathbb{R}^2 \times S^1, \tag{2.2}$$

the periodicity of the vector field $f_1(x,t)$ being reflected in the phase space $\mathbb{R}^2 \times S^1$. Here $S^1 = \mathbb{R}/T$, the circle of length T. For $\varepsilon = 0$ or $\neq 0$, (2.2) possesses a Poincaré map

$$P_\varepsilon^{t_0} : \Sigma_{t_0} \to \Sigma_{t_0}, \tag{2.3}$$

where $\Sigma_{t_0} = \{(x,\theta) \in \mathbb{R}^2 \times S^1 \mid \theta = t_0\}$ is a global cross section of the flow. $P_\varepsilon^{t_0}$ is obtained by projecting the three dimensional flow $F_t^\varepsilon(x_0, t_0)$ of (2.2) onto the first factor

$$P_\varepsilon^{t_0}(x) = \pi_1 . F_T^\varepsilon(x, t_0). \tag{2.4}$$

By hypothesis, when $\varepsilon = 0$, (2.2) has a circular periodic orbit $\gamma_0 = p_0 \times S^1$ which has stable and unstable manifolds $W^s_0(\gamma_0)$, $W^u_0(\gamma_0)$ which intersect to form the "cylinder" $\Gamma_0 \times S^1$. Such non-transversal intersections are known to be non-generic (Smale [1967]) and would therefore be expected to break under small perturbations. The unperturbed Poincaré map, $P_0^{t_0}$, inherits this degeneracy in the tangency of $W^s_0(p_0) = W^s_0(\gamma_0) \cap \Sigma_{t_0}$ and $W^u_0(p_0) = W^u_0(\gamma_0) \cap \Sigma_{t_0}$, since, when $\varepsilon = 0$, the $\theta$ component of (2.2) is trivial and all cross sections $\Sigma_{t_0}$ are identical. However, although the degenerate intersection of the manifolds in general breaks, the hyperbolic periodic orbit $\gamma_0$ persists as a slightly perturbed orbit $\gamma_\varepsilon$, thus giving a perturbed saddle point $p_\varepsilon \in \Sigma_{t_0}$ which possesses local stable and unstable manifolds $W^s_\varepsilon(p_\varepsilon)$, $W^u_\varepsilon(p_\varepsilon)$ $C^r$ close to $W^s_0(p_0)$ and $W^u_0(p_0)$. Note that the position of $p_\varepsilon$ will, in general, vary from section to section as $t_0$ varies.

In general $F_T^\varepsilon$ and hence $P_\varepsilon^{t_0}$ cannot be calculated, but if we assume that (2.1) can be solved (in some sense) for $\varepsilon = 0$, then regular perturbation methods may be used to approximate $F_T^\varepsilon$ for $\varepsilon \neq 0$, small. Let us consider the case of a single homoclinic cycle

$\Gamma_0$ (Figure 1a), on which we suppose that we know the solution $x_0(t-t_0, \hat{x}_0)$ based at a point $(\hat{x}_0, t_0) \in \mathbb{R}^2 \times S^1$. From this global knowledge of the flow $F_t^0$ we are able to approximate the positions of the perturbed stable and unstable manifolds of the Poincaré map $P_\varepsilon^{t_0}$ on different sections $\Sigma_{t_0}$, simply by assuming power series expansions of the forms

$$x_\varepsilon^s(t,t_0) = x_0(t-t_0,\hat{x}_0) + \varepsilon x_1^s(t,t_0) + 0(\varepsilon^2) \tag{2.5a}$$

$$x_\varepsilon^u(t,t_0) = x_0(t-t_0,\hat{x}_0) + \varepsilon x_1^u(t,t_0) + 0(\varepsilon^2) \tag{2.5b}$$

for orbits lying close to the stable and unstable manifolds of the unperturbed circular periodic orbit $\gamma_0 = p_0 \times S^1$. The first variations $x_1^s$, $x_1^u$ are obtained from the linearisation in $\varepsilon$ of (2.1) about the homoclinic orbit $\Gamma_0$; for example

$$\dot{x}_1^s(t,t_0) = Df_0(x_0(t-t_0,\hat{x}_0))x_1^s(t,t_0) + f_1(x_0(t-t_0,\hat{x}_0),t) , \tag{2.6}$$

for solution curves $x_1^s(t,t_0)$ $W_\varepsilon^s(\gamma_\varepsilon)$. Solutions of (2.6) are valid in the time interval $[t_0,\infty)$, since $x_\varepsilon^s(t,t_0) \to \gamma_\varepsilon$ as $t \to +\infty$. A similar expression holds for $x_1^u(t,t_0)$, with validity in the interval $t\varepsilon(-\infty,t_0]$. Piecing these two solutions together permits us to extimate the distance $d(t_0)$ between the perturbed manifolds $W_\varepsilon^s(p_\varepsilon)$ and $W_\varepsilon^u(p_\varepsilon)$ at the point $\hat{x}_0 \varepsilon \Gamma_0$ on the section $\Sigma_{t_0}$.

Measuring $d(t_0)$ normal to the vector field $f_0(\hat{x}_0)$ at $\hat{x}_0$, and neglecting the normalisation factor $|f_0(\hat{x}_0)|$, we have

$$d(t_0) = f_0(\hat{x}_0) \wedge (\hat{x}_\varepsilon^s - \hat{x}_\varepsilon^u) , \tag{2.7}$$

where $\hat{x}_\varepsilon^s$(resp. $\hat{x}_\varepsilon^u$) are points lying on $W_\varepsilon^s(\gamma_\varepsilon) \cap \Sigma_{t_0}$ (resp. $W_\varepsilon^u(\gamma_\varepsilon) \cap \Sigma_{t_0}$) and the normal to $\Gamma_0$ at $\hat{x}_0$, $f_0^\perp(\hat{x}_0)$. Here the wedge product is defined as $x \wedge y = x_1y_2 - x_2y_1$. We cannot calculate $\hat{x}_\varepsilon^s$ and $\hat{x}_\varepsilon^u$ directly but can use (2.6) to find $d(t_0)$ in the following manner. We define a moving distance function

$$d(t,t_0) = f_0(x_0(t-t_0,\hat{x}_0)) \wedge (x_\varepsilon^s(t,t_0) - x_\varepsilon^u(t,t_0)) , \tag{2.8}$$

and note that $d(t_0,t_0) = d(t_0)$, provided that $x_\varepsilon^s$, $x_\varepsilon^u$ are the solutions of (2.6) with initial conditions $\hat{x}_\varepsilon^s$, $\hat{x}_\varepsilon^u$. This we cannot ensure, but as we shall see, it does not matter, since the manifolds are $C^r$ close and the distance $d(t_0)$ projected onto $f_0^\perp(\hat{x}_0)$ is a good measure (to $0(\varepsilon^2)$) of the true distance between $\hat{x}^s_\varepsilon$ and $\hat{x}^u_\varepsilon$. (If we could find $\hat{x}^s_\varepsilon$ and $\hat{x}^u_\varepsilon$ directly then no projection would be necessary).

We now write

$$\begin{aligned} d(t,t_0) &= \varepsilon(f_0(x_0) \wedge x^s_1 - f_0(x_0) \wedge x^u_1) + 0(\varepsilon^2) \\ &\triangleq \varepsilon(d^s_1(t,t_0) - d^u_1(t,t_0)) + 0(\varepsilon^2) , \end{aligned} \tag{2.9}$$

so that $d(t_0) = \varepsilon(d^s_1(t_0,t_0) = d^u_1(t_0,t_0)) + 0(\varepsilon^2)$ . (2.10)

To calculate $d^s_1(t_0,t_0)$ we differentiate $d^s_1(t,t_0) = f_0(x_0(t-t_0,\hat{x}_0)) \wedge x^s_1(t,t_0)$ with respect to t, and use (2.6) and the definition of $d^s_1(t,t_0)$ to obtain

$$\dot{d}^s_1(t,t_0) = [\text{trace } Df_0(x_0(t-t_0,\hat{x}_0))]d^s_1(t,t_0) + f_0(x_0(t-t_0,\hat{x}_0)) \wedge f_1(x_0(t-t_0,\hat{x}_0),t) \tag{2.11}$$

Integrating (2.11) from $t_0$ to $\infty$ we obtain

$$d^s_1(\infty,t_0) = d^s_1(t_0,t_0) + \int_{t_0}^{\infty}\{f_0(x_0(t-t_0,\hat{x}_0)) \wedge f_1(x_0(t-t_0,\hat{x}_0),t)$$

$$\exp[-\int_0^{t-t_0} \text{trace } Df_0(x_0(s,\hat{x}_0))ds]\}dt \tag{2.12}$$

Noting that $d^s_1(\infty,t_0) = 0$, since $f_0(x_0(\infty,\hat{x}_0)) = f_0(p_0) = 0$, and performing a similar calculation for $d^u_1(t_0,t_0)$, we obtain the Melnikov function $M(t_0)$:

$$M(t_0) = \int_{-\infty}^{\infty}\{f_0(x_0(t-t_0,\hat{x}_0)) \wedge f_1(x_0(t-t_0,\hat{x}_0),t) \quad .$$

$$\exp[\int_0^{t-t_0} \text{trace } Df_0(x_0(s,\hat{x}_0))ds]dt \tag{2.13}$$

As we show in Holmes [1980] and Holmes-Marsden [1980], the true distance between the manifolds $\hat{x}^s_\varepsilon - \hat{x}^u_\varepsilon$ measured normal to $f_0(\hat{x}_0)$ is given to leading order by $M(t_0)$

$$d(t_0) = \varepsilon M(t_0) + 0(\varepsilon^2) \quad . \tag{2.14}$$

Now if, as $t_0$ varies, $d(t_0)$ passes through zero then it is easy to see that the points $\hat{x}^s_\varepsilon$, $\hat{x}^u_\varepsilon$ change orientation with respect to $f_0(\hat{x}_0)$ and hence the manifolds $W^s_\varepsilon(p_\varepsilon)$, $W^u_\varepsilon(p_\varepsilon)$

intersect. In fact we have

Theorem 1. Let the system (2.1) for $\varepsilon = 0$ possess a homoclinic orbit $\Gamma_0$ to a non-degenerate saddle point $p_0$, given by $x_0(t-t_0, \hat{x}_0)$. Then the Poincaré map $P_0^{t_0}: \Sigma_{t_0} \to \Sigma_{t_0}$ associated with (2.1) for $\varepsilon \neq 0$ small possesses a hyperbolic saddle point $p_\varepsilon$ with stable and unstable manifolds $W_\varepsilon^s(p_\varepsilon)$, $W_\varepsilon^u(p_\varepsilon)$. Moreover, if the Melnikov function $M(t_0)$ given in (2.13) has simple zeros, then $W_\varepsilon^s(p_\varepsilon)$ and $W_\varepsilon^u(p_\varepsilon)$ intersect transversally. If $M(t_0)$ has no zeros in $t_0 \in [0, T]$, then $W_\varepsilon^s(p_\varepsilon) \cap W_\varepsilon^u(p_\varepsilon) = \emptyset$.

Transversal intersection of one dimensional stable and unstable manifolds in $\mathbb{R}^2$ implies that some power of the map $P_\varepsilon^{t_0}$ possesses Smale horseshoes (see Corollary 2 in §2.2 below).

Corollary 1. (cf. Chow, Hale and Mallet-Paret [1980]. Consider a parametrised family of perturbations with parameter $\mu$. Suppose that $M_\mu(t_0)$ has a quadratic zero ($M_{\mu_0}(t_0) = M'_{\mu_0}(t_0) = 0$ but $M''_{\mu_0}(t_0) \neq 0$ at $t_0 = \tau$.) Then for some $\mu$ near $\mu_0$ the map $P^\tau_{\varepsilon,\mu}: \Sigma_\tau \to \Sigma_\tau$ possesses a hyperbolic saddle $p_{\varepsilon,\mu}$ whose stable and unstable manifolds have quadratic tangencies.

This corollary allows us to apply a result due to Newhouse [1974, 1977, 1980], which implies that there are maps near $P^\tau_{\varepsilon,\mu_0}$ which possess infinitely many asymptotically stable periodic orbits, provided that $P^\tau_{\varepsilon,\mu_0}$ is a contraction, (i.e. the eigenvalues $\lambda_1, \lambda_2$ of $DP^\tau_{\varepsilon,\mu}(p_{\varepsilon,\mu})$ satisfy $\lambda_1\lambda_2 < 1$).

## 2.2 Periodic Perturbations of Infinite Dimensional Evolution Equations.

Here the situation is essentially the same as the above, with the phase-space $\mathbb{R}^2$ replaced by a Banach space X. We shall in addition require that the unperturbed system $\dot{x} = f_0(x)$ be Hamiltonian, so that the analogue of trace $Df(x_0(t-t_0, \hat{x}_0))$ is identically zero and the calculations become considerably simpler. This restriction does not, however, appear essential. Full details can be found in Holmes-Marsden [1980], from which we excerpt the relevant results.

We consider an evolution in a Banach space X of the form

$$\dot{x} = f_0(x) + \varepsilon f_1(x, t) \quad , \tag{2.15}$$

where $f_1$ is periodic of period T in t. Our hypotheses on (2.15) are as follows :

1. (a) Assume $f_0(x) = Ax + B(x)$ where A is an (unbounded) linear operator that generates a $C^o$ one parameter group of transformations on X and where $B:X \to X$ is $C^\infty$, and has bounded derivatives on bounded sets.

(b) Assume $f_1 : X \times S^1 \to X$ is $C^\infty$ and has bounded derivatives on bounded sets, where $S^1 = \mathbb{R}/(T)$, the circle of length T.

(c) Assume that $F_t^\varepsilon$ is defined for all $t \in \mathbb{R}$ for $\varepsilon > 0$ sufficiently small and $F_t^\varepsilon$ maps bounded sets in $X \times S^1$ to bounded sets in $X \times S^1$ uniformly for small $\varepsilon \geq 0$ and t in bounded time-intervals.

(Cf. Segal [1962], Holmes-Marsden [1978]).

Assumption 1 implies that the associated suspended autonomous system on $X \times S^1$,

$$\begin{cases} \dot{x} = f_0(x) + \varepsilon f_1(x, \theta) , \\ \dot{\theta} = 1 , \end{cases} \tag{2.16}$$

has a smooth local flow, $F_t^\varepsilon$, which can be extended globally in time, i.e. solutions do not escape to infinity in finite time. Energy estimates often suffice to prove the latter, as in the present application (§4 below).

2. (a) Assume that the system $\dot{x} = f_0(x)$ (the unperturbed system) is Hamiltonian with energy $H_0 : X \to \mathbb{R}$ .

(b) Assume there is a symplectic 2-manifold $N \subset X$ invariant under the flow $F_t^0$ and that on N there is a fixed point $p_0$ and a homoclinic orbit $x_0(t)$ i.e.,

$$f_0(p_0) = 0, \dot{x}_0(t) = f_0(x_0(t))$$

and

$$p_0 = \lim_{t \to +\infty} x_0(t) = \lim_{t \to -\infty} x_0(t)$$

This means that X carries a skew symmetric continuous bilinear map $\Omega : X \times X \to \mathbb{R}$ that is weakly non-degenerate (i.e. $\Omega(u, v) = 0$ for all v implies $u = 0$) called the symplectic form and there is a smooth function $H_0 : X \to \mathbb{R}$ such that

$$\Omega(f_0(x), u) = dH_0(x).u \tag{2.17}$$

for all x in $D_A$, the domain of A. Consult Abraham and Marsden [1978] and Chernoff and Marsden [1974] for details about Hamiltonian systems. The symplectic form will replace the wedge product in the two dimensional analysis above.

Remarks.

(a) The condition that N be symplectic means that $\Omega$ restricted to vectors tangent to N defines a non-degenerate bilinear form.

(b) Assumption 2 can be replaced by a similar assumption on the existence of heteroclinic orbits connecting two saddle points and the existence of transverse heteroclinic orbits can then be proven using the methods below. In this way homoclinic cycles can be dealt with (figure 1b).

The next assumption states that the homoclinic orbit through $p_0$ arises from a hyperbolic saddle.

3. Assume that $\sigma(Df_0(p_0))$, the spectrum of $Df_0(p_0)$, consists of two nonzero real eigenvalues $\pm\lambda$, with the remainder of the spectrum on the imaginary axis, strictly bounded away from 0. Assume that, if $\pm i\lambda_0 \in \sigma(Df_0(p_0))$, then $\lambda_n T$ is strictly bounded away from $2m\pi$, $\forall m$. Assume that $\sigma(e^{tDf_0(p_0)})$, the spectrum of $e^{tDf_0(p_0)}$, equals the closure of $e^{t\sigma(Df_0(p_0))}$.

Note that, if $Df_0(p_0)$ and $e^{tDf_0(p_0)}$ have only point spectra, then $\sigma(e^{tDf_0(p_0)}) = e^{t\sigma(Df_0(p_0))}$.

Consider the suspended system (2.16) with its flow $F_t^\varepsilon : X \times S^1 \to X \times S^1$. Let the Poincaré map $P_\varepsilon : X \to X$ be defined by

$$P_\varepsilon(x) = \pi_1 . (F_T^\varepsilon(x, 0)) \tag{2.18}$$

where $\pi_1 : X \times S^1 \to X$ is the projection onto the first factor. As in the planar case, $P_0(p_0) = p_0$, and fixed points of $P_\varepsilon$ correspond to periodic orbits of $F_t^\varepsilon$. Hypothesis 3 ensures that $\sigma(DP_0(p_0))$ does not contain 1 and we can now prove :

Lemma 1. For $\varepsilon > 0$ small, there is a unique fixed point $p_\varepsilon$ for $P_\varepsilon$ near $p_0$; moreover $p_\varepsilon$ is a smooth function of $\varepsilon$.

For ordinary differential equations, lemma 1 is a standard fact about persistence of fixed points. For general partial differential equations, its validity can be a delicate matter; see Holmes-Marsden [1980] for more details. In the application below, hypothesis 3 cannot be satisfied directly and we must prove lemma 1 by alternative methods, since simple application of the inverse function theorem will not suffice.

Our final hypothesis implies that the perturbation $f_1(x,t)$ is "Hamiltonian plus damping". Using assumption 3, above, this condition can be stated either in terms of the spectrum of the linearisation of equation (4) or in terms of the Poincaré map.

4. Assume that for $\varepsilon > 0$ the spectrum of $DP_\varepsilon(p_\varepsilon)$ lies strictly inside the unit circle with the exception of a single real eigenvalue $e^{T\lambda_\varepsilon^+} > 1$.

In lemma 1 we saw that the fixed point $p_0$ perturbs to another fixed point $p_\varepsilon$ for the perturbed system. The same is true for the local invariant manifolds of the map $P_\varepsilon, W_\varepsilon^{ss}(p_\varepsilon)$ and $W_\varepsilon^u(p_\varepsilon)$, which remain $C^r$ close to the unperturbed manifolds $W_0^s(p_0)$ and $W_0^u(p_0)$. Here $W_\varepsilon^{ss}(p_\varepsilon) \subset W_\varepsilon^s(p_\varepsilon)$ and the superscript ss denotes the strong stable manifold. Assumptions 3 and 4 guarantee that the centre-stable manifold $W_0^{sc}(p_0)$ of the unperturbed system and the perturbed stable manifold $W_\varepsilon^s(p_\varepsilon)$ are codimension one, while the unstable manifold is one dimensional for $\varepsilon = 0$ and $\neq 0$. The flow in $X \times S^1$ similarly has a periodic orbit $\gamma_\varepsilon$ $C^r$ close to $\{p_0\} \times S^1$, with invariant manifolds close to $W_0^s(p_0) \times S^1$, etc :

Lemma 2.

(a) For $\varepsilon > 0$ sufficiently small there are two unique simple real eigenvalues $e^{T\lambda_\varepsilon^\pm}$ of $DP_\varepsilon(p_\varepsilon)$ such that $\lambda_\varepsilon^\pm$ vary smoothly with $\varepsilon$ and $\lambda_0^+ = \lambda, \lambda_0^- = -\lambda$ (see assumption 3).

(b) Corresponding to the eigenvalues $\lambda_\varepsilon^\pm$ there are unique invariant manifolds $W_\varepsilon^{ss}(p_\varepsilon)$ ( the strong stable manifold) and $W_\varepsilon^u(p_\varepsilon)$ (the unstable manifold) of $p_\varepsilon$ for the map $P_\varepsilon$ such that

i. $W_\varepsilon^{ss}(p_\varepsilon)$ and $W_\varepsilon^u(p_\varepsilon)$ are tangent to the eigenspaces of $e^{T\lambda_\varepsilon^\pm}$ respectively at $p_\varepsilon$

ii. they are invariant under $P_\varepsilon$

iii. if $x \in W^{ss}_\varepsilon(p_\varepsilon)$ then

$$\lim_{n \to \infty} (P^\varepsilon)^n(x) = p_\varepsilon$$

and if $x \in W^u_\varepsilon(P_\varepsilon)$ then

$$\lim_{n \to -\infty} (P^\varepsilon)^n(x) = p_\varepsilon$$

iv. For any finite $t^*$, $W^{ss}(p_\varepsilon)$ is $C^r$ close as $\varepsilon \to 0$ to the homoclinic orbit $x_0(t)$, $t^* \leq t < \infty$ and for any finite $t_*$, $W^u_\varepsilon(p_\varepsilon)$ is $C^r$ close to $x_0(t)$, $-\infty < t \leq t_*$ as $\varepsilon \to 0$. Here, r is any fixed integer, $0 \leq r < \infty$.

Equipped with these assumptions and preliminary results, we now proceed to calculate the separation of the perturbed manifolds $W^s_\varepsilon(p_\varepsilon)$ and $W^u_\varepsilon(p_\varepsilon)$, by calculating the $0(\varepsilon)$ compon of perturbed solution curves of equation (2.15) from the first variation equation. As in the planar case, we expand solution curves in $W^s_\varepsilon(\gamma_\varepsilon)$ and $W^u_\varepsilon(\gamma_\varepsilon)$. Points in $W^s_\varepsilon(p_\varepsilon)$ (resp. $W^u_\varepsilon(p_\varepsilon)$) are obtained by intersecting $W^s_\varepsilon(\gamma_\varepsilon)$ (resp. $W^u_\varepsilon(\gamma_\varepsilon)$) with the section $X \times \{0\}$. This can also be done on general sections $X \times \{t_0\}$, as above.

It is then possible to compute the Melnikov function $M(t_0)$ which acts as a measure of the separation of the perturbed manifolds $W^s_\varepsilon(p_\varepsilon)$, $W^u_\varepsilon(p_\varepsilon)$ on different Poincaré sections $X \times \{t_0\}$. The formal conputation is identical to that in §2.1, except that trace $Df_0 \equiv 0$ and the two dimensional wedge product is replaced by the symplectic form $\Omega$. Again $M(t_0)$ is periodic of period T in $t_0$ and we have

Theorem 2. Let hypotheses 1-4 hold. Let

$$M(t_0) = \int_{-\infty}^{\infty} \Omega(f_0(x_0(t-t_0)), f_1(x_0(t-t_0), t))dt \tag{2.19}$$

Suppose that $M(t_0)$ has a simple zero as a function of $t_0$. Then for $\varepsilon > 0$ sufficiently small, the stable manifold $W^s_\varepsilon(p_\varepsilon(t_0))$ of $p_\varepsilon$ for $P^{t_0}_\varepsilon$ and the unstable manifold $W^u_\varepsilon(p_\varepsilon(t_0))$ intersect transversally.

The main idea of the extension of the two dimensional Melnikov result lies in the use of a projected distance function $d^N_\varepsilon(t_0)$, projected from $X \times \{t_0\}$ into the tangent space $T_{\hat{x}_0}N$ to N at a specified point $\hat{x}_0(0)$, lying on the unperturbed homoclinic loop.

The $C^r$ closeness of $W^s_\varepsilon(p_\varepsilon)$ and $W^{sc}_0(p_0)$ then guarantees that $d^N_\varepsilon(t_0)$ is a good measure of the actual separation of the manifolds $W^s_\varepsilon(p_\varepsilon(t_0))$ and $W^u_\varepsilon(p_\varepsilon(t_0))$ near $\hat{x}_0$. The function $M(t_0)$ in theorem 1 is the leading non-zero term $d^N_\varepsilon(t_0)$. The power of the method rests on the fact that $M(t_0)$ is easily calculated in specific cases, just as for the planar problem.

The second major result required is an extension of the Smale-Birkhoff homoclinic theorem (Smale [1967]) to infinite dimensions :

Theorem 3. If the diffeomorphism $P^{t_0}_\varepsilon : X \to X$ possesses a hyperbolic saddle point $p_\varepsilon$ and an associated transverse homoclinic point $q \in W^u_\varepsilon(p_\varepsilon) \pitchfork W^s_\varepsilon(p_\varepsilon)$, with $W^u_\varepsilon(p_\varepsilon)$ of dimension 1 and $W^s_\varepsilon(p_\varepsilon)$ of codimension 1, then some power of $P^{t_0}_\varepsilon$ possesses an invariant zero dimensional hyperbolic set $\Lambda$, homeomorphic to a Cantor set on which a power of $P^{t_0}_\varepsilon$ is conjugate to a shift on two symbols.

As in the finite dimensional case, this implies

Corollary 2. A power of $P^{t_0}_\varepsilon$ restricted to $\Lambda$ possesses a dense set of periodic points, there are points of arbitrary high period and there is a non-periodic orbit dense in $\Lambda$.

The hyperbolicity of $\Lambda$ under a power of $P^{t_0}_\varepsilon$ and the structural stability theorem of Robbin implies that the situation of Theorem 2 persists under perturbations :
so that the complex dynamics cannot be removed by making small (lower order, bounded) changes in equation (2.15) :

Corollary 3. If $\bar{P}:X \to X$ is a diffeomorphism that is sufficiently close to $P^{t_0}_\varepsilon$ in $C^1$ norm, then a power of $\bar{P}$ has in invariant set $\bar{\Lambda}$ and there is a homeomorphism $h:\bar{\Lambda} \to \Lambda$ such that $(P^{t_0}_\varepsilon)^N \circ h = h \circ \bar{P}^N$ for a suitable power N.

We also have a result analogous to Corollary 1 of §2.1.

## 3. Spatially Periodic Perturbations.

We consider certain stationary solutions of the system

$$\begin{aligned} &\varphi_{tt} - \varphi_{zz} + \sin\varphi = \varepsilon f(z), \quad x \in \mathbb{R}; \\ &\varphi_z = 0 \quad \text{at } z = \pm\infty , \end{aligned} \tag{3.1}$$

where $f(z) = f(z+L)$ is periodic of period L in z. Setting $\varphi_t = \varphi_{tt} \equiv 0$ we obtain the "phase-plane" problem

$$\varphi_{zz} = \sin\varphi - \varepsilon f(z),$$

or

$$\left.\begin{aligned} \varphi_z &= v \\ v_z &= \sin\varphi - \varepsilon f(z);\ v \to 0 \text{ as } x \to \pm\infty . \end{aligned}\right\} \tag{3.2}$$

Note that only when $\varepsilon = 0$ is (3.2) a true autonomous phase plane problem. When $\varepsilon \neq 0$ it falls into the class of problems discussed in §2.1 above, with t replaced by the spatial variable z. Travelling wave problems with wave frame coordinate $\xi = (z \pm ct)/\sqrt{1-c^2}$ might be treated in a similar manner, and the functions f(z) could be replaced by a first order differential operator $F(z,\varphi)$ or $F(z,\varphi,v)$. For simplicity, we only consider the spatially periodic field f(z) in this section.

When $\varepsilon = 0$ the phase plane of (3.2) is simply that of the pendulum (Figure 2). Note that there are equilibrium solutions $\varphi(z,t) \equiv 0,\ \pi,\ 2\pi$ etc. We shall concentrate on the stationary "kink" solutions $\Gamma_0^{\pm} = \bar{\varphi}(z)$ connecting $(\varphi, v) = (0,0)$ and $(\varphi, v) = (2\pi, 0)$, which may be obtained by integration of the Hamiltonian

$$H_0(\varphi, v) = \frac{v^2(z)}{2} + (\cos\varphi(z) - 1) = \text{const}, \tag{3.3}$$

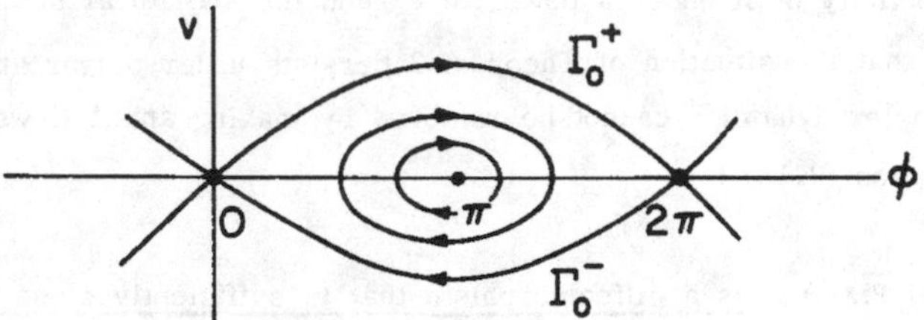


Figure 2. The phase-plane of (3.2) with $\varepsilon = 0$.

with $H_0(\varphi, v) = 0$. $\Gamma_0^+$ is given by

$$\begin{aligned} \bar{\varphi}(z) &= 2\text{arcot}(-\sinh z) \\ \bar{v}(z) &= 2\text{sech } z , \end{aligned} \tag{3.4}$$

where we have taken $(\varphi, v) = (\pi, 2)$ at $x = 0$. Changing the signs yields an expression for the solution $\Gamma_0^-$ connecting $(2\pi, 0)$ and $(0,0)$ and passing through $(\pi, -2)$ at $z = 0$. Note that the curves $\Gamma_0^{\pm}$ are also the invariant manifolds $W_0^s(0,0)$, $W_0^u(0,0)$, $W_0^s(2\pi,0)$, $W_0^u(2\pi,0)$

of the unperturbed Poincaré map $P_0^{z_0}: \Sigma_{z_0} \to \Sigma_{z_0}$ defined on any section

$$\Sigma_{z_0} = \{(\varphi, v, z) \mid z = z_0\}. \tag{3.5}$$

We now calculate the Melnikov function. Since trace $Df_0 \equiv 0$ in this case, we have

$$\begin{aligned} M(z_0) &= \int_{-\infty}^{\infty} f_0 \wedge f_1 \, dx = \int_{-\infty}^{\infty} \{\bar{v}(x-z_0).(-f(x)) - \sin(\bar{\varphi}(x-z_0).0\} dx \\ &= -\int_{-\infty}^{\infty} 2\operatorname{sech}(x-z_0) f(x) dx \end{aligned} \tag{3.6}$$

Expressing f(z) as a Fourier series

$$f(z) = \alpha_0 + \sum_{j=1}^{\infty} \{\alpha_j \cos(j\pi z/L) + \beta_j \sin(j\pi z/L)\}, \tag{3.7}$$

and letting $y = z - z_0$, we find that

$$\begin{aligned} M(z_0) = -[2\alpha_0 \int_{-\infty}^{\infty} \operatorname{sech} y dy + 2 \sum_{j=1}^{\infty} \{(\alpha_j \cos(\frac{j\pi z_0}{L}) \\ + \beta_j \sin(\frac{j\pi z_0}{L})) \int_{-\infty}^{\infty} \operatorname{sech} y \cos(j\pi y/L) dy]. \end{aligned} \tag{3.8}$$

The integrals may be evaluated by the method of residues to yield

$$\begin{aligned} M(z_0) = -[2\pi\alpha_0 + 2 \sum_{j=1}^{\infty} (\alpha_j \cos(j\pi z_0/L) \\ + \beta_j \sin(j\pi z_0/L))\pi \operatorname{sech}(j\pi^2/2L)] \end{aligned} \tag{3.9}$$

Depending upon the relative values of the Fourier coefficients, $M(t_0)$ may or may not have simple zeros. For example, if $f(z) = \alpha_0 + \alpha_1 \cos(\pi z/L)$ then we require

$$\alpha_1 > \alpha_0 \cosh(\pi^2/2L) \tag{3.10}$$

for simple zeros. In this case the Poincaré map $P_\varepsilon^{z_0}$ associated with (3.2) will appear somewhat as shown in figure 3.

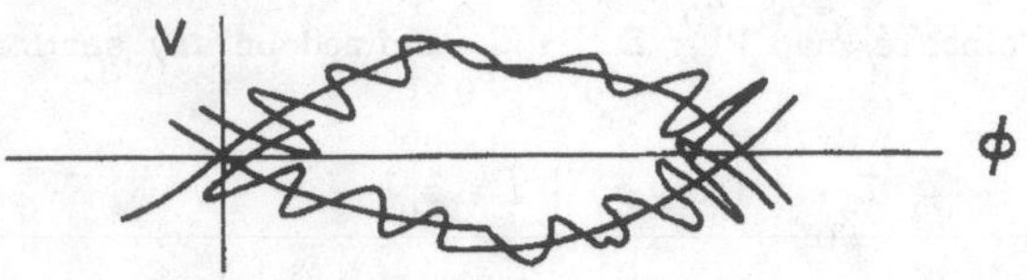


Figure 3. The perturbed map $P_\varepsilon^{z_0}$, showing transversal intersections.

We can now use topological arguments to prove the following

<u>Theorem 4.</u> <u>If the Melnikov function</u> $M(z_0)$ <u>given in</u> (3.9) <u>has simple zeros, then</u> (3.1) <u>has a countable set of stationary solutions of arbitrarily high spatial periods</u> NkL, $k \to \infty$ <u>for some</u> $N < \infty$ <u>in addition to an uncountable set of non periodic solutions.</u>

<u>Remark</u> : These solutions look like spatial translates and superpositions of the kink $\bar{\varphi}^+(z) = 2\text{arcot}(-\sinh z)$ and the antikink $\bar{\varphi}^-(z) = -2\text{arcot}(\sinh z)$.

<u>Sketch of Proof.</u> We shall only indicate the geometrical ideas behind the proof of theorem 4, since the estimates necessary to prove hyperbolicity are essentially similar to those in the case of a single homoclinic orbit (cf. Moser [1973], Smale [1963]).

Consider two rectangles $R_1$ and $R_2$ containing the two saddle points $p_1, p_2$ and portions of their local stable manifolds in their boundaries (Figure 4). Since $R_1$ sits astride

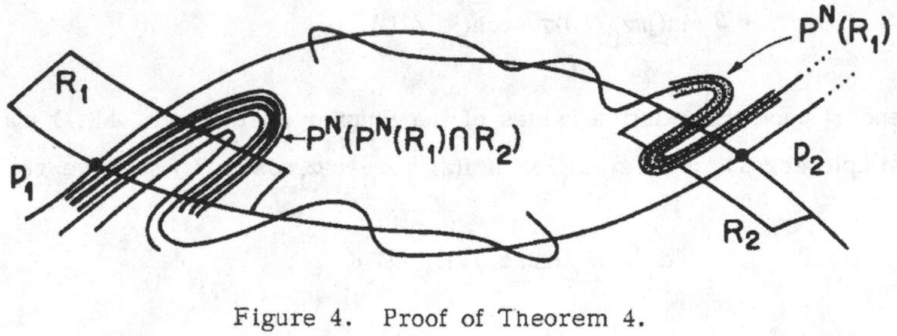


Figure 4. Proof of Theorem 4.

the unstable manifold $W_\varepsilon^u(p_1)$, and the flow is contracting transverse to $W_\varepsilon^u(p_1)$ and expanding along $W_\varepsilon^u(p_1)$*, one can choose $R_1, R_2$ and an integer N such that $P_\varepsilon^N(R_1)$ lies

* In fact $P_\varepsilon$ is an area preserving map in this case.

across $R_2$ in the manner shown and $P^N(R_1) \cap R_2$ has two components. By symmetry, $P^N(R_2) \cap R_1$ will likewise have two components and hence $S \triangleq P^N(P^N(R_1) \cap R_2) \cap R_1$ will have four components as indicated. Carrying out the inverse mapping, one see that $P^{-2N}(S) \cap R_1$ will consist of four strips transverse to those of S. Denoting these latter strips as $H_j$ and their images $P_\varepsilon^{2N}(H_j) \subset S$ as $V_j$, j = 1,2,3,4, we have a map

$$P_\varepsilon^{2N} : H_j \to V_j \tag{3.11}$$

which can be shown to be equivalent to a shift on four symbols in a manner similar to that employed by Moser [1973]. $P_\varepsilon^{2N}$ has an invariant set $\Lambda \subset \bigcup_{i,j} (H_i \cap V_j)$ which is homeomorphic to a Cantor set and contains a countable set of periodic points in addition to an uncountable set of nonperiodic orbits and a dense orbit. Moreover, it follows from the hyperbolicity of the saddle points and the transversality of the intersections $W_\varepsilon^u(p_1) \cap W_\varepsilon^s(p_2)$ and $W_\varepsilon^u(p_2) \cap W_\varepsilon^s(p_1)$ that $\Lambda$ is a hyperbolic set, and thus stable to further small perturbations in the map $P_\varepsilon$. □

By a slight extension of the geometric methods used above we can extend theorem 4 as follows :

Theorem 5. If $M(z_0)$ has simple zeros, then (3.1) has stationary solutions in which kinks and antikinks follow one another in any order. More precisely, let the kink $\bar{\varphi}^+$ be denoted by the symbol "+" and the antikink $\bar{\varphi}^-$ by "-". Then, given any sequence of these two symbols, there exists a stationary solution corresponding to that sequence.

Sketch Proof. This result follows from the periodicity of the Poincaré map $P_\varepsilon$ in $\varphi$. We define a countable set of rectangles $R_{11}, R_{12}; R_{21}, R_{22}; \ldots; R_{n1}, R_{n2}; \ldots$ enclosing the saddle points $\ldots, p_1, p_2, \ldots, p_n$ which lie near $(\varphi, v) = (0,0), (2\pi, 0), \ldots, (2n\pi, 0), \ldots,$ as shown in Figure 5. Letting $F_{ij} = P_\varepsilon^N$, where N is a suitably chosen integer, it is easy to see that the maps

$$F_{11}:R_{11} \to R_{21} \cup R_{22}, \quad F_{21}:R_{21} \to R_{31} \cup R_{32}, \quad F_{22}:R_{22} \to R_{11} \cup R_{12}$$

etc. are well defined. Application of the maps $F_{i1}$ corresponds to kinks and of the maps $F_{i2}$ to antikinks and simple topological arguments suffice to show that points in $R_{11}$ (say) can be chosen such that repeated applications of F ($F_{i1}$ or $F_{i2}$ being selected as required) lead to any desired kink-antikink sequence. □

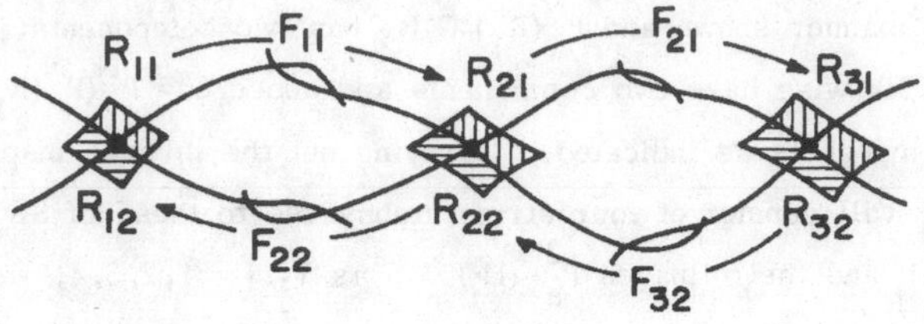


Figure 5. Kink and antikink maps.

4. Time-Periodic Perturbations.

We consider the system

$$\varphi_{tt} - \varphi_{zz} + \sin\varphi = -\varepsilon\sigma\varphi_t ,$$
$$\varphi_{z|z=0} = \varepsilon H, \ \varphi_{z|z=1} = \varepsilon(H+I(t)), \ 0 \leq \varepsilon \ll 1 \tag{4.1}$$

where I(t) is a bounded periodic perturbation of period T with bounded derivatives. This system models a Josephson junction in a weak magnetic field εH and subject to a weak periodic current εI(t). See Matisoo [1969] for physical details and Levi, Hoppensteadt and Miranker [1978] for additional mathematical details on equation (4.1). Levi considers the case of constant current and ε of order 1.

We first employ the transformation

$$\psi(z,t) = \varphi(z,t) - \varepsilon(zH + \frac{z^2}{2} I(t)) , \tag{4.2}$$

under which (4.1) becomes a problem with homogeneous boundary conditions :

$$\psi_{tt} - \psi_{zz} + \sin\psi = \varepsilon\{I(t) - \frac{z^2}{2} I_{tt}(t) - \sigma\psi_t - (zH + \frac{z^2}{2} I(t))\cos\psi\} + 0(\varepsilon^2),$$
$$\psi_{z|z=0} = \psi_{z|z=1} = 0 \ . \tag{4.3}$$

We shall consider the evolution equation arising from (4.3)

$$\frac{dx}{dt} = f_0(x) + \varepsilon f_1(x,t); \ x = \{\psi, \psi_t\} \ \varepsilon \ X = H_0^1(0,1) \times L^2(0,1) , \tag{4.4}$$

$$\text{where } f_0(x) = Ax + B(x) = \begin{bmatrix} 0 & I \\ \frac{\partial^2}{\partial z^2} & 0 \end{bmatrix} \begin{pmatrix} \psi \\ \psi_t \end{pmatrix} + \begin{pmatrix} 0 \\ -\sin\psi \end{pmatrix} \tag{4.5a}$$

$$f_1(x,t) = \begin{pmatrix} 0 \\ -\sigma\psi_t + I(t) - \frac{z^2}{2} I_{tt}(t) - (zH + \frac{z^2}{2} I(t))\cos\psi \end{pmatrix} \tag{4.5b}$$

and $H_0^1$ denotes the Sobolev space of square integrable functions $\psi(z)$ whose first derivatives are square integrable and $\psi_z = 0$ at $z = 0, 1$. The domain of A is $H_0^2 \subset H_0^1$ and it can be shown that A generates a $C^0$ semigroup on X. Moreover $B{:}H^1 \to H^1$ is a $C^\infty$ operator and the pertubation $f_1(x,t)$ is of lower order than A.

To check the remaining condition of hypothesis 1 of §2.2 (global existence) we consider the Hamiltonian energy function

$$H_0(\psi, \psi_t) = \frac{|\psi_t|^2}{2} + \frac{|\psi_z|^2}{2} + \int_0^1 (1 - \cos\psi(z))dz \tag{4.6}$$

associated with the unperturbed ($\varepsilon$=0) system. Here $|.|$ denotes the $L^2$ norm $|f| = (\int_0^1 f(z)^2 dz)^{1/2}$.

Differentiating (4.6) along solution curves of (4.3) we obtain

$$\begin{aligned} \frac{d}{dt} H_0(\psi, \psi_t) &= \varepsilon(\psi_t, -\sigma\psi_t + I(t) - \frac{z^2}{2} I_{tt}(t) - (zH + \frac{z^2}{2} I(t))\cos\psi)) \\ &= -\varepsilon\sigma|\psi_t|^2 + \int_0^1 \{I(t) - \frac{z^2}{2} I_{tt}(t) - (zH + \frac{z^2}{2} I(t))\cos\psi(z)\}\psi_t(z)dz \\ &\le -\varepsilon\sigma|\psi_t|^2 + [I(t) - \frac{I_{tt}(t)}{6} - (\frac{H}{2} + \frac{I(t)}{6})]\int_0^1 |\psi_t(z)|dz \end{aligned} \tag{4.7}$$

where we have used the Schwartz inequality. Thus, if $|\psi_t|$ is sufficiently large, we have $\frac{d}{dt} H_0(\psi, \psi_t) < 0$ and energy (and hence solutions) remain bounded for all time in $X = H_0^1(0,1) \times L^2(0,1)$ with respect to the "energy" norm $||x|| = (|\psi_x|^2 + |\psi_t|^2)^{1/2}$. This and the preceeding observations show that hypothesis 1 holds.

We now prove the following Lemma, which guarantees that hypothesis 2 of §2.2 is likewise satisfied.

Lemma 3 : For $\varepsilon = 0$ the solutions $(\psi, \psi_t) = (\pm\pi, 0)$ are saddle points with one dimensional stable and unstable manifolds. Moreover, the subspace N = span $\{1,1\} \in H_0^1 \times L^2$ is a symplectic two-manifold containing a pair of heteroclinic orbits $\{\bar{\psi}, \bar{\psi}_t\} = \{\pm 4 \arctan(\tanh t/2), \pm 2 \operatorname{sech} t\}$ connecting $(\psi, \psi_t) = (-\pi, 0)$ and $(\psi, \psi_t) = (+\pi, 0)$.

Proof. Linearising (4.3) about $\psi = \pm\pi$ for $\varepsilon = 0$ we obtain

$$u_{tt} - u_{zz} - u = 0; \quad u_z = 0 \text{ at } z = 0,1 \tag{4.8}$$

Letting $u = u_0 e^{\lambda t}$ and solving for the eigenvalues and eigenfunctions we obtain

$$u_{zz} + (1-\lambda^2)u = 0 , \tag{4.9}$$

so that $\lambda_0 = \pm 1, \; i\lambda_j = \pm i\sqrt{j^2\pi^2 - 1}, \; j = 1,2,\ldots$ (4.10a)

$$u_0(z) = 1, \quad u_j(z) = \cos j\pi z \tag{4.10b}$$

This proves the first assertion. To prove the second, we substitute $\{\psi, \psi_t\} = \{1.a(t), 1.a_t(t)\}$ into (4.3) with $\varepsilon = 0$ to obtain

$$a_{tt} + \sin a = 0 \quad . \tag{4.11}$$

The plane N = span$\{1,1\}$ is thus invariant under the flow. It is now a simple computation to check that the solutions

$$\begin{aligned} a &= \pm 4 \arctan(-\tanh(t/2)) \\ a_t &= \pm 2 \operatorname{sech} t \end{aligned} \tag{4.12}$$

of (4.11) satisfy the boundary conditions $(a(t), a_t(t)) \to (\pm\pi, 0)$ as $t \to \pm\infty$ . □

Note that we cannot satisfy hypothesis 3, since the pure imaginary eigenvalues of $Df_0(p_0)$ are given by $i\lambda_j = \pm i\sqrt{j^2\pi^2 - 1}, \; j = 1,2,\ldots$ (eq. 4.10) and the condition that $\lambda_j T$ be bounded away from $2m\pi$ becomes

$$\sqrt{j^2\pi^2 - 1}\; T \neq 2m\pi \tag{4.13}$$

or $T \neq \frac{2m\pi}{\sqrt{j^2\pi^2-1}} \simeq \frac{2m}{j}$ as $j \to \infty$. One can therefore find (possibly large) integers j, m such that for <u>any</u> excitation period T one is arbitrarily close to resonance and thus $\sigma(DP_0(p_0))$ cannot be bounded away from 1. We therefore resort to the following argument.

We linearise (4.3) at the saddle points $\{\bar{\psi}, \bar{\psi}_t\} = \{\pm\pi, 0\}$ :

$$u_{tt} - u_{zz} - u = \varepsilon\{I(t) - \frac{z^2}{2} I_{tt}(t) - \sigma u_t - (zH + \frac{z^2}{2} I(t))\} + O(\varepsilon^2) . \tag{4.14}$$

Using Galerkin's method with the set of basis functions $\{\cos j\pi z, \cos j\pi z\}$ (of (4.10b)), we obtain an infinite set of uncoupled second order equations for the modal components :

$$\ddot{a}_j + \varepsilon\sigma\dot{a}_j + (j^2\pi^2-1)a_j = 2\varepsilon\int_0^1 [(I(t) - \frac{z^2}{2} I_{tt}(t) - (zH + \frac{z^2}{2} I(t))]\cos j\pi z dz \tag{4.15}$$

where $u(z,t) = \sum_{j=0}^{\infty} a_j(t)\cos j\pi z$. It is easy to check that, provided the frequency component $\omega_j = \sqrt{j^2\pi^2-1}$, j = 1, 2, ... is not present in the forcing term I(t), then the response amplitudes $a_j(t)$ will all be $O(\varepsilon)$. For general T-periodic excitations, components $\omega_m = 2m\pi/T$, n = 1, 2, ... will be present and thus the non-resonance requirement

$$\omega_m = \frac{2m\pi}{T} \neq \sqrt{j^2\pi^2-1} \Rightarrow T \neq \frac{2m\pi}{\sqrt{j^2\pi^2-1}} \simeq \frac{2m}{j} \tag{4.16}$$

is, of course, the same as (4.13), above. However, for single (or finite component) frequency excitations we <u>can</u> ensure that the non-resonance conditions are met. In particular, if $I = I_0 \cos \omega t$, as assumed below, we merely require that $\omega$ be bounded away from $\sqrt{j^2\pi^2-1}$, j = 1, 2, ..., which can be easily achieved. This linear analysis ensures that the perturbed periodic orbit of the linearised problem has amplitude of $O(\varepsilon)$. A contraction mapping argument using the variation of constants formulation then shows that the perturbed orbit $\gamma_\varepsilon$ of the full non-linear problem is also of $O(\varepsilon)$, provided that the semigroup $e^{(A+A_1)T}$, defined below, has all eigenvalues bounded away from 1 by $O(\varepsilon)$ and the nonlinear terms are of $O(\varepsilon^3)$ for x of $O(\varepsilon)$. See Holmes-Marsden [1980] for details. We check these conditions for the sine-Gordon system in our verification of hypothesis 4, below.

It is convenient to rewrite equation (4.4) as

$$\frac{dx}{dt} = (A+A_1)x + G(x,t) , \tag{4.17}$$

with $A_1 = \begin{pmatrix} 0 & 0 \\ 1 & -\varepsilon\sigma \end{pmatrix}$, $G(x,t) = \begin{pmatrix} 0 \\ -(\sin\psi+\psi)+\varepsilon\{I(t)-\frac{z^2}{2}I_{tt}(t)-(zH+\frac{z^2}{2}I(t))\cos\psi\} \end{pmatrix}$

The Poincaré map $P_\varepsilon : \Sigma_0 \to \Sigma_0$ associated with (4.13) is obtained by integration of (4.13), i.e.

$$x(T) = e^{(A+A_1)T}x(0) + \int_0^T e^{(A+A_1)(T-s)} G(x(s),s)ds . \tag{4.18}$$

We require the linearised map $DP_\varepsilon(p_\varepsilon)$. Linearising (4.14) about the perturbed saddle solutions $p_\varepsilon(t) = \pm\pi + 0(\varepsilon)$ we obtain

$$u(T) = e^{(A+A_1)T}u(0) + \int_0^T e^{(A+A_1)(T-s)} DG(p_\varepsilon(s),s)ds \tag{4.19}$$

where $DG(p_\varepsilon(s),s) = \begin{pmatrix} 0 \\ -(\cos(\pm\pi+0(\varepsilon))\pm 1)+\varepsilon(zH+\frac{z}{2}I(t))\sin(\pm\pi+0(\varepsilon)) \end{pmatrix}$

$$= 0(\varepsilon^2) .$$

Thus, for $\varepsilon$ sufficiently small, the map $DP_\varepsilon(p_\varepsilon)$ is dominated by the linear semigroup $e^{(A+A_1)T}$. It remains to check that this semigroup has all eigenvalues inside the unit circle but one. Since

$$A + A_1 = \begin{pmatrix} 0 & I \\ \frac{\partial^2}{\partial z^2}+1 & -\varepsilon\sigma \end{pmatrix} \tag{4.20}$$

the dynamical problem assumes the form

$$\frac{d}{dt}\begin{pmatrix} u \\ u_t \end{pmatrix} = \begin{pmatrix} u_t \\ u_{zz}+u-\varepsilon\sigma\, u_t \end{pmatrix} \tag{4.21}$$

Using Galerkin's method again with the basis functions $\{\cos j\pi z, \cos j\pi z\}$ we obtain an infinite set of uncoupled second order equations

$$\ddot{a}_j + \varepsilon\sigma\dot{a}_j + (j^2\pi^2 - 1)a_j = 0; \quad j = 0,1,2,\ldots \tag{4.22}$$

where $u(z,t) = \sum_{j=0}^{\infty} a_j(t)\cos j\pi z$. The spectrum of $(A+A_1)$ is thus

$$\lambda_j = \frac{-\varepsilon\sigma}{2} \pm \sqrt{\frac{\varepsilon^2\sigma^2}{4} + (1-j^2\pi^2)}\ , \tag{4.23}$$

and hypothesis 4 therefore holds.

We can now compute the Melnikov function $M(t_0)$ of equation (2.19). The symplectic form is given by $\Omega(\{u,\dot{u}\},\{v,\dot{v}\}) = \int_0^1[\dot{v}(z)u(z) - \dot{u}(z)v(z)]dz$ and hence

$$M(t_0) = \int_{-\infty}^{\infty}[\int_0^1 \bar{\psi}_t(z,t-t_0)\{-\sigma\bar{\psi}(z,t-t_0) + I(t) - \frac{z^2}{2}I_{tt}(t)$$

$$- (zH + \frac{z^2}{2}I(t))\cos\bar{\psi}(z,t-t_0)\}dz]dt\ , \tag{4.24}$$

where $\{\bar{\psi},\bar{\psi}_t\} = \{4\arctan(-\tanh(t-t_0)),\ 2\,\mathrm{sech}(t-t_0)\}$ is the heteroclinic orbit connecting $(-\pi,0)$ to $(\pi,0)$. (Since the heteroclinic pair is symmetric we need only calculate one branch.) Carrying out the spatial integration we have

$$M(t_0) = \int_{-\infty}^{\infty}\bar{\psi}_t(t-t_0)\{-\sigma\bar{\psi}(t-t_0) + I(t) - \frac{I_{tt}(t)}{6} - (\frac{H}{2} + \frac{I(t)}{\sigma})\cos\bar{\psi}(t-t_0)\}dt \tag{4.25}$$

Letting $\tau = t - t_0$ and using $\cos\bar{\psi} = 2\,\mathrm{sech}^2(t-t_0) - 1 = 2\,\mathrm{sech}^2\tau - 1$ we have

$$M(t_0) = \int_{-\infty}^{\infty} 2\,\mathrm{sech}\,\tau\{\sigma\mathrm{sech}\,\tau + I(\tau+t_0) - \frac{I_{\tau\tau}(\tau+t_0)}{6}$$

$$- (\frac{H}{2} + \frac{I(\tau+t_0)}{6})[2\,\mathrm{sech}^2\tau - 1]\}d\tau$$

$$= -4\sigma\int_{-\infty}^{\infty}\mathrm{sech}^2\tau d\tau + H\int_{-\infty}^{\infty}(\mathrm{sech}\tau - 2\,\mathrm{sech}^3\tau)d\tau$$

$$+ \int_{-\infty}^{\infty}[(2I(\tau+t_0) - \frac{I_{\tau\tau}(\tau+t_0)}{3} + \frac{I(\tau+t_0)}{3})\mathrm{sech}\,\tau d\tau$$

$$+ \int_{-\infty}^{\infty}\frac{2I(\tau+t_0)}{3}\,\mathrm{sech}^3\,\tau d\tau$$

$$\triangleq -8\sigma + H(\pi - 2(\frac{\pi}{2})) + I_1 + I_2\ , \tag{4.26}$$

$$= -8\sigma + I_1 + I_2$$

where the first two integrals were evaluated directly and by the method of residues respectively.

We now take a specific forcing function $I(t) = I_0\cos\omega t$, so that $I(\tau+t_0) = I_0(\cos\omega\tau\cos\omega t_0 - \sin\omega\tau\sin\omega t_0)$ and $I_{\tau\tau}(\tau+t_0) = -\omega^2 I_0(\tau+t_0)$. Thus, using the fact that $\mathrm{sech}\,\tau$ and $\mathrm{sech}^3\tau$ are even functions, we have

$$I_1 = \int_{-\infty}^{\infty} \frac{1}{3}(7+\omega^2)\mathrm{sech}\,\tau\cos\omega\tau d\tau . I_0\cos\omega t_0 \tag{4.27a}$$

$$I_2 = \int_{-\infty}^{\infty} \frac{2}{3}\,\mathrm{sech}^3\tau\cos\omega\tau d\tau . I_0\cos\omega\, t_0 \ . \tag{4.27b}$$

Evaluating these integrals by the method of residues yields

$$I_1 = \frac{1}{3}(7+\omega^2)\pi I_0\mathrm{sech}(\frac{\omega\pi}{2})\cos\omega t_0,\quad I_2 = \frac{2}{3}\cdot\frac{\pi}{2}\, I_0(1+\omega^2)\mathrm{sech}(\frac{\omega\pi}{2})\cos\omega t_0 \ ,$$

so that the Melnikov function becomes

$$M(t_0) = -8\sigma + [\frac{7+\omega^2}{3}+\frac{1+\omega^2}{3}]\pi I_0\mathrm{sech}(\frac{\pi\omega}{2})\cos\omega t_0$$

$$M(t_0) = -8\sigma + \frac{(8+2\omega^2)\pi I_0}{3}\,\mathrm{sech}(\frac{\pi\omega}{2})\cos\omega t_0 \ . \tag{4.28}$$

Thus, we have transversal (heteroclinic) intersections of the manifolds $W^s_\varepsilon(p_\varepsilon)$ and $W^u_\varepsilon(p_\varepsilon)$ if $\varepsilon$ is sufficiently small and

$$(4+\omega^2)\pi I_0 > 12\sigma\cosh(\frac{\pi\omega}{2}) \ . \tag{4.29}$$

Note that the magnetic field strength ($\varepsilon H$) does not appear in (4.29), this is as one would expect, since it is applied equally at both boundaries $z = 0$ and $z = 1$.

Also, if the damping ($\sigma$) is relatively high then (4.29) is not satisfied and no intersections occur.

Finally, arguments similar to those of the proof of theorem 4, adapted for infinite dimensional maps as in Holmes and Marsden [1980] suffice to prove that if (4.29) is satisfied then theorem 3 and Corollaries 2 and 3 of section 2.2 hold for equation (4.3) and hence for equation (4.1).

Remark. The analysis developed in this section can be applied directly to a related, but not full integrable problem, the $\varphi^4$ system :

$$\varphi_{tt} - \varphi_{xx} - \varphi + \varphi^3 = 0 , \tag{4.30}$$

and it can be shown that the presence of weak time-periodic perturbations and weak dissipation can lead to chaotic motions in this case also.

## 5. Bifurcation, Stability and the Problem of Infinite Domains.

In this section we make some comments on possible extensions to the work on time periodic motions described above.

### 5.1 Bifurcation to horseshoes : Newhouse Sinks.

In a series of papers ([1974, 1979, 1980]) Newhouse has shown that, given a finite dimensional map possessing a homoclinic orbit with quadratic tangencies between stable and unstable manifolds, there is a residual set of nearby maps possessing infinitely many attracting periodic orbits of arbitrarily long period. As the stable and unstable manifolds are "pulled through each other" under the increase of, say force amplitude, saddle-node bifurcations occur in which pairs of periodic orbits are created. If the map is (overall) a contraction, then the nodes are stable (periodic) sinks. As the force increases further these sinks apparently loose their stability in a succession of period doubling bifurcations until ultimately one has a full horseshoe in the neighbourhood of the (now transversal) intersections of the manifolds. Thus, associated with the infinite sequence of bifurcations into horseshoes, one has short lived periodic attractors of arbitrarily long period.

In our case, the verification of hypothesis 4, above, shows that the product of the eigenvalues of the perturbed map $DP_\varepsilon^{t_0}(p_\varepsilon)$ is less than one and thus that we have a contracting map, or, more precisely, the saddle point $p_\varepsilon$ is dissipative in Newhouse's terminology (Newhouse, [1979]). It should therefore follow, by arguments similar to those of Newhouse (cf. Chow, Hale and Mallet-Paret [1980]) that near parameter values for which the Melnikov function has quadratic zeros and we have quadratic heteroclinic tangencies (cf. Corollary 1), the Poincaré map $P_\varepsilon^{t_0}$ will possess attracting periodic orbits of arbitrarily high period in the neighbourhood of the unperturbed heteroclinic orbits $\{\overline{\psi}, \overline{\psi}_t\} \subset H_0^1 \times L^2$. Thus, while the invariant set $\Lambda$ of the horseshoe is not an attractor, infinitely many periodic attractors are created in a countable sequence of bifurcations as I increases relative to $\sigma$ and $\Lambda$ appears.

## 5.2 Other Perturbed Orbits.

Following the analysis of section 4, it is easy to check that the symplectic manifold $N = \text{span}\,\{1,1\} \subset H_0^1 \times L^2$ contains a region D, bounded by the heteroclinic orbits, which is filled with a continuous family of time periodic orbits whose periods vary from $2\pi$ to $\infty$. In fact, the flow of the unperturbed problem ((4.3), with $\varepsilon = 0$) restricted to N is just that of the pendulum, equation (4.11). Thus, if periodic forcing of period $2\pi/\omega$ is applied there is a countable set of orbits $O_m \subset D$ with periods $2\pi m/\omega n$ $m = m_0, m_0+1, \ldots, n = 1,2,\ldots$, in resonance with the external force. Here $m_0$, the lowest order, is chosen such that $\frac{2\pi m_0}{\omega} > 2\pi$. The second part of Melnikov's [1963] paper is devoted to the study of such motions and their perturbations in the two dimensional case. Chow, Hale and Mallet-Paret [1980] also study such motions, and discuss the bifurcation sets in force-damping space on which they appear. The two dimensional analysis should again extend to the infinite dimensional case, but we have not yet carried out all the details. The subharmonic orbits thus found occur in pairs for each period $\frac{2\pi m}{\omega n}$, one being of saddle-type and the other stable (provided that the damping, $\sigma$, is positive). Thus in addition to the horseshoes and the attracting periodic motions associated with their creation, one may have additional infinite sets of attracting periodic motions.

The equilibrium solution $\{\psi, \psi_t\} \equiv \{0,0\}$ (of centre type), in the unperturbed problem also perturbs into a stable periodic orbit of period $2\pi/\omega$. Stability for this and the other subharmonics follows from arguments similar to those employed in the validation of hypothesis 4 in §4 above.

## 5.3 Infinite Domains.

Consider equation (4.3) recast on an infinite domain, for example

$$\psi_{tt} - \psi_{zz} + \sin\psi = \varepsilon(\gamma f(z)\cos\omega t - \sigma\psi_t) \tag{5.1}$$
$$\psi_z \to 0 \text{ as } z \to \pm\infty .$$

This is one of the examples considered by Kaup and Newell [1978 a,b]. It is well known that the unperturbed problem ($\varepsilon=0$) possesses travelling kink (and antikink) solutions

$$\psi(z,t) = 4\arctan(\exp Z) = 2\ \mathrm{arcctg}(-\sinh Z) \tag{5.2}$$

in addition to breather modes :

$$\psi(z,t) = 4\arctan\left\{\frac{2a}{\Omega}\cos(\Omega T)\ \mathrm{sech}(2aZ)\right\} \quad . \tag{5.3}$$

Here $\Omega^2 = 1 - 4a^2$; $T = (t-cz)/b$; $Z = (z-ct)/b$; $b^2 = 1 - c^2$. Note that the breathers are time (T) periodic in the Z, T coordinates, in terms of which the unperturbed equation becomes

$$\psi_{TT} - \psi_{ZZ} + \sin\psi = \varepsilon\left(f\left(\frac{Z+ct}{b}\right)\cos\omega\left(\frac{cZ+T}{b}\right) - \frac{\sigma}{b}(\psi_T - c\psi_Z)\right) \tag{5.4}$$

Attempting to carry through the method of §(2.2) in this case we find that, while the kink (5.2) is analogous to that on a finite domain, in the infinite domain case the spectrum of $DP_0(p_0)$ contains zero. Linearising (5.4) with $\varepsilon = 0$ about (5.2) we obtain

$$u_{TT} - u_{ZZ} + (1-2\mathrm{sech}^2 Z)u = 0 \ . \tag{5.5}$$

Letting $u(Z,T) = \tilde{u}(Z)e^{\lambda T}$ and seeking eigenvalues and eigenfunctions we have

$$\tilde{u}_{ZZ} + (2\mathrm{sech}^2 Z - 1)\tilde{u} = \lambda^2\tilde{u} \tag{5.6}$$

and it is easy to check that zero is an eigenvalue with eigenfunction $\tilde{u}_0 = \mathrm{sech}\ z$. The assumptions of hypothesis 3 are therefore not satisfied and our method cannot be applied without modification. In general the continuous spectrum of $DP_0(p_0)$, which is a consequence of the infinite spatial domain, may require more delicate analytical techniques than those described in the present paper.

## References.

R. Abraham & J. Marsden [1978] "Foundations of Mechanics," 2nd Edition, Addison-Wesley.

A.R. Bishop, J.A. Krumhansl & S.E. Trullinger [1980] Physica D : Nonlinear Phenomena 1, 1-44. Solitons in condensed matter : a paradigm.

P. Chernoff & J. Marsden [1974] "Properties of Infinite Dimensional Hamiltonian Systems," Springer Lecture Notes in Math., no. 425.

S.N. Chow, J. Hale & J. Mallet-Paret [1980] J. Diff. Eqn's 37, 351-373, "An example of bifurcation to homoclinic orbits ".

P. Hartman [1964] "Ordinary Differential Equations", Wiley.

B. Hassard [1980] Computation of invariant manifolds, in Proc. "New Approaches to Nonlinear Problems in Dynamics", ed. P. Holmes, SIAM publications, Philadelphia.

E. Hille & R. Phillips [1957] "Functional Analysis and Semigroups," A.M.S. Colloq. Publ.

M. Hirsch, C. Pugh & M. Shub [1977] "Invariant Manifolds," Springer Lecture Notes in Math., 583.

P. Holmes [1979a] Proc. 1978 IEEE Conf. on Decision and Control, San Diego, CA., 181-185, Global bifurcations and chaos in the forced oscillations of buckled structures.

[1979b] Phil. Trans. Roy. Soc. A292. 419-448. A nonlinear oscillator with a strange attractor.

[1980] SIAM J. on Appl. Math. 38 (1), 65-80, Averaging and chaotic motions in forced oscillations.

P. Holmes & J. Marsden [1978], Automatica 14, 367-384. Bifurcation to divergence and flutter in flow induced oscillations; an infinite dimensional analysis.

[1980] Arch. Rat. Mech. Anal. (in press), A partial differential equation with infinitely many periodic orbits : chaotic oscillations of a forced beam.

T. Kato [1977] "Perturbation Theory for Linear Operators" 2nd Ed., Springer.

D.J. Kaup & A. Newell [1978a] Proc. Roy. Soc. Lond. A361 413-446, Solitons as particles, oscillators, and in slowly changing media : a singular perturbation theory.

[1978b] Phys. Rev. B. 18(10) 5162-5167. Theory of nonlinear oscillating dipolar excitations in one dimensional condensates.

M. Levi, F.C. Hoppensteadt & W.C. Miranker [1978] Quart. of Appl. Math. 36, 167-198. Dynamics of the Josephson junction.

J. Matisoo [1969] IEEE Trans. on Magnetics MAG-5(4), 848-873. Josephson-type superconductive tunnel junctions and applications.

V.K. Melnikov [1963] Trans. Moscow Math. Soc. 12, 1-57. On the stability of the centre for time periodic perturbations.

J. Moser [1973] "Stable and Random Motions in Dynamical Systems," Ann. of Math. Studies no. 77, Princeton Univ. Press.

A. Newell [1980] The inverse scattering transform (to appear in "Topics of Modern Physics : Solitons", ed. R. Bullough & P. Caudrey, Springer Verlag, Berlin.

S. Newhouse [1974] Topology 12, 9-18. Diffeomorphisms with infinitely many sinks.

[1979] Publ. IHES 50, 101-151, The abundance of wild hyperbolic sets and non-smooth stable sets for diffeomorphisms.

[1980] Proc. CIME summer school, Bressanone, Italy, 1978, (to appear). Lectures on Dynamical Systems.

J. Robbin [1971] Ann. of Math. 94, 447-493. A structural stability theorem.

I. Segal [1962] Ann. of Math. 78, 334-362. Nonlinear Semigroups.

S. Smale [1963] Diffeomorphisms with many periodic points, in "Differential and combinatorial Topology," (ed. S.S. Cairns), Princeton Univ. Press., 63-80.

[1967] Bull. Am. Math. Soc. 73, 747-817. Differentiable dynamical systems.

P. Holmes : Department of Theoretical and Applied Mechanics, Cornell University, Ithaca, N.Y. 14853, USA.

Simple computation of bifurcating invariant circles for mappings.

G. Iooss*, A. Arneodo**, P. Coullet***, C. Tresser***.

## I. Introduction.

Let us consider a mapping $X \mapsto \Phi(\mu, X)$ in a Banach space E, sufficiently smooth in $(\mu, X)$, where $\mu$ is a real parameter. We assume that 0 is a fixed point :

(1) $$\Phi(\mu, 0) = 0 ,$$

and that the linear bounded operator defined by

(2) $$A_0 = D_x \Phi(0, 0)$$

is such that it has two simple eigenvalues $\lambda_0$ and $\bar{\lambda}_0$ on the unit circle, while the remainder of its spectrum lies at a non-zero distance from the unit circle (see fig. 1).

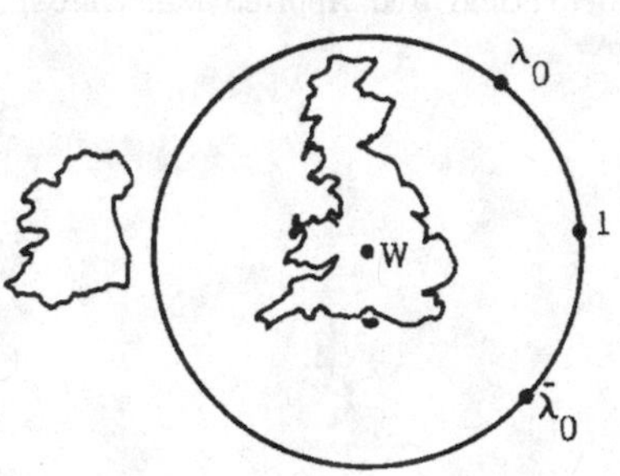


We add the Hopf transversality condition which says that, if we denote by $\lambda(\mu)$ the eigenvalue of $D_x\Phi(\mu, 0)$ perturbing $\lambda_0$ for $\mu \neq 0$, then we assume

(3) $$\frac{d}{d\mu}|\lambda(\mu)|\Big|_{\mu=0} > 0.$$

It is known that if $\lambda_0^n \neq 1$ for n = 1, 2, 3, 4, then in general there bifurcates, from the origin, an invariant simple closed curve for $\mu$ close to 0, on one side of $\mu = 0$.

**Physique Théorique, ***Mécanique Statistique, *Mathématiques, Université de Nice, Parc Valrose, 06034 Nice, France.

The usual analysis uses (i) a centre manifold theorem technique to reduce the problem into a two-dimensional one, and (ii) changes of variables to put the trace of the mapping into a normal form. Then (iii) one can prove the existence and compute the principal part of the "circle" [9, 10,5, 6, 3]. The trouble comes when somebody asks you a more economic way to compute it, and not only the principal part for which there are formulas [2, 3].

The aim of this paper is to give an economic way (save energy!) to compute the invariant bifurcating circle, in the spirit of the elementary way known for computing Hopf bifurcations for vector fields [4]. The same type of computation is possible for invariant bifurcating two-tori for vector fields (see Appendixes X.2 and X.3 of [4]) using two independent times. In fact, these computations strongly influenced this paper.

## II Main Idea.

The mapping restricted to the invariant "circle" satisfies

$$\Phi\,[\mu, X(\theta)] = X[f_\mu(\theta)] \tag{4}$$

where the map $\theta \mapsto f_\mu(\theta)$ is a diffeomorphism of $\mathbb{R}$ satisfying

$$f_\mu(\theta) = \theta + g_\mu(\theta) \ ,$$

with a $2\pi$-periodic function $g_\mu$. It is known [3, 4] that, for any integer N, there is a suitable parametrisation of the "circle" such that

$$f_\mu(\theta) = \theta + \omega(\mu) + O(|\mu|^N) \ , \tag{5}$$

where the dependence in $\theta$ of $g_\mu$ is only at order $(|\mu|^N)$. This means that $f_\mu$ is "asymptotically" conjugate to a family of rotations, depending smoothly on the amplitude of the bifurcating "circle". From (5) we can also derive an estimate for the rotation number of $f_\mu$ :

$$\rho(\mu) = \frac{\omega(\mu)}{2\pi} + O(|\mu|^N) \ . \tag{6}$$

A general result is that at each rational value, $\rho$ stays constant on a closed interval (in general not reduced to a point), so its graph, which is continuous, contains infinitely many steps corresponding to these rational values. The estimate (6) leads to the result about the size of these steps (eventual lockings) [4].

This result allows us to look for a solution satisfying

$$\Phi[\mu, X(\theta)] = X(\theta+\omega) , \tag{7}$$

where we pose

$$\begin{cases} X(\theta) = \sum_{p\geq 1} \varepsilon^p X_p(\theta) \\ \mu = \sum_{p\geq 1} \varepsilon^p \mu_p \\ \omega = \sum_{p\geq 0} \varepsilon^p \omega_p \quad . \end{cases} \tag{8}$$

Even when $\Phi$ is analytic, the series (8) are in general divergent, moreover (7) is not correct in general, but identifications of different powers of $\varepsilon$ will lead to a correct asymptotic estimate, which is actually what we need.

In fact the formula (7) is strictly correct in the cases when a one parameter compact group operates on the mapping (for instance an invariance under a rotation about an axis) [7, 8, 4].

We shall consider two different cases : 1) the case when $\lambda_0$ is not a root of unity, 2) the case when $\lambda_0$ is a root of unity of order $\geq 5$.

## III. Case when $\lambda_0$ is not a root of unity.

### III.1 Fredholm alternative.

In the identification of powers of $\varepsilon$ in (7) we have to solve at each step an equation of the form

$$A_0 Y(\theta) - Y(\theta + \omega_0) = Z(\theta) . \tag{9}$$

In (9) Z is a known $C^\infty$ function, $2\pi$-periodic in $\theta$, taking values in E, and the $2\pi$-periodic Y is unknown. An identification of the Fourier coefficients in (9) leads to

$$(A_0 - e^{ki\omega_0}) Y_k = Z_k, \quad k \in \mathbb{Z} . \tag{10}$$

Now the term of first order in ε in (7) gives

$$A_0X_1(\theta) - X_1(\theta + \omega_0) = 0 . \tag{11}$$

So, we need a non trivial solution for (11), and thanks to (10) this leads to the fact that, for some k, $e^{ki\omega_0}$ is an eigenvalue of $A_0$. This, in fact, leads to the following choice of $\omega_0$ :

$$e^{i\omega_0} = \lambda_0 . \tag{12}$$

Remark : Another choice would give the same solution, but parametrised poorly : we should obtain $\omega' = \omega/k$, $\theta' = \theta/k$ in (7), hence one rotation of $2\pi$ in $\theta'$ would correspond to k turns in the $\theta$ coordinate.

Once $\omega_0$ is chosen, we immediately see in (10) that we need to have $Z_1$ and $Z_{-1}$ respectively in the range of $A_0 - \lambda_0$, and $A_0 - \bar{\lambda}_0$. Let us define the eigenvectors $\zeta_0$, $\zeta_0^*$ such that

$$A_0\zeta_0 = \lambda_0\zeta_0, \qquad A_0^*\zeta_0^* = \bar{\lambda}_0\zeta_0^*, \qquad (\zeta_0, \zeta_0^*) = 1 , \tag{13}$$

where $A_0^*$ is the adjoint of $A_0$, and $(.,.)$ is the duality product between E and its dual E*. A necessary condition for the existence of a solution Y of (9) is then

$$\int_0^{2\pi} (Z(\theta), \zeta_0^*)e^{-i\theta}d\theta = \int_0^{2\pi} (Z(\theta), \bar{\zeta}_0^*)e^{i\theta}d\theta = 0 . \tag{14}$$

If (14) is realised we are able to compute all $Y_k$ in (10), but the problem is now the convergence of the Fourier series. In fact if we project (10) on the invariant one dimensional space generated by $\zeta_0$, we obtain

$$(\lambda_0 - \lambda_0^k)y_k = z_k, \quad k \neq 1, \tag{15}$$

where $\quad y_k = (Y_k, \zeta_0^*), \quad z_k = (Z_k, \zeta_0^*) \quad .$

The fact $\lambda_0 - \lambda_0^k$, $k \neq 1$ can be as close as we wish to zero provided k is large enough, hence there are small divisors in the problem. This leads to the fact that the linear operator on the left hand side of (9) considered for instance in $C^k(T^1;E)$ has non closed range. If we really wish to invert (9), we would have to make diophantine assumptions on $\omega_0$ as in [1] and then we loose some differentiability at each step. In fact, in what follows, we do not need to solve (9) for general Z. We only consider Z with finite Fourier expansions, so (10) gives the solution, unique if we impose, as for Z :

(16) $$(Y_1, \zeta_0^*) = (Y_{-1}, \bar{\zeta}_0^*) = 0 \ .$$

III.2 Computation of the bifurcated circle.

We first explicitly give the Taylor expansion of $\Phi$ :

(17) $$\Phi(\mu, X) = A_0 X + \sum_{\substack{p+q\geq 2 \\ q\geq 1}} \mu^p \Phi_{pq}(X, \ldots, X) \ ,$$

where $\Phi_{pq}$ is bounded q-linear symmetric in E. We write the expansion (17) without saying where it stops up to higher order terms. In fact, once required coefficients are computed, it is easy to say a posteriori the required smoothness on $\Phi$, to justify these computations.

Now we identify powers of $\varepsilon$ in (7), using (8) and (17). The order $\varepsilon$ gives

(18) $$A_0 X_1(\theta) - X_1(\theta + \omega_0) = 0 \ ,$$

the general solution of which is

(19) $$X_1(\theta) = a\zeta_0 e^{i\theta} + \bar{a}\bar{\zeta}_0 e^{-i\theta} \ ,$$

because all $X_n(\theta)$ are real. We have then to fix the determination of the "amplitude" $\varepsilon$. For this we impose

(20) $$\frac{1}{2\pi}\int_0^{2\pi} (X_n(\theta), \zeta_0^*) e^{-i\theta} d\theta = \begin{cases} 1 & \text{for } n = 1 \\ 0 & \text{for } n \geq 2 \end{cases} \ .$$

Hence we have

(21) $$X_1(\theta) = \zeta_0 e^{i\theta} + \bar{\zeta}_0 e^{-i\theta} \ .$$

The order $\varepsilon^2$ in (7) gives :

(22) $$A_0X_2(\theta) - X_2(\theta+\omega_0) + \mu_1\Phi_{11}[X_1(\theta)] - \omega_1X_1'(\theta+\omega_0) + \Phi_{02}[X_1(\theta),X_1(\theta)] = 0 \; *$$

So we use the results of §III.1, and the compatibility condition gives :

(23) $$\mu_1\lambda_1 - i\omega_1 e^{i\omega_0} = 0 \quad ,$$

(24) $$\lambda_1 = (\Phi_{11}(\zeta_0),\zeta_0^*) \quad .$$

In fact a classical result of perturbation theory is that for $\mu$ near to 0, the eigenvalue $\lambda(\mu)$ has the form

(25) $$\lambda(\mu) = \lambda_0 + \mu\lambda_1 + 0(\mu^2) \quad ,$$

where $\lambda_1$ is given by (24). Now, thanks to (12), the strict crossing condition (3) can be expressed

(26) $$\text{Re}\,(\lambda_1 e^{-i\omega_0}) > 0 \; .$$

Hence the equation (23) leads to

(27) $$\mu_1 = \omega_1 = 0 \quad ,$$

and (22) finally gives, thanks to (20) :

(28) $$\begin{cases} X_2(\theta) = X_2^{(2)}e^{2i\theta} + X_2^{(0)} + \bar{X}_2^{(2)}e^{-2i\theta} \quad , \\ X_2^{(2)} = (\lambda_0^2 - A_0)^{-1}\Phi_{02}(\zeta_0,\zeta_0) \quad , \\ X_2^{(0)} = 2(1-A_0)^{-1}\Phi_{02}(\zeta_0,\bar{\zeta}_0) \; . \end{cases}$$

Let us now consider the order $\varepsilon^3$ in (7) :

(29) $$A_0X_3(\theta) - X_3(\theta+\omega_0) + \mu_2\Phi_{11}[X_1(\theta)] - \omega_2X_1'(\theta+\omega_0) + 2\Phi_{02}[X_1(\theta),X_2(\theta)] + \\ + \Phi_{03}[X_1(\theta),X_1(\theta),X_1(\theta)] = 0.$$

---

* we denote by X' the derivative of X with respect to $\theta$.

The compatibility condition gives :

$$\mu_2\lambda_1 - i\omega_2\lambda_0 + \lambda_0\Lambda_2 = 0 \quad , \tag{30}$$

where

$$\Lambda_2 = \bar{\lambda}_0 \left(2\Phi_{02}(\zeta_0, X_2^{(0)}) + 2\Phi_{02}(\bar{\zeta}_0, X_2^{(2)}) + 3\Phi_{03}(\zeta_0,\zeta_0,\bar{\zeta}_0), \zeta_0^*\right) . \tag{31}$$

Hence,

$$\mu_2 = -\frac{\mathrm{Re}\,\Lambda_2}{\mathrm{Re}(\lambda_1\bar{\lambda}_0)}, \quad \omega_2 = \frac{\mathrm{Re}(\lambda_1\bar{\lambda}_0)\,\mathrm{Im}\,\Lambda_2 - \mathrm{Im}(\lambda_1\bar{\lambda}_0)\,\mathrm{Re}\,\Lambda_2}{\mathrm{Re}\,(\lambda_1\bar{\lambda}_0)} \quad . \tag{32}$$

Now $X_3$ is uniquely determined thanks to (20).

Identification at order $\varepsilon^n$ in (7) gives

$$A_0X_n(\theta) - X_n(\theta+\omega_0) + \mu_{n-1}\Phi_{11}[X_1(\theta)] - \omega_{n-1}X_1'(\theta+\omega_0) + F_n(\theta) = 0 \quad , \tag{33}$$

where $F_n$ is a known function, easy to compute from the $X_k$'s, $k \le n-1$, and $\mu_k$, $\omega_k$, $k \le n-2$. So, more generally it can be easily shown that

$$\begin{cases} \omega_{2p+1} = \mu_{2p+1} = 0 \text{ for any } p \; , \\ X_{2p}(\theta) = \sum_{k=1}^{p} [X_{2p}^{(2k)} e^{2ki\theta} + \bar{X}_{2p}^{(2k)} e^{-2ki\theta}] + X_{2p}^{(0)} \quad , \\ X_{2p+1}(\theta) = \sum_{k=0}^{p} [X_{2p+1}^{(2k+1)} e^{(2k+1)i\theta} + \bar{X}_{2p+1}^{(2k+1)} e^{-(2k+1)i\theta}] \quad , \end{cases} \tag{34}$$

where all coefficients are easy to compute, thanks to the conditions (26) and $\lambda_0^n \neq 1$ for any n.

Remark. Let us note that at each step we have to solve an equation of the form (9) for $X_n$. The second member has a finite Fourier series : $|k| \le n$, hence it is possible to have explicitly the result at each step, as for $X_2(\theta)$. When n increases, we have more and more possibilities to have small denominators $\lambda_0 - \lambda_0^k$ occuring via equation (15) in the solution of (9). This is why, in general, the series we just computed are divergent.

## IV Case when $\lambda_0$ is a root of unity.

### IV.1 Fredholm alternative.

Now, we assume that $\lambda_0^n = 1$, and adapt the method of §III. We pose

$$\lambda_0 = e^{i\omega_0} \; , \quad \omega_0 = 2\pi m/n, \quad n \ge 5 \; . \tag{35}$$

We first look for $X \in C^{\infty}(T^1;E)$ (i.e. smooth $2\pi$-periodic functions with values in E) solutions of

(36) $$A_0X(\theta) - X(\theta+\omega_0) = 0 .$$

Let us decompose any smooth $2\pi$-periodic function into the following form :

(37) $$X(\theta) = \sum_{k=0}^{n-1} X^{(k)}(n\theta)e^{ik\theta}, \quad X^{(k)} \in C^{\infty}(T^1;E) ,$$

where the n coefficients $X^{(k)}$ are $2\pi$-periodic functions of $n\theta$, defined by :

(38) $$X^{(k)}(n\theta) = \frac{1}{n}\sum_{p=0}^{n-1} [e^{-ik(\theta+p\omega_0)} X(\theta+p\omega_0)] .$$

Now (36) leads to the system

(39) $$A_0X^{(k)}(n\theta) - e^{ik\omega_0} X^{(k)}(n\theta) = 0, \quad k = 0,1,\ldots,n-1 .$$

Hence we obtain the general solution of (36) :

(40) $$X(\theta) = \alpha(n\theta)\zeta_0e^{i\theta} + \beta(n\theta)\bar{\zeta}_0e^{-i\theta} ,$$

where $\alpha$ and $\beta$ are smooth arbitrary scalar $2\pi$-periodic functions of $n\theta$, and where $X^{(1)}(n\theta) = \alpha(n\theta)\zeta_0$, $X^{(n-1)}(n\theta) = \beta(n\theta)e^{-ni\theta}\bar{\zeta}_0$, and other $X^{(k)}(n\theta) \equiv 0$. We observe that this kernel is infinite dimensional, which gives rise to one of the difficulties of this case. Let us now consider the inhomogeneous problem

(41) $$A_0Y(\theta) - Y(\theta+\omega_0) = Z(\theta)$$

where Z is known in $C^{\infty}(T^1;E)$. Using the decomposition (37), this leads to the system

(42) $$A_0Y^{(k)}(n\theta) - e^{ik\omega_0}Y^{(k)}(n\theta) = Z^{(k)}(n\theta), \quad k = 0,1,\ldots,n-1 .$$

Hence, for $k \neq 1,\ n-1$, we have

(43) $$Y^{(k)}(n\theta) = (A_0 - \lambda_0^k)^{-1}Z^{(k)}(n\theta) ,$$

and for k = 1 and n - 1 we have a solution if and only if

$$(Z^{(1)}(n\theta), \zeta_0^*) = (Z^{(n-1)}(n\theta), \bar{\zeta}_0^*) = 0 \quad , \tag{44}$$

and the solution Y of (41) is unique if we impose

$$(Y^{(1)}(n\theta), \zeta_0^*) = (Y^{(n-1)}(n\theta), \bar{\zeta}_0^*) = 0 \quad , \tag{45}$$

otherwise we should have to add an arbitrary element of the form (40).

IV.2 Computation of the bifurcated circle.

As in §III, we identify powers of ε in (7), using (8) and (17). Order ε gives (18) as before, and now

$$X_1(\theta) = \alpha_1(n\theta)\zeta_0 e^{i\theta} + \bar{\alpha}_1(n\theta)\bar{\zeta}_0 e^{-i\theta} \quad , \tag{46}$$

where $\alpha_1(n\theta)$ is an unknown scalar function .

In fact we shall show in §IV.3, that one can choose $\alpha_1 \equiv 1$, as in (21). Let us then construct the series (8) by starting with

$$X_1(\theta) = \zeta_0 e^{i\theta} + \bar{\zeta}_0 e^{-i\theta} \quad , \tag{47}$$

and consider the following normalisation condition to fix the determination of ε :

$$\frac{1}{2\pi}\int_0^{2\pi}(X_p^{(1)}(n\theta), \zeta_0^*)d\theta = \begin{cases} 1 \text{ for } p = 1 \\ 0 \text{ for } p \geq 2 \end{cases} \quad , \tag{48}$$

recalling that all $X_p$ are real, hence

$$\bar{X}_p^{(k)}(n\theta) = X_p^{(n-k)}(n\theta)e^{in\theta}, \quad k = 0,1,\ldots,n-1 \quad . \tag{49}$$

The order $\varepsilon^2$ in (7) gives (22) as previously, and the compatibility condition (44) leads again to (23), because n > 3. So, we obtain at this order

$$\mu_1 = \omega_1 = 0 \quad , \tag{50}$$

and

(51) $$X_2(\theta) = X_2^{(2)} e^{2i\theta} + X_2^{(0)} + \bar{X}_2^{(2)} e^{-2i\theta} + \alpha_2(n\theta)\zeta_0 e^{i\theta} + \bar{\alpha}_2(n\theta)\bar{\zeta}_0 e^{-i\theta} \quad ,$$

where $X_2^{(2)}, X_2^{(0)}$ are defined by (28) and where $\alpha_2$ is unknown, but satisfies, thanks to (48) :

(52) $$\int_0^{2\pi} \alpha_2(s)ds = 0 \quad .$$

Let us consider the order $\varepsilon^3$ in (7), written in equation (29); then the compatibility condition, using (51) for the form of $X_2$ leads again to (30) because $n > 4$. Hence $\mu_2$ and $\omega_2$ are given by (32) as before. Moreover we have

(53) $$\begin{aligned} X_3(\theta) = {} & X_3^{(3)} e^{3i\theta} + \bar{X}_3^{(3)} e^{-3i\theta} + X_3^{(1)} e^{i\theta} + \bar{X}_3^{(1)} e^{-i\theta} + \alpha_2(n\theta) X_3^{(2)} e^{2i\theta} + \\ & + \bar{\alpha}_2(n\theta)\bar{X}_3^{(2)} e^{-2i\theta} + (\alpha_2(n\theta) + \bar{\alpha}_2(n\theta))X_3^{(0)} + \alpha_3(n\theta)\zeta_0 e^{i\theta} + \\ & + \bar{\alpha}_3(n\theta)\bar{\zeta}_0 e^{-i\theta} \quad , \end{aligned}$$

where we have the relations

(54) $$\begin{cases} X_3^{(2)} = 2X_2^{(2)} \ , \ X_3^{(0)} = X_2^{(0)} \quad , \\ (X_3^{(1)}, \zeta_0^*) = 0 \ , \ \int_0^{2\pi} \alpha_3(s)ds = 0 \quad , \end{cases}$$

and all coefficients except $\alpha_2$ and $\alpha_3$ are known.

To show how the construction works, it is necessary to go to higher orders. The order $\varepsilon^4$ leads to :

(55) $$\begin{aligned} & A_0 X_4(\theta) - X_4(\theta+\omega_0) + \mu_3 \Phi_{11}[X_1(\theta)] - \omega_3 X_1'(\theta+\omega_0) + \mu_2 \Phi_{11}[X_2(\theta)] - \omega_2 X_2'(\theta+\omega_0) \\ & + 2\Phi_{02}[X_1(\theta), X_3(\theta)] + 3\Phi_{03}[X_1(\theta), X_1(\theta), X_2(\theta)] \\ & + \Phi_{02}[X_2(\theta), X_2(\theta)] + \Phi_{04}[X_1(\theta), X_1(\theta), X_1(\theta), X_1(\theta)] \\ & + \mu_2 \Phi_{12}[X_1(\theta), X_1(\theta)] = 0 \ . \end{aligned}$$

The compatibility condition (44) leads to

(56) $$\mu_3\lambda_1 - i\omega_3\lambda_0 + \mu_2\lambda_1\alpha_2(n\theta) - i\omega_2\lambda_0\alpha_2(n\theta) - \omega_2\lambda_0 n\alpha_2'(n\theta) + \lambda_0\Lambda_2[2\alpha_2(n\theta) + \bar{\alpha}_2(n\theta)] + \lambda_0\Lambda_3 e^{-5i\theta} = 0 ,$$

where the term in $e^{-5i\theta}$ only exists if n = 5, and

(57) $$\Lambda_3 = \bar{\lambda}_0(2\Phi_{02}(\bar{\zeta}_0, \bar{X}_3^{(3)}) + 3\Phi_{03}(\bar{\zeta}_0, \bar{\zeta}_0, \bar{X}_2^{(2)}) + \Phi_{02}(\bar{X}_2^{(2)}, \bar{X}_2^{(2)}) + \Phi_{04}(\bar{\zeta}_0, \bar{\zeta}_0, \bar{\zeta}_0, \bar{\zeta}_0), \zeta_0^*) .$$

The equation (56) allows us to determine $\alpha_2$ satisfying (52) : the zero mean value condition leads to

(58) $$\mu_3\lambda_1 - i\omega_3\lambda_0 = 0 ,$$

hence

(59) $$\mu_3 = \omega_3 = 0 ,$$

and using (30), (56) may now be written

(60) $$\Lambda_2[\alpha_2(n\theta) + \bar{\alpha}_2(n\theta)] - \omega_2 n\alpha_2'(n\theta) + \Lambda_3 e^{-5i\theta} = 0 .$$

Assuming now that

(61) $$\omega_2 \neq 0 ,$$

it is easy to prove that the only $2\pi$-periodic solution $\alpha_2$ of (60), which satisfies (52) has the form :

(62) $$\begin{cases} \alpha_2(5\theta) = ae^{5i\theta} + be^{-5i\theta} & \text{if } n = 5 , \\ \alpha_2(n\theta) = 0 & \text{if } n \geq 6 , \end{cases}$$

where a and b are easy to compute from (60) .

Remark. For n = 5, the condition $\omega_2 \neq 0$ corresponds to the fact that we have to avoid weak-resonance, which corresponds to a bifurcation from the origin of a family of periodic points of period 5 ( see [3,4] for computations of weak resonant solutions). For $n \geq 6$ the condition (61) is strictly speaking not necessary, but we should have to assume for instance Re $\Lambda_2 \neq 0$ to go to higher orders and then a non-weak resonant condition analogous to (61) would arise later [3]. To simplify the analysis we assume from now on (61) to be realised. It is clear that in the case of weak resonance, it is unreasonable to hope to compute the invariant "circle" using (7), because this formula would give all points of the circle as periodic points for the map, which is not true in general.

We now completely determine $X_2(\theta)$. For $X_3(\theta)$, the scalar function $\alpha_3(n\theta)$ is still unknown, and we have*

$$(63) \qquad X_4(\theta) = X_4^{(4)} e^{4i\theta} + X_4^{(3)}(5\theta)e^{3i\theta} + X_4^{(2)} e^{2i\theta} + X_4^{(1)}(5\theta)e^{i\theta} + X_4^{(0)} + \\ + \bar{X}_4^{(1)}(5\theta)e^{-i\theta} + \bar{X}_4^{(2)} e^{-2i\theta} + \bar{X}_4^{(3)}(5\theta)e^{-3i\theta} + \bar{X}_4^{(4)} e^{-4i\theta} + \\ + \alpha_3(n\theta)2X_2^{(2)}e^{2i\theta} + [\alpha_3(n\theta) + \bar{\alpha}_3(n\theta)]X_2^{(0)} + \bar{\alpha}_3(n\theta)2\bar{X}_2^{(2)}e^{-2i\theta} + \\ + \alpha_4(n\theta)\zeta_0 e^{i\theta} + \bar{\alpha}_4(n\theta)\bar{\zeta}_0 e^{-i\theta} ,$$

where all coefficients are known, except $\alpha_3$ and $\alpha_4$, and where $X_4^{(3)}$ and $X_4^{(1)}$ cancel if $n \geq 6$, and $X_4^{(4)}$ is orthogonal to $\bar{\zeta}_0^*$ if $n = 5$ .

The order $\varepsilon^p$ in (7), for $p \geq 5$ leads to the equation :

$$(64) \qquad A_0X_p(\theta) - X_p(\theta+\omega_0) + \mu_{p-1}\Phi_{11}[X_1(\theta)] - \omega_{p-1}X_1'(\theta+\omega_0) + \mu_2\Phi_{11}[X_{p-2}(\theta)] - \\ - \omega_2X'_{p-2}(\theta+\omega_0) + 2\Phi_{02}[X_1(\theta), X_{p-1}(\theta)] + 2\Phi_{02}[X_2(\theta), X_{p-2}(\theta)] + \\ + 3\Phi_{03}[X_1(\theta), X_1(\theta), X_{p-2}(\theta)] + F_p(\theta) = 0 ,$$

where $F_p(\theta)$ is known at this step, and the unknowns are $\mu_{p-1}$, $\omega_{p-1}$, $X_p(\theta)$, $\alpha_{p-1}(n\theta)$ and $\alpha_{p-2}(n\theta)$ (which appear in $X_{p-1}(\theta)$ and $X_{p-2}(\theta)$).

The compatibility condition (44) leads to

$$(65) \qquad \mu_{p-1}\lambda_1 - i\omega_{p-1}\lambda_0 + \lambda_0\Lambda_2[\alpha_{p-2}(n\theta) + \bar{\alpha}_{p-2}(n\theta)] - \lambda_0\omega_2 n\alpha'_{p-2}(n\theta) = f_p(n\theta) ,$$

---

* In (63), the notation of coefficients does not follow the definition (37).

where $f_p$ is a known $2\pi$-periodic function in $n\theta$. Thanks to the assumption (61), the equation (65) has a unique solution $\alpha_{p-2}(n\theta)$ of zero mean value, provided that

$$(66) \qquad \mu_{p-1}\lambda_1 - i\omega_{p-1}\lambda_0 = \frac{1}{2\pi}\int_0^{2\pi} f_p(s)ds \ .$$

So, $\mu_{p-1}$, $\omega_{p-1}$ and $\alpha_{p-2}$ are now determined, and equation (64) leads to an expression of $X_p(\theta)$ where the only unknowns are $\alpha_{p-1}(n\theta)$ and $\alpha_p(n\theta)$ .

<u>Remark.</u> In general, the series (8) obtained by this way do not converge. Even though there are no small divisors in the construction, there is a loss of differentiability at each step, due to the expansion of $X(\theta+\omega)$ near $\omega = \omega_0$. An alternative way would be to keep $\omega$ in $X(\theta+\omega)$ without expanding it, but then we should obtain small divisors in solving linear problems such as (41), because $\omega$ depends on the amplitude $\varepsilon$.

<u>Summary of results of IV.</u>

We obtained $X_1(\theta)$ at (47), $\mu_1 = \omega_1 = \mu_3 = \omega_3 = 0$, and $X_2(\theta)$ is known thanks to (51) and (62). We note that if $\lambda_0^5 = 1$, $X_2(\theta)$ differs from the one given in (28), while $\mu_2$ and $\omega_2$ stay the same. More generally if $\lambda_0^n = 1$, $n \geq 5$, formulas giving $\mu_p, \omega_p, X_p(\theta)$ are the same as in the case when $\lambda_0$ is not a root of unity until $X_{n-3}(\theta)$, $\mu_{n-2}$, $\omega_{n-2}$ (excluded), taking account that $\mu_{2p+1} = \omega_{2p+1} = 0$ for $2p \leq n - 3$ if n is odd, and for $2p \leq n - 4$ if n is even.

IV.3 <u>About the choice of $\alpha_1$.</u>

Let us come back to the expression (46) of $X_1(\theta)$ where we do not choose $\alpha_1(n\theta)$, and consider order $\varepsilon^2$ in (7), which leads to (22). The compatibility condition leads to :

$$(67) \qquad \alpha_1(n\theta)(\mu_1\lambda_1 - i\omega_1\lambda_0) - n\omega_1\alpha_1'(n\theta)\lambda_0 = 0 \ .$$

Hence there exists $k \in \mathbb{Z}$ such that

$$(68) \qquad \mu_1\lambda_1\bar{\lambda}_0 - (nk+1)i\omega_1 = 0 \ ,$$

and $\mu_1 = 0$ , $\omega_1 = 0$ . Moreover we have

$$(69) \qquad X_2(\theta) = \alpha_1^2(n\theta)X_2^{(2)}e^{2i\theta} + \alpha_1(n\theta)\bar{\alpha}_1(n\theta)X_2^{(0)} + \bar{\alpha}_1^2(n\theta)\bar{X}_2^{(2)}e^{-2i\theta} + \alpha_2(n\theta)\zeta_0 e^{i\theta} + \bar{\alpha}_2(n\theta)\bar{\zeta}_0 e^{-i\theta} \ .$$

Now the order $\varepsilon^3$ in (7) gives (29) and the compatibility condition is :

(70) $$-n\omega_2\alpha_1'(n\theta) + \alpha_1(n\theta)[\mu_2\lambda_1\bar{\lambda}_0 - i\omega_2] + \Lambda_2\alpha_1(n\theta)|\alpha_1(n\theta)|^2 = 0 ,$$

which is a non-linear differential equation in $\alpha_1$, $\omega_2$ and $\mu_2$ being unknown. Moreover $\alpha_1$ has to be $2\pi$-periodic. Equation (70) leads to

(71) $$-\frac{n\omega_2}{2}\frac{d}{ds}|\alpha_1(s)|^2 + \mu_2 \operatorname{Re}(\lambda_1\bar{\lambda}_0)|\alpha_1(s)|^2 + \operatorname{Re}\Lambda_2|\alpha_1(s)|^4 = 0 ,$$

and it is easy to see that if $\operatorname{Re}\Lambda_2 \neq 0$, the only possibility for us to obtain a periodic solution $|\alpha_1(s)|^2$ of (71) is when $|\alpha_1(s)|^2$ is constant. Now to fix the scale of the amplitude $\varepsilon$ in (8), we choose

(72) $$|\alpha_1(n\theta)| \equiv 1 ,$$

hence

(73) $$\mu_2 \operatorname{Re}(\lambda_1\bar{\lambda}_0) + \operatorname{Re}\Lambda_2 = 0 ,$$

which is one of the relations (32). Now (70) gives a linear equation in $\alpha_1$ :

(74) $$-n\omega_2\alpha_1' + (\mu_2\lambda_1\bar{\lambda}_0 + \Lambda_2 - i\omega_2)\alpha_1 = 0 ,$$

hence there is $k \in \mathbb{Z}$ such that

(75) $$\mu_2 \operatorname{Im}(\lambda_1\bar{\lambda}_0) + \operatorname{Im}\Lambda_2 = (1+kn)\omega_2 ,$$

which is the second part of (32) if $k = 0$ .

We claim now that we can choose k, hence

(76) $$\alpha_1(n\theta) = e^{i(kn\theta+\varphi_0)} ,$$

and go on for the construction as in §IV.2 with an adaptation of (48) which would be

(77) $$\frac{1}{2\pi}\int_0^{2\pi}(X_p^{(1)}(n\theta), \zeta_0^*)e^{-ikn\theta}d\theta = \begin{cases} e^{i\varphi_0} & \text{for } p = 1 \\ 0 & \text{for } p \geq 2 \end{cases} .$$

In fact this choice would give the same "circle" as before parametrised in another way by $\theta'$ such that $(1+kn)\theta' = \theta$, and with another rotation $\omega'$ such that $(1+kn)\omega' = \omega$ (see remark in §III.1).

It remains to remark that we took $\varphi_0 = 0$ in §IV.2, which is a deliberate choice of the origin on the circle. In fact we could here modify the definition of $\varepsilon$, i.e. choose something else than (48). This could be useful if we wish to obtain the same circle for $\varepsilon$ and $-\varepsilon$, which is not the case with our previous choice. For this to be true we should have to pose

(78) $$X(\theta) = \varepsilon(e^{i\varphi(\varepsilon)}\zeta_0 e^{i\theta} + e^{-i\varphi(\varepsilon)}\bar{\zeta}_0 e^{-i\theta}) + \sum_{p\geq 2} \varepsilon^p X_p(\theta) ,$$

where

(79) $$\varphi(\varepsilon) = \sum_{p\geq 1} \varepsilon^p \varphi_p$$

is unknown, but could be computed thanks to the normalisation condition (48) for the new $X_p$ and the condition that $\mu$ and $\omega$ are even in $\varepsilon$, all this leading to

(80) $$X(\theta, -\varepsilon) = X(\theta+\pi, \varepsilon) .$$

We have to notice that a change in the origin on the circle, also changes the parametrisation $\theta$ (so it is not just a translation).

Remark about the case $\lambda_0^4 = 1$ $(\lambda_0 = i)$ .

In the case $\lambda_0 = i$, it is known that when there are no bifurcating periodic points of period 4, there exists a bifurcating circle as in cases studied above [11]. This case would give an equation like (70) for $\alpha_1(4\theta)$, but with an additional term in $[\overline{\alpha_1(4\theta)}]^3$ which leads to a difficulty because we cannot compute explicitly now the solution of the differential equation. A qualitative study as in [11] is required leading to the determination of $\mu_2$ and $\omega_2$, and a choice of $\alpha_1$. Further steps are as before, but with a more complicated $\alpha_1$ (no more constant).

## V. Example*.

Let us consider the following map in $\mathbb{R}^2$:$(x', y') \mapsto (y', \mu' y'(1-x'))$ already studied by

* We thank the referee for posing us this challenge.

Aronson et.al. in [12]. Fixed points are $0 : x' = y' = 0$, and $0' = x' = y' = 1 - 1/\mu'$. The fixed point 0' is stable for $1 < \mu' < 2$ and looses its attractivity for $\mu' > 2$ leading to a bifurcated circle.

Let us pose
$$\begin{cases} x' = x + 1 - 1/\mu' \\ y' = y + 1 - 1/\mu' \\ \mu' = \mu + 2 \end{cases}$$
the new map takes the form :

$$(81) \qquad (x, y) \mapsto (y, -(1+\mu)x + y - (\mu+2)xy) \quad ,$$

So, with the notations of this paper we have here :

$$(82) \qquad A_0 = \begin{pmatrix} 0 & 1 \\ -1 & 1 \end{pmatrix} \quad , \quad \Phi_{11} = \begin{pmatrix} 0 & 0 \\ -1 & 0 \end{pmatrix} \quad ,$$

$$(83) \qquad \Phi_{02}(Y_1, Y_2) = \begin{pmatrix} 0 \\ -x_1y_2 - x_2y_1 \end{pmatrix} \quad \text{where} \quad Y_i = \begin{pmatrix} x_i \\ y_i \end{pmatrix}$$

$$(84) \qquad \Phi_{12} = \Phi_{02} \text{ and other } \Phi_{pq} = 0 \quad .$$

In this case we observe that $\lambda_0 = \frac{1}{2}(1+i\sqrt{3})$ satisfies $\lambda_0^6 = 1$ , and we choose

$$(85) \qquad \zeta_0 = \begin{pmatrix} 1 \\ \lambda_0 \end{pmatrix} \quad , \quad \zeta_0^* = \frac{1}{3}\begin{pmatrix} 2-\lambda_0 \\ 1-2\bar{\lambda}_0 \end{pmatrix} \quad .$$

Formula (24) now gives

$$(86) \qquad \lambda_1 = \frac{i}{\sqrt{3}} \quad ,$$

and we can check the transversality condition

$$\lambda_1\bar{\lambda}_0 = \frac{1}{2} + \frac{i}{2\sqrt{3}} \quad , \text{ hence } \operatorname{Re}(\lambda_1\bar{\lambda}_0) = \frac{1}{2} > 0 \quad .$$

Results of §IV lead to the a priori expansion :

$$(87) \qquad \begin{aligned} X(\theta) &= \varepsilon X_1(\theta) + \varepsilon^2 X_2(\theta) + \varepsilon^3 X_3(\theta) + \varepsilon^4 X_4(\theta) + 0(\varepsilon^5) \\ \mu &= \varepsilon^2\mu_2 + \varepsilon^4\mu_4 + 0(\varepsilon^5) \\ \omega &= \omega_0 + \varepsilon^2\omega_2 + \varepsilon^4\omega_4 + 0(\varepsilon^5) \end{aligned}$$

where $\omega_0 = \pi/3$ ,

(88) $$X_1(\theta) = \zeta_0 e^{i\theta} + \bar{\zeta}_0 e^{-i\theta}$$

(89) $$X_2(\theta) = X_2^{(2)} e^{2i\theta} + X_2^{(0)} + \bar{X}_2^{(2)} e^{-2i\theta}$$

(90) $$X_3(\theta) = X_3^{(3)} e^{3i\theta} + \bar{X}_3^{(3)} e^{-3i\theta} + X_3^{(1)} e^{i\theta} + \bar{X}_3^{(1)} e^{-i\theta} + \alpha_3(6\theta)\zeta_0 e^{i\theta} + \\ + \bar{\alpha}_3(6\theta)\bar{\zeta}_0 e^{-i\theta}$$

(91) $$X_4(\theta) = X_4^{(4)} e^{4i\theta} + \bar{X}_4^{(4)} e^{-4i\theta} + X_4^{(2)} e^{2i\theta} + \bar{X}_4^{(2)} e^{-2i\theta} + X_4^{(0)} + 2\alpha_3(6\theta)X_2^{(2)} e^{2i\theta} \\ + 2\bar{\alpha}_3(6\theta)\bar{X}_2^{(2)} e^{-2i\theta} + (\alpha_3(6\theta) + \bar{\alpha}_3(6\theta))X_2^{(0)} + \alpha_4(6\theta)e^{i\theta}\zeta_0 \\ + \bar{\alpha}_4(6\theta)e^{-i\theta}\bar{\zeta}_0 \ .$$

To compute all coefficients we need :

(92) $$\Phi_{02}(\zeta_0,\zeta_0) = \begin{pmatrix} 0 \\ -2\lambda_0 \end{pmatrix} \quad , \quad 2\Phi_{02}(\zeta_0,\bar{\zeta}_0) = \begin{pmatrix} 0 \\ -2 \end{pmatrix} ,$$

which, thanks to (28), leads to :

(93) $$X_2^{(0)} = \begin{pmatrix} -2 \\ -2 \end{pmatrix} \quad , \quad X_2^{(2)} = \begin{pmatrix} \bar{\lambda}_0 \\ \lambda_0 \end{pmatrix} .$$

Now we obtain :

(94) $$2\Phi_{02}(\zeta_0, X_2^{(0)}) = 4 \begin{pmatrix} 0 \\ 1+\lambda_0 \end{pmatrix}$$
$$2\Phi_{02}(\bar{\zeta}_0, X_2^{(2)}) = \begin{pmatrix} 0 \\ 0 \end{pmatrix}$$
$$2\Phi_{02}(\zeta_0, X_2^{(2)}) = -2 \begin{pmatrix} 0 \\ 1+\lambda_0 \end{pmatrix} ,$$

and formula (31), (32) give :

(95) $$\Lambda_2 = -4\lambda_0 \ ,$$

(96) $$\mu_2 = 4 \quad , \quad \omega_2 = -4/\sqrt{3} \ .$$

Now we can compute from (29) :

$$X_3^{(3)} = \tfrac{2}{3}(1+\lambda_0)\begin{pmatrix}-1\\1\end{pmatrix}, \quad X_3^{(1)} = \frac{4i}{3\sqrt{3}}\begin{pmatrix}1+\bar{\lambda}_0\\1-2\lambda_0\end{pmatrix}, \tag{97}$$

where $X_3^{(1)}$ is orthogonal to $\boldsymbol{\zeta}_0^*$ .

We are now able to determine $\mu_4$, $\omega_4$ and $\alpha_3(6\theta)$ in using formulas :

$$\begin{aligned}\mu_4\lambda_1 - i\omega_4\lambda_0 + \lambda_0\omega_2^2/2 + (\mu_2\Phi_{11}X_3^{(1)} + 2\Phi_{02}(\bar{\boldsymbol{\zeta}}_0, X_4^{(2)}) + 2\Phi_{02}(\boldsymbol{\zeta}_0, X_4^{(0)}) +\\ + 2\Phi_{02}(X_2^{(2)}, \bar{X}_3^{(1)}) + 2\Phi_{02}(X_2^{(0)}, X_3^{(1)}) + 2\Phi_{02}(\bar{X}_2^{(2)}, X_3^{(3)}) +\\ + 2\mu_2\Phi_{12}(\boldsymbol{\zeta}_0, X_2^{(0)}) + 2\mu_2\Phi_{12}(\bar{\boldsymbol{\zeta}}_0, X_2^{(2)}), \boldsymbol{\zeta}_0^*) = 0 ,\end{aligned} \tag{98}$$

$$\Lambda_2[\alpha_3(6\theta) + \bar{\alpha}_3(6\theta)] - 6\omega_2\alpha_3'(6\theta) + \bar{\lambda}_0(2\Phi_{02}(\bar{\boldsymbol{\zeta}}_0, \bar{X}_4^{(4)}) + 2\Phi_{02}(\bar{X}_2^{(2)}, \bar{X}_3^{(3)}), \boldsymbol{\zeta}_0^*)e^{-6i\theta} = 0 , \tag{99}$$

$$(A_0 - \lambda_0^4)X_4^{(4)} + 2\Phi_{02}(\boldsymbol{\zeta}_0, X_3^{(3)}) + \Phi_{02}(X_2^{(2)}, X_2^{(2)}) = 0 , \tag{100}$$

$$\begin{aligned}(A_0 - \lambda_0^2)X_4^{(2)} + \mu_2\Phi_{11}(X_2^{(2)}) - 2i\omega_2\lambda_0^2X_2^{(2)} + 2\Phi_{02}(\boldsymbol{\zeta}_0, X_3^{(1)}) + 2\Phi_{02}(\bar{\boldsymbol{\zeta}}_0, X_3^{(3)}) +\\ + 2\Phi_{02}(X_2^{(2)}, X_2^{(0)}) + \mu_2\Phi_{12}(\boldsymbol{\zeta}_0; \boldsymbol{\zeta}_0) = 0 ,\end{aligned} \tag{101}$$

$$\begin{aligned}(A_0 - 1)X_4^{(0)} + \mu_2\Phi_{11}(X_2^{(0)}) + 2\Phi_{02}(\boldsymbol{\zeta}_0, \bar{X}_3^{(1)}) + 2\Phi_{02}(\bar{\boldsymbol{\zeta}}_0, X_3^{(1)}) + 2\Phi_{02}(X_2^{(2)}, \bar{X}_2^{(2)}) +\\ + \Phi_{02}(X_2^{(0)}, X_2^{(0)}) + 2\mu_2\Phi_{12}(\boldsymbol{\zeta}_0, \bar{\boldsymbol{\zeta}}_0) = 0 .\end{aligned} \tag{102}$$

So, we need

$$2\Phi_{02}(\boldsymbol{\zeta}_0, X_3^{(3)}) = -\tfrac{4}{3}(1+\bar{\lambda}_0)\begin{pmatrix}0\\1\end{pmatrix}, \qquad \Phi_{02}(X_2^{(2)}, X_2^{(2)}) = \begin{pmatrix}0\\-2\end{pmatrix},$$

$$\mu_2\Phi_{11}(X_2^{(2)}) = -4\bar{\lambda}_0\begin{pmatrix}0\\1\end{pmatrix}, \qquad -2i\omega_2\lambda_0^2X_2^{(2)} = \frac{-8i}{\sqrt{3}}\begin{pmatrix}\lambda_0\\-1\end{pmatrix},$$

$$2\Phi_{02}(\boldsymbol{\zeta}_0, X_3^{(1)}) = -\frac{8i}{3\sqrt{3}}\begin{pmatrix}0\\2-\lambda_0\end{pmatrix}, \qquad 2\Phi_{02}(\bar{\boldsymbol{\zeta}}_0, X_3^{(3)}) = -\frac{4i}{\sqrt{3}}\begin{pmatrix}0\\1\end{pmatrix},$$

$$2\Phi_{02}(X_2^{(2)}, X_2^{(0)}) = 4\begin{pmatrix}0\\1\end{pmatrix}, \quad \mu_2\Phi_{12}(\boldsymbol{\zeta}_0, \boldsymbol{\zeta}_0) = -8\lambda_0\begin{pmatrix}0\\1\end{pmatrix},$$

$$\mu_2\Phi_{11}(X_2^{(0)}) = \begin{pmatrix}0\\8\end{pmatrix}, \qquad 2\Phi_{02}(\boldsymbol{\zeta}_0, \bar{X}_3^{(1)}) = \frac{16i}{3\sqrt{3}}\begin{pmatrix}0\\1-2\bar{\lambda}_0\end{pmatrix},$$

$$2\Phi_{02}(\bar{\boldsymbol{\zeta}}_0, X_3^{(1)}) = -\frac{16i}{3\sqrt{3}}\begin{pmatrix}0\\1-2\lambda_0\end{pmatrix}, \quad 2\Phi_{02}(X_2^{(2)}, \bar{X}_2^{(2)}) = \begin{pmatrix}0\\1\end{pmatrix}$$

$$\Phi_{02}(X_2^{(0)}, X_2^{(0)}) = -8\begin{pmatrix}0\\1\end{pmatrix}, \quad 2\mu_2\Phi_{12}(\boldsymbol{\zeta}_0, \bar{\boldsymbol{\zeta}}_0) = \begin{pmatrix}0\\-8\end{pmatrix}.$$

We then obtain

$$(103)\qquad X_4^{(4)} = \begin{pmatrix} -\frac{1}{2}+\frac{7i}{2\sqrt{3}} \\ 2-\frac{i}{\sqrt{3}} \end{pmatrix} , \quad X_4^{(2)} = \frac{1}{3}\begin{pmatrix} -10 - 4i\sqrt{3} \\ -1 + i\sqrt{3} \end{pmatrix} , \quad X_4^{(0)} = \frac{50}{3}\begin{pmatrix}1\\1\end{pmatrix} ,$$

and

$$2\Phi_{02}(\bar{\zeta}_0, \bar{X}_4^{(4)}) = \begin{pmatrix}0\\0\end{pmatrix} , \quad 2\Phi_{02}(\bar{X}_2^{(2)}, \bar{X}_3^{(3)}) = -\frac{4i}{\sqrt{3}}(1+\bar{\lambda}_0)\begin{pmatrix}0\\1\end{pmatrix} ,$$

$$\mu_2\Phi_{11}X_3^{(1)} = \begin{pmatrix} 0 \\ -\frac{16i}{3\sqrt{3}}(1+\bar{\lambda}_0) \end{pmatrix} , \quad 2\Phi_{02}(\bar{\zeta}_0, X_4^{(2)}) = 8(1 - i/\sqrt{3})\begin{pmatrix}0\\1\end{pmatrix} ,$$

$$2\Phi_{02}(\zeta_0, X_4^{(0)}) = -\frac{100}{3}(1+\lambda_0)\begin{pmatrix}0\\1\end{pmatrix} , \quad 2\Phi_{02}(X_2^{(2)}, \bar{X}_3^{(1)}) = \frac{8i}{\sqrt{3}}\lambda_0\begin{pmatrix}0\\1\end{pmatrix} ,$$

$$2\Phi_{02}(X_2^{(0)}, X_3^{(1)}) = \frac{16i}{\sqrt{3}}\bar{\lambda}_0\begin{pmatrix}0\\1\end{pmatrix} , \quad 2\Phi_{02}(\bar{X}_2^{(2)}, X_3^{(3)}) = -\frac{4i}{\sqrt{3}}(i+\lambda_0)\begin{pmatrix}0\\1\end{pmatrix} ,$$

$$2\mu_2\Phi_{12}(\zeta_0, X_2^{(0)}) = 16(1+\lambda_0)\begin{pmatrix}0\\1\end{pmatrix} , \quad 2\mu_2\Phi_{12}(\bar{\zeta}_0, X_2^{(2)}) = \begin{pmatrix}0\\0\end{pmatrix} .$$

Now, equation (99) becomes :

$$(104)\qquad -\lambda_0[\alpha_3(6\theta) + \bar{\alpha}_3(6\theta)] + 2\sqrt{3}\,\alpha_3'(6\theta) + \frac{i}{\sqrt{3}} e^{-6i\theta} = 0$$

with $\int_0^{2\pi}\alpha_3(\theta)d\theta = 0$ ,

and the solution is

$$(105)\qquad \alpha_3(6\theta) = \left(\frac{5}{156} - \frac{i\sqrt{3}}{52}\right)e^{6i\theta} + \left(\frac{19}{156} + \frac{i}{52\sqrt{3}}\right)e^{-6i\theta} .$$

Finally, equation (98) gives :

$$\mu_4\lambda_1 - i\omega_4\lambda_0 + \lambda_0\frac{\omega_2^2}{2} - 12 + \frac{44i}{3\sqrt{3}} = 0 ,$$

hence

$$(106)\qquad \mu_4 = -8 \quad , \quad \omega_4 = \frac{64}{3\sqrt{3}} .$$

At this step of the computation, we completely know $X(\theta)$ up to order $\varepsilon^3$ and $\mu(\varepsilon)$, $\omega(\varepsilon)$

up to order $\varepsilon^4$ included. To determine $X_4(\theta)$, it remains to compute $\alpha_4(\theta)$ which comes from identification at order $\varepsilon^6$ in (7). If the reader wishes to eliminate $\varepsilon$ in the expression of $X(\theta)$, he will observe that we know the solution completely up to order $\mu^{3/2}$ and the rotation number up to order $\mu^2$.

We draw the reader's attention to the fact that this example is two dimensional, hence very simple even with other methods because there is no need to compute the centre manifold. The real efficiency of our method would appear on an example in more than 2 dimensions and the authors wish to check this on some existing example of any solution computed further than the first term if such one exists.

## References.

1. Chenciner A. & G. Iooss, Bifurcations de tores invariants. Arch. Rational Mech. Anal. 69, 2, p.109-198, 1979.

2. Hassard P. & Y.H. Wan, Bifurcation formulae derived from centre manifold theory. J. Math. Anal. Appl. 63, 1, p.297-312, 1978.

3. Iooss G., Bifurcation of maps and Applications. Math. Studies 36, North-Holland, 1979.

4. Iooss G. & D.D. Joseph, Elementary Stability and Bifurcation Theory. Springer Verlag, Berlin-Heidelberg-New York, 1980.

5. Lanford O.E. III, Bifurcation of Periodic Solutions into invariant Tori. Lecture Notes in Math. 322, p.159-192, Springer Verlag, Berlin-Heidelberg New York, 1973.

6. Marsden J.E. & M. McCracken, The Hopf bifurcation and its applications. Applied Math. Sciences 19, Springer Verlag, 1976.

7. Rand D. The pre-turbulent transitions and flows of a viscous fluid between concentric rotating cylinders. Preprint, University of Warwick, 1980.

8. Renardy M., Bifurcation from rotating waves, to appear.

9. Ruelle D. & Takens F., On the nature of turbulence. Comm. Math. Phys. 20, 167-192, 1971.

10. Sacker R.J., On invariant surfaces and bifurcation of periodic solutions of ordinary differential equations. New York University, IMM-NYU, 333, 1964.

11. Wan Y.H., Bifurcation into invariant tori at points of resonance. Arch. Rational Mech. Anal. 68, 343-357, 1978.

12. Holmes P., New approaches to nonlinear problems in dynamics, SIAM, 1980.

# Families of Vector Fields with Finite Modulus of Stability.

I.P. Malta & J. Palis *

## 1. Introduction.

The purpose of this paper is to show that a large class of one-parameter families of vector fields (flows) on surfaces have finite modulus of stability. The key point is that many of these families are not stable and yet we can describe all equivalence classes of nearby families using a finite number of (real) parameters. In particular, for the sphere or the projective plane, we can approximate any given one-parameter family of vector fields by one with finite modulus of stability.

These results were motivated by Sotomayor [4] where the class of families we deal with was introduced and the question whether they were stable was posed and by Newhouse, Palis, Takens [2] where it was shown that in general this question has a negative answer (correcting an early attempt to the contrary by Guckenheimer [1]). We are now able to understand these families from the stability point of view by exhibiting a complete set of parameters needed to describe the equivalence classes.

We now state our results in a more formal way. First we briefly go through a number of concepts most of which will be presented in detail in the next section.

Let M be a $C^\infty$ compact 2-manifold without boundary. Let $\mathfrak{B} = \mathfrak{B}(M)$ be the space of $C^\infty$ arcs of vector fields on M with the usual $C^\infty$ topology. That is, $\mathfrak{B}$ consists of $C^\infty$ mappings $\xi: I \to \mathfrak{X}(M)$ of the interval $I = [-1,1]$ into the space of $C^\infty$ vector fields on M. Such an arc will be frequently denoted by $\xi = \{X_\mu\}$, $\mu \in I$ and $X_\mu = \xi(u)$.

We consider the following equivalence relation on $\mathfrak{B}$. We say that two arcs of $C^\infty$ vector fields on M are topologically equivalent if, modulo an orientation preserving homeomorphism of the interval I, each element of the first family is topologically equivalent to the corresponding element of the second family and the equivalence varies continuously with the parameter. That is,

<u>Definition</u> : Let $\{X_\mu\}$ and $\{Y_\mu\}$ be two arcs of $C^\infty$-vector fields on M. We say that they are topologically equivalent if there exists a homeomorphism $H = (h, \eta): M \times I \to M \times I$,

*The second author gratefully acknowledges the financial support of Stiftung Volkswagenwerk for a visit to the IHES during which this work was partly developed.

with $\eta$ an orientation preserving homeomorphism of the interval I, such that, for each $\mu \in I$, the homeomorphism $h_\mu$ of M, defined by $h_\mu(x) = h(x,\mu)$ is a topological equivalence between $X_\mu$ and $Y_{\eta(\mu)}$.

We call $\{X_\mu\} \in \mathfrak{B}$ <u>stable</u>, if it is equivalent to any nearby family, that is, if $\{X_\mu\}$ belongs to the interior of its equivalence class. If we can describe the equivalence classes of all families close enough to $\{X_\mu\}$ using a finite number of (real) parameters, we say that $\{X_\mu\}$ has a finite modulus of stability.

Through a classical result of Peixoto [3], it is possible to describe the stable vector fields in $\mathfrak{X}(M)$, at least when M is orientable : they are the so-called Morse-Smale vector fields. They form an open subset of $\mathfrak{X}(M)$ which we denote by $\Sigma$ . If a family $\{X_\mu\}$ is such that $X_\mu \in \Sigma$ for all $\mu \in I$ then $\{X_\mu\}$ is clearly stable in $\mathfrak{B}$. Thus from the stability point of view it is relevant to consider families $\{X_\mu\}$ such that $X_\mu \notin \Sigma$ for some $\mu \in I$. To this effect, it is natural to search for a subset $\Sigma_1 \subset \mathfrak{X}(M)-\Sigma$ that plays a role similar to that of $\Sigma$ in $\mathfrak{X}(M)$. The construction of such subset was performed by Sotomayor [4], and is presented in section 2. The elements of $\Sigma_1$ are almost as simple as those of $\Sigma$; because of this, they are called quasi-Morse-Smale. In particular, if $X \in \Sigma_1$ then either all critical elements (singularities or closed orbits) of X are hyperbolic and there is a saddle-connection or there is one and only one critical element that is not hyperbolic (in this case we require it to be quasi-hyperbolic) and X has no saddle-connections. In all cases, the $\alpha$ and $\omega$-limit sets of the orbits consist of critical elements or a loop (self-saddle-connection).

To explain out results we need to pay special attention to the elements $X \in \Sigma_1$ that exhibit a saddle-node closed orbit $\gamma$. In this case we say that X has a cycle if $\gamma$ is both the $\alpha$ and $\omega$-limit set of some orbit of X distinct from $\gamma$. Moreover, let us associate to X the following characteristic number K(X). Let $m \geq 0$ and $n \geq 0$, be the number of saddle-separatrices that have $\gamma$ as $\alpha$-limit set and $\omega$-limit set, respectively. We define K(X) by :

$$K(X) = 0 \text{ if } m = n = 0$$

$$K(X) = n-1 \text{ if } m = 0 \text{ and } n \geq 1$$

$$K(X) = m-1 \text{ if } n = 0 \text{ and } m \geq 1$$

and

$$K(X) = m+n-2 \text{ if } m \geq 1 \text{ and } n \geq 1.$$

Finally we say that $\mu \in I$ is a regular value of $\{X_\mu\} \in \mathfrak{B}$, if $X_\mu \in \Sigma$. Otherwise we say that $\mu$ is a bifurcation value.

We can now define the set $\mathcal{A}$ of one-parameter families that we consider here : $\{X_\mu\} \in \mathcal{A}$ if

(i) $X_\mu \in \Sigma \cup \Sigma_1$ for each $\mu \in I$

(ii) $\{X_\mu\}$ unfolds generically at the bifurcation values

(iii) if $\mu_0$ is a bifurcation value such that $X_{\mu_0}$ has a saddle-node closed orbit, then $X_{\mu_0}$ has no cycles.

Our results can be stated as follows :

Theorem : $\mathcal{A}$ is open in $\mathfrak{B}$.

Main Theorem : If $\{X_\mu\} \in \mathcal{A}$, then $\{X_\mu\}$ has finite modulus of stability. More precisely, the modulus of $\{X_\mu\}$ is given by the sum of the characteristic numbers associated to the vector fields in $\{X_\mu\}$ having a saddle-node closed orbit.

Corollary : On the sphere or the projective plane the one-parameter families of vector fields with finite modulus of stability are dense.

Sketches of the key parts of the main theorem, are in Sections 3 and 4 (see also the comments below). Full proofs will appear in a forthcoming paper. In Section 2 we recall all necessary concepts.

Several comments are in order.

In [5], Sotomayor announced results concerning the stability of families in $\mathcal{A}$ but with a weaker version of the notion of stability (which we called mild stability in [2]) : the topological equivalence between two arcs is not required to vary continuously with the parameter.

Let us now explain how we get the full set of invariants (parameters) that define the modulus of stability of $\{X_\mu\} \in \mathcal{a}$ near $\mu_0 \in I$, where $X_{\mu_0}$ has a saddle-node closed orbit $\gamma$. We take a cross-section S to $X_{\mu_0}$ at some $x \in \gamma$ and let f be the corresponding Poincaré transformation. Then there is a unique $C^\infty$ vector field Z on S such that $f = Z_{t=1}$, where $Z_t$ with $t \in \mathbb{R}$ is the flow induced by Z. On each side of x in S, we consider fundamental domains $I_1 = [y, f(y)]$ and $I_2 = [z, f(z)]$. If $s_1, s_2, \ldots, s_n$ are the points where saddle-separatrices of $X_{\mu_0}$ cut $I_1$, we define $t_2, \ldots, t_n$ as the real numbers such that $Z_{t_i}(s_1) = s_i$, $2 \leq i \leq n$. Similarly we define $\bar{t}_2, \ldots, \bar{t}_m$ corresponding to $I_2$. These are the full set of invariants for the equivalence classes of arcs $\{Y_\mu\}$ close to $\{X_\mu\}$ near $\mu = \mu_0$. The proof of this fact is one of the key points of our main theorem. We construct conjugacies for the Poincaré mappings on cross-sections from which we get local topological equivalences near $\gamma \times \{\mu_0\}$ in $M \times I$. This construction is sketched in Section 3. The globalisation of the equivalences to all of $M \times I$ is done via arc length of orbits.

The construction of the topological equivalences between arcs of vector fields near loops is also quite delicate. We present the ideas for doing so in Section 4. Near the other bifurcation values the situation is much simpler and it will not be mentioned here.

Finally, let us consider a family $\{X_\mu\}$ such that $X_{\mu_0}$ has an attracting or repelling loop (see Sections 2 and 4) at the singularity $x_0$. In general the unstable (stable) separatrix of $x_0$ that does not form the loop can have the loop as its $\omega$-limit ($\alpha$-limit) set. However, this is not the case if $\{X_\mu\} \in \mathcal{a}$ since, for some $\mu$ near $\mu_0$, $X_\mu$ must have a non-critical recurrence. To prove this fact, one defines a transversal circle to $X_\mu$, $\mu$ near $\mu_0$, through a point of the loop and consider the corresponding Poincaré mappings. These mappings are discontinuous but strictly increasing. Then one can use the theory of rotation number as usually defined for homeomorphisms of the circle.

## 2. Basic concepts on paramatrised families of vector fields.

Let X be a $C^\infty$ vector field on M, and p be a singularity of X. We call p quasi-hyperbolic if one of the following two conditions holds :

(i) $D_pH$ has rank one and the second derivative of X at p restricted to centre manifold tangent to the zero-eigenspace is non zero. In this case p is called a saddle-node .

(ii) $D_pX$ has non-zero eigenvalues on the imaginary axis and the third-order jet of X at p differs from $D_pX$. We call p a Hopf point.

A closed orbit $\gamma$, of X, is called quasi-generic if one of the two conditions below holds :

Let $x \in \gamma$ and $\pi$ a Poincaré map on $\gamma$.

(i) $\pi'(x) = 1$ and $\pi^{(2)}(x) \neq 0$. In this case $\gamma$ is called a saddle-node closed orbit.

(ii) $\pi'(x) = -1$ and $(\pi^2)^{(3)}(x) \neq 0$. In this case $\gamma$ is called a flip closed orbit.

Let $\Sigma_1$ be the set of $C^\infty$ vector fields of M, X, which satisfy condition (a) or (b) below.

(a) (i) X has one and only quasi-generic critical element (singularity or closed orbit). The others are all generic.

(ii) X does not have saddle-connections.

(iii) The $\alpha$ and $\omega$-limit sets of every orbit of X are critical elements.

(b) (i) X has only generic (i.e.hyperbolic) critical elements.

(ii) X has one and only one saddle connection. If the orbit joins a saddle point p to itself, we require the trace of $D_pX$ to be non-zero. In this case we say that X has a loop at p.

(iii) The $\alpha$ and $\omega$-limit sets of every trajectory of X are critical elements or a loop, if it exists.

Let $\Sigma$ be the set of Morse-Smale vector fields, i.e. $X \in \Sigma$ if

(i) X has only generic critical elements;

(ii) X has no saddle-connections; and

(iii) the $\alpha$ and $\omega$-limit sets of every orbit of X are critical elements.

Suppose $\{X_\mu\}$ is an arc such that $X_\mu \in \Sigma \cup \Sigma_1$ for every $\mu \in I$. If $X_\mu$ is a bifurcation point of this arc, then $X_\mu \in \Sigma_1$, and so it has a quasi-generic critical element or a saddle connection. In both cases we will impose some generic conditions on the dependence upon $\mu$, to get then the subset $\mathcal{Q}$ we are interested in.

(a) Saddle-node closed orbits.

If $\mu_0$ is a bifurcation value such that $X_{\mu_0}$ has a saddle-node closed orbit, $\gamma$, taking a transversal section at $x \in \gamma$, we obtain a saddle-node arc of diffeomorphisms $\{\varphi_u\}$, defined on an interval around $\mu_0$ (see [2]). We impose then, the condition $\frac{d}{du}\varphi(x,\mu) \neq 0$.

(b) Flip closed orbit.

If $\gamma$ is a flip closed orbit of $X_{\mu_0}$, taking a transversal section S, at $x \in \gamma$, and a diffeomorphism $\alpha$ of R onto a neighbourhood of x in S, such that $\alpha(0) = x$, we can obtain an arc $\{\varphi_\mu\}$ of diffeomorphisms on R, such that $\varphi_0(0) = 0$, $(d\varphi_0)_0 = -1$ and such that the 3-jet of $(\varphi_0)^2$ in the origin differs from the 3-jet of the identity. The origin is either a source or a sink of $\varphi_0$; in the following we shall assume it to be a sink of $\varphi_0$; the other case is completely analogous. With a coordinate change of the form

$$\tilde{t} = t(t,\mu)$$
$$\tilde{\mu} = \mu$$

we can bring $\varphi_\mu$ in the form

$$\varphi_{\tilde{\mu}}(\tilde{t}) = -\tilde{t} + \tilde{t}^3 + \lambda(\mu).\tilde{t} + 0(\tilde{t}^4) + 0(|\tilde{u}|.\tilde{t}^2)$$

where $\lambda$ is a real function and $\lambda(0) = 0$. We shall impose the condition $\frac{d\lambda}{d\tilde{\mu}}(0) \neq 0$, which is, of course, a generic condition.

(c) Saddle-node.

If p is a saddle-node of $X_{\mu_0}$, then we restrict $X_\mu$ to a centre-manifold (tangent to the zero-eigenspace of $D_pX_{\mu_0}$). We obtain an arc $Y = \{Y_\mu\}$ defined in a neighbourhood of $\mu_0$ of one dimensional vector fields such that $D_pY_{\mu_0} = 0$ and the second derivative of $Y_{\mu_0}$

at p is non-zero. We impose the condition $\frac{d}{d\mu} Y(p,\mu_0) \neq 0$.

(d) Hopf points.

Suppose $X_{\mu_0}$ has a Hopf point p. Without loss of generality we may assume that p is a singularity of $X_\mu$ for $\mu$ near $\mu_0$. Let $\lambda(\mu)$ be the absolute value of the eigenvalues of $D_pX_\mu$. We require $\frac{d}{d\mu}\lambda(\mu_0) \neq 0$.

(e) Saddle-connections.

Let $X_{\mu_0} \in \Sigma_1$ be such that $X_{\mu_0}$ has a saddle-connection. Let p and q be the saddles of $X_{\mu_0}$ such that one unstable separatrix of p, $W^u(p)$ equals one stable separatrix of q, $W^s(q)$ (the case p = q is included). Take a cross-section S to $X_{\mu_0}$ at $x \in W^u(p)$. For each $\mu$ sufficiently close to $\mu_0$, let $p_\mu$ and $q_\mu$ be the saddles of $X_\mu$ corresponding to p and q respectively. Using the properties of the invariant manifolds of hyperbolic singularities, we can define differentiable functions $g_s(\mu)$ and $g_u(\mu)$, defined in a neighbourhood of $\mu_0$, such that $g_u(\mu) \in S \cap W^u(p_\mu)$, $g_s(\mu) \in S \cap W^s(q_\mu)$ and $g_u(\mu_0) = g_s(\mu_0)$. Let $g(\mu) = g_s(\mu) - g_u(\mu)$. We will say that $\{X_\mu\}$ unfolds generically at $\mu = \mu_0$ if $\frac{dg}{d\mu}(\mu_0) \neq 0$.

In general, we say that an arc $\{X_\mu\} \in \mathfrak{B}$ unfolds generically at a bifurcation value $\mu \in I$, such that $X_\mu \in \Sigma_1$, if $X_\mu$ satisfies one of the conditions given above.

Finally, if $X \in \Sigma_1$ is a vector field having a saddle-node closed orbit $\gamma$, we say that X has a cycle if there exists $x \in M-\gamma$ such that $\alpha(x) = \gamma = \omega(x)$.

We can now define the set $\mathfrak{A}$ we are interested in. Let $\mathfrak{A}$ be the set of arcs $\{X_\mu\}$ in $\mathfrak{B}$ which satisfy the following conditions :

(i) $X_\mu \in \Sigma \cup \Sigma_1$ for each $\mu \in I$

(ii) $\{X_\mu\}$ unfolds generically at every bifurcation value

(iii) if $\mu_0$ is a bifurcation value such that $X_{\mu_0}$ has a saddle-node closed orbit, then $X_{\mu_0}$ has no cycles.

## 3. Local structure of a saddle-node arc.

Here we consider arcs of $C^\infty$ local diffeomorphisms $\{\varphi_\mu\}$ defined in an interval $(-\delta, \delta) \subset R$ such that $\varphi_0(0) = 0$, $\frac{\partial}{\partial x}\varphi_0(0) = 1$, $\frac{\partial^2}{\partial x^2}\varphi_0(x) > 0$ and $\frac{\partial}{\partial \mu}\varphi_0(0) > 0$. Following [2], these arcs will be called saddle-node arcs. We say that two saddle-node arcs, $\{\varphi_\mu\}$ and $\{\psi_\mu\}$ are locally conjugate at $(0,0)$ if there exists a local homeomorphism $H = (h, \eta)$ defined in a neighbourhood of $(0,0)$ in $\mathbb{R}^2$ such that

$$h_\mu \circ \varphi_\mu = \psi_{\eta(\mu)} \circ h_\mu \text{ whenever defined}$$

where $\eta: \mathbb{R} \to \mathbb{R}$ is a local homeomorphism defined in a neighbourhood of zero such that $\eta(0) = 0$, and $h_\mu$ is defined by $h_\mu(x) = h(x, \mu)$.

We are interested in answering the following question:
"If $\{\varphi_\mu\}$ and $\{\psi_\mu\}$ are two saddle-node arcs and $x_1, \ldots, x_k$, $\bar{x}_1, \ldots, \bar{x}_k$ are given real numbers close to $x = 0$, when does there exist a local conjugacy $H = (h, \eta)$ between $\{\varphi_\mu\}$ and $\{\psi_\mu\}$ such that $h_\mu(x_i) = \bar{x}_i$ for $i = 1, \ldots, k$ ?"

This question was motivated by the fact that if one wants to construct a topological equivalence between two vector fields having saddle-node closed orbits from conjugacies between Poincaré transformations in cross-sections, then necessarily these conjugacies need to preserve those saddle separatrices that intersect the cross sections.

This question is answered by Theorem 1 below, which is the main result of this section.

First we recall that if $\{\varphi_\mu\}$ is a saddle-node arc, then there is a unique $C^\infty$ vector field Z, defined on a neighbourhood of zero in $\mathbb{R}$, such that the time one map $Z_1$ of Z equals $\varphi_0$, see [6].

<u>Theorem 1</u> : Let $\{\varphi_\mu\}$ and $\{\bar{\varphi}_\mu\}$ be two saddle-node arcs with corresponding vector fields Z and $\bar{Z}$, that is, $Z_1 = \varphi_0$, $\bar{Z}_1 = \bar{\varphi}_0$. If $x_1 < \ldots < x_k < \varphi_0(x_1) < 0$ and $0 < y_1 < \ldots < y_\ell < \varphi_0(y_1)$ are small, let $t_i$ and $\bar{t}_j$, $i = 1, \ldots, k$, $j = 1, \ldots, \ell$, be defined by

$$Z_{t_i}(x_1) = x_i \qquad Z_{\bar{t}_j}(y_1) = y_j \quad .$$

Then if $|t_i - t_m| \neq |\bar{t}_j - \bar{t}_r|$ and $1 - |t_i - t_m| \neq |\bar{t}_j - \bar{t}_\ell|$ for i, m = 1,...,k and j, r = 1,...,$\ell$, there exists a local homeomorphism $H = (h, \eta)$, such that H is a conjugacy between $\{\varphi_\mu\}$ and $\{\bar{\varphi}_\mu\}$ in a neighbourhood of (0,0),

$$h_\mu(x_i, \mu) = (\bar{x}_i, \eta(\mu)), \quad \text{and} \quad h_\mu(y_j, \mu) = (\bar{y}_j, \eta(\mu))$$

where $\bar{x}_i = \bar{Z}_{t_i}(h_0(x_1))$, and $\bar{y}_j = \bar{Z}_{\bar{t}_j}(h_0(y_1))$.

Consider now, $C^2$ vector fields $Z = Z(x,\mu)\frac{\partial}{\partial x}$ on $\mathbb{R}^2$ with $Z(0,0) = 0$, $\frac{\partial}{\partial x} Z(0,0) = 0$, $\frac{\partial^2}{\partial x^2} Z(0,0) > 0$, $\frac{\partial}{\partial \mu} Z(0,0) > 0$. These vector fields are called saddle-node fields. Two saddle-node fields Z and $\bar{Z}$ are called locally conjugate if there is a homeomorphism $H = (h, \eta)$ from a neighbourhood of (0,0) in $\mathbb{R}^2$ to another such neighbourhood such that

$$h_\mu \circ Z_t(x,\mu) = \bar{Z}_t(h_\mu(x), \eta(\mu))$$

whenever both sides are defined and $\eta : \mathbb{R} \to \mathbb{R}$ is a local homeomorphism such that $\eta(0) =$

We say that a $C^4$ saddle-node field is adapted to a saddle-node arc $\{\varphi_\mu\}$ if the function $g(x,\mu)$, defined by

$$(\varphi_\mu(x) + g(x,\mu), \mu) = Z_1(x,\mu)$$

vanishes along $\mu = 0$ and has a zero 4-jet at (0,0). [2].

The results that we present in this section are based on the following theorems, which are proved in [2].

<u>Theorem 2</u> : Let $\{\varphi_\mu\}$ and $\{\bar{\varphi}_\mu\}$ be two saddle-node arcs with corresponding vector fields Z and $\bar{Z}$, i.e., such that $Z_1 = \varphi_0$ and $\bar{Z}_1 = \bar{\varphi}_0$. Let $H = (h, \eta)$ be a local conjugacy between $\{\varphi_\mu\}$ and $\{\bar{\varphi}_\mu\}$. Then $h_0 Z_t = \bar{Z}_t h_0$, where $h_0(x) = h(x,0)$ and $t \in \mathbb{R}$.

<u>Remark</u> : The above theorem implies that the choice of the conjugacy along $\{u=0\}$ is extremely restricted. Instead of the usual freedom to fix the conjugacy arbitrarily on a fundamental domain, we are here only free to fix $h_0$ in two points : one in $\{x < 0\}$ and one in $\{x > 0\}$.

Theorem 3 : Let $\{\varphi_\mu\}$ be a saddle-node arc. Then there exists an adapted saddle-node field Z for $\{\varphi_\mu\}$, which is at least $C^5$, and positive numbers, $\varepsilon$, a, and C such that if

(i) $(x,\mu) \in U = \{(x,\mu) | 0 \leq \mu \leq \varepsilon, \ -a \leq x \leq a\}$, and (ii) $i \in \mathbb{N}$ is such that $(\varphi^i_\mu(x),\mu) \in U$, then $Z_\alpha(x,\mu) = (\varphi^i_\mu(x),\mu)$ for some $\alpha \in \mathbb{R}$ with

$$|i-\alpha| \leq \mu . C \ .$$

We present below the ideas used to prove Theorem 1.

Lemma 1 : Let $\{\varphi_\mu\}$ be a saddle-node arc. There exist $\varepsilon > 0$, and $a > 0$ such that if $-a < x < 0$, then for each $n \in \mathbb{N}$ there exists $0 < \varepsilon_n < \varepsilon$ such that the function

$$g_n : [0,\varepsilon] \to \mathbb{R},$$

defined by

$$g_n(\mu) = \varphi^n(x,\mu)$$

satisfies :

i) $\varepsilon_{n+1} < \varepsilon_n$

ii) $\frac{d}{d\mu} g_n(\mu) > 0 \qquad \forall \mu \in [0,\varepsilon_n]$ .

Remark : Suppose $\{X_\mu\} \in \mathfrak{B}$ is such that for $\mu = 0$ we have :

(i) $X_0 \in \Sigma_1$ has a saddle-node orbit $\gamma$ ;

(ii) $X_0$ has no cycles ;

(iii) $\gamma$ is the $\alpha$ and $\omega$-limit sets of saddle separatrices ; and

(iv) $\{X_\mu\}$ unfolds generically at $\mu = 0$ .

Then the previous lemma implies that there exists $\varepsilon > 0$, and a sequence $\mu_n \to 0$ as $n \to \infty$, with $0 < \mu_n < \varepsilon$, such that for each $|\mu| \leq \varepsilon$ we have

(a) $X_\mu \in \Sigma$ if $\mu \neq \mu_n$ for all n.

(b) $\mu_n$ is a bifurcation value such that $X_{\mu_n}$ has one saddle-connection.

(c) $\{X_\mu\}_{|\mu|\leq\varepsilon}$ belongs to $\mathfrak{A}$.

This is one of the basic facts that we use to prove that $\mathfrak{A}$ is open in $\mathfrak{B}$, as stated in the introduction.

Sketch of the proof of Lemma 1.

Let V be a neighbourhood of (0,0) such that if $(x,\mu) \in V$ then $\frac{\partial\varphi}{\partial x}(x,\mu) > 0$ and $\frac{\partial\varphi}{\partial\mu}(x,\mu) > 0$. Let Z be an adapted saddle-node field for $\{\varphi_\mu\}$, with a, ε and C, given by Theorem 3. It is easy to see that we can take a, ε so small that $\varepsilon C < \frac{1}{2}$ and if $-a < x < 0$, $0 < \delta \le \varepsilon$ and $\varphi^n(x,\delta) \in U = \{(x,\mu) | 0 \le \mu \le \varepsilon, -a \le x \le a\}$, then $\varphi^k(x,\mu) \in V$, for all $0 \le k \le n+1$ and $0 \le \mu \le \delta$. Let $-a < x < 0$ and $0 < y < a$ be fixed. For each $n \in \mathbb{N}$, let $\varepsilon_n = \min\{\max\{0 \le \mu \le \varepsilon | \varphi^{n-1}(x,\mu) = y\}, \varepsilon\}$. Then,

(i) $\qquad \varepsilon_{n+1} \le \varepsilon_n$ .

In fact, if $0 < \mu, \mu' < \varepsilon$ and $\varphi^n(x,\mu) = y = \varphi^{n+1}(x,\mu')$ then $\mu > \mu'$ . Indeed

if $\quad X_t(x,\mu) = \varphi^n(x,\mu) = y$

and $\quad X_{t'}(x,\mu') = \varphi^{n+1}(x,\mu')$, then by Theorem 3

$$|t-n| < \mu C < \frac{1}{2} \quad \text{and} \quad |t'-(n+1)| < \mu' C < \frac{1}{2}$$

and so $\quad t < t'$, which implies $\mu > \mu'$ .

(ii) $\qquad$ We can prove that $\frac{dg_n}{d\mu}(\mu) > 0 \quad \forall \mu \in [0,\varepsilon_n]$ by induction on n, using the fact that

$$\frac{dg_n}{du}(\mu) = \frac{\partial\varphi}{\partial x}(\varphi_\mu^{n-1}(x), \mu)\,\frac{dg_{n-1}}{du}(\mu) + \frac{\partial\varphi}{\partial u}(\varphi^{n-1}(x,\mu),\mu) \quad \text{and (i)} .$$

Lemma 2 : Let $\{\varphi_\mu\}$ be a saddle-node arc. Let Z be an adapted saddle-node field for $\{\varphi_\mu\}$, a, ε and C as in Lemma 1. If $-a < x_1 < x_2 < \varphi_0(x_1) < 0$ and $0 < y_1 < y_2 < \varphi_0(y_1) < a$, let t and $\bar{t}$ defined by

$$Z_t(x_1,0) = (x_2,0), \qquad\qquad Z_{\bar{t}}(y_1,0) = (y_2,0)$$

and $\mu^i_{j,n}$, for i, j = 1, 2 defined by $\varphi^n(x_j,\mu^i_{j,n}) = y_i$. Then if $t \ne \bar{t}$ and $1-t \ne \bar{t}$, the order in which the $\mu^i_{j,n}$'s occur, for n sufficiently large, is determined by the order relation between $t, \bar{t}$ and $1-t, \bar{t}$.

Sketch of the proof. We will consider one case, say $t > \bar{t}$ and $1-t > \bar{t}$. The other cases are similar. We will prove that

$$\mu^1_{1,n+1} < \mu^2_{1,n+1} < \mu^1_{2,n} < \mu^2_{2,n} < \mu^1_{1,n} .$$

Let $\delta = \min\{|t-\bar{t}|, |t+\bar{t}-1|\}$. Let $\varepsilon' < \varepsilon$ such that $\varepsilon' C < \frac{\delta}{4}$ and if

$0 \le \mu \le \varepsilon'$ then $|t_\mu - \bar{t}_\mu| > \frac{\delta}{2}$ and $|\alpha_\mu - \bar{t}_\mu| > \frac{\delta}{2}$, where $t_\mu, \bar{t}_\mu$ and $\alpha_\mu$ are defined by

$$Z_{t_\mu}(x_1, \mu) = (x_2, \mu)$$
$$Z_{\bar{t}_\mu}(y_1, \mu) = (y_2, \mu)$$
$$Z_{t_\mu + \alpha_\mu}(x_1, \mu) = (\varphi_\mu(x_1), \mu).$$

The fact that $\mu^1_{i,n+1} < \mu^2_{i,n+1} < \mu^1_{i,n} < \varepsilon'$ for $i = 1,2$ and $n \ge n_0$ is a consequence of the fact that the function $g_{i,n}(\mu) = \varphi^n(x_i, \mu)$ $i = 1,2$, given in Lemma 1, are strictly increasing continuous functions. On the other hand, since $\varphi^n_\mu(x_1) < \varphi^n_\mu(x_2) < \varphi^{n+1}_\mu(x_1)$ we conclude that $\mu^i_{1,n+1} < \mu^i_{2,n} < \mu^i_{1,n}$. So we need to prove that $\mu^2_{2,n} < \mu^1_{1,n}$ and $\mu^2_{1,n+1} < \mu^1_{2,n}$.

First we observe that if $\varphi^n(x_2, \mu^1_{1,n}) > y_2$, then necessarily we will have $\mu^2_{2,n} < \mu^1_{1,n}$ since $g_{2,n}(\mu^2_{2,n}) = y_2$, $g_{2,n}(\mu^1_{1,n}) = \varphi^n(x^2, \mu^1_{1,n})$ and $g_{2,n}$ is an increasing function. But the fact that $t > \bar{t}$ implies that if $0 < \mu \le \varepsilon'$ and $\varphi^n(x_1, \mu) = y_1$ then $\varphi^n(x_2, \mu) > y_2$. To see this, let $t, t'$ and $\alpha$ defined by

$$Z_t(x_2, \mu) = (y_2, \mu), \quad \alpha = t_\mu + t, \quad Z_{t'}(x_2, \mu) = (\varphi^n_u(x_2), \mu).$$

Then $Z_\alpha(x_1, \mu) = (y_1, \mu)$
and $Z_{t + \bar{t}_\mu}(x_2, \mu) = Z_{\bar{t}_\mu}(y_1, \mu) = (y_2, \mu)$.

From the estimates in Theorem 3, we have that $|n - \alpha| < \frac{\delta}{4}$ and $|n - t'| < \frac{\delta}{4}$, and so, $t' > t + t_\mu$, which implies that $\varphi^n(x_2, \mu) > y_2$. Similarly, using that $1 - t > \bar{t}$ we prove that $\mu^2_{1,n+1} < \mu^1_{2,n}$.

As a corollary of this lemma, we have the following result

<u>Corollary</u> : Let $-a < x_0 < x_1 < \ldots < x_k < \varphi_0(x_0) < 0$ and $0 < y_0 < y_1 < \ldots < y_\ell < \varphi_0(y_0) < a$. Let $t_i, \bar{t}_j, \mu^j_{i,n}$ for $i = 0, \ldots, k$, $j = 0, \ldots, \ell$ and $n \in \mathbb{N}$ defined by

$$Z_{t_i}(x_0, 0) = (x_i, 0)$$
$$Z_{\bar{t}_j}(y_0, 0) = (y_j, 0) \quad \text{and}$$
$$\varphi^n(x_i, \mu^j_{i,n}) = (y_j, \mu^j_{i,n}) .$$

Then the order of the numbers $(\mu^j_{i,n})$ is determined by the order relations between $|t_m - t_i|$ and $|\bar{t}_p - \bar{t}_j|$ and between $1 - |t_m - t_i|$ and $|\bar{t}_p - \bar{t}_j|$ for $i, m = 0, 1, \ldots, k$ and $p, j = 0, 1, \ldots, \ell$.

To prove Theorem 1, we first consider the case where $\{\tilde{\varphi}_\mu\}$ is the time-one map of Z, a saddle-node field adapted to $\{\varphi_\mu\}$. That is

Lemma : Let $\{\varphi_\mu\}$ be a saddle-node arc, and Z a saddle-node field adapted to $\{\varphi_\mu\}$ as in Theorem 3. Let $x_1 < x_2 < \ldots < x_k < \varphi_0(x_1) < 0$ and $0 < y_1 < y_2 \ldots < < y_\ell < \varphi_0(y_1)$ such that if $Z_{t_i}(x_1, 0) = (x_i, 0)$ and $Z_{\bar{t}_i}(y_1, 0) = (y_j, 0)$, then $|t_i - t_m| \neq |\bar{t}_j - \bar{t}_p|$ and $1 - |t_i - t_m| \neq |\bar{t}_j - \bar{t}_p|$ for $i, m = 1, \ldots, k$; $j, p = 1, \ldots, \ell$. Then, if $x_1$ and $y_1$ are sufficiently small, there exists a local conjugacy $H = (h, \eta)$ between $\{\varphi_\mu\}$ and the time-one map of Z, such that :

(i) $h_\mu(x_i) = x_i$, $\quad h_\mu(y_i) = y_i$

(ii) $\varphi^n_\mu(x_i) = y_j \Leftrightarrow Z^n_1(x_i, \eta(\mu)) = y_j$

for $i = 1, \ldots, k$; $j = 1, \ldots, \ell$, $n \in \mathbb{N}$.

Sketch of the proof.

First we observe that we can assume, without loss of generality that $Z_{t_i}(x_1, \mu) = (x_i, \mu)$ and $Z_{\bar{t}_j}(y_1, \mu) = (y_1, \mu)$ for $i = 1, \ldots, k$, $j = 1, \ldots, \ell$ and $\mu$ small.

(i) Construction of $\eta$.

Let $a > 0$, $\varepsilon > 0$, and $C > 0$, given by Theorem 3. Let $n_0 \in \mathbb{N}$ be the first number such that

$$\varphi^n_\varepsilon(x_1) \geq y_\ell \text{ and } Z^n_1(x_1, \varepsilon) \geq y_\ell. \text{ For each } n \geq n_0 \text{ let } \mu^j_{i,n} \text{ and } \alpha^j_{i,n}$$

defined by

$$\varphi^n(x_i, \mu^j_{i,n}) = y_j \text{ and } Z^n(x_i, \alpha^j_{i,n}) = y_j; \; i = 1, \ldots, k; \; j = 1, \ldots, \ell .$$

Let $f:(0, \varepsilon] \to \mathbb{R}$ be defined by

$$Z_{f(\mu)}(x_1, \mu) = (y_1, \mu) .$$

We know, from [2], that $\lim_{\mu \to 0} f(\mu) = \infty$ and the derivative of f is less than zero in $(0, \varepsilon]$.

Let $\varepsilon_1 = \mu^1_{1,n_0}$ and $g:(0,\varepsilon_1] \to \mathbb{R}$ a strictly decreasing differentiable function such that

$$g(\mu^j_{i,n}) = n + t_1 - \overline{t}_j \quad \text{and} \quad |g(\mu) - f(\mu)| \to 0 \quad \text{as} \quad \mu \to 0 \ .$$

This function exists, since g is strictly increasing in $\mu^j_{i,n}$'s and $|g(\mu^j_{i,n}) - f(\mu^j_{i,n})| = |n-t^j_i| < \mu^j_{1,n}C$.

Let $\varepsilon'_1 = \alpha^1_{1,n_0}$, that is $Z_{n_0}(x_1,\varepsilon'_1) = (y_1,\varepsilon'_1)$ and so $f(\varepsilon'_1) = n_0 = g(\varepsilon_1)$. Let $\eta:[0,\varepsilon_1] \to [0,\varepsilon'_1]$ a homeomorphism such that $\eta(0) = 0$ and $g(\mu) = f(\eta(\mu))$ for $0 \le \mu \le \varepsilon_1$.

(ii) Construction of h.

Let
$$D = \{(x,\mu)|x_1 \le x \le \varphi_\mu(x_1),\ 0 \le \mu \le \varepsilon_1\} \text{ and}$$
$$D' = \{(x,\mu)|x_1 \le x \le Z_1(x_1,\mu),\ 0 \le \mu \le \varepsilon'_1\} \ .$$

Let

$$I_i = \{(\mu, x_i)|0 \le \mu \le \varepsilon_1\}$$
$$J^n_j = \{(z,\mu) \in D|0 \le \mu \le \varepsilon_1,\ \varphi^n_\mu(z) = y_j\}$$
$$\tilde{J}^n_j = \{(z,\mu) \in D'|0 \le \mu \le \varepsilon'_1,\ Z_n(z,\mu) = y_j\}$$
$$R = (\bigcup_{i=1}^k I_i) \cup (\bigcup_{j,n} J^n_j)$$
$$\tilde{R} = (\bigcup_{i=1}^k I_i) \cup (\bigcup_{j,n} \tilde{J}^n_j) \ .$$

The idea is to define h in R, with the properties we want, and then extend it to D.

To define h in R, we put $h(z,\mu) = Z_1^{-n}(y_j,\eta(\mu))$ if $(z,\mu) \in J^n_j$, and $h(x_i,\mu) = (x_i,\eta(\mu))$ for $i = 1,\dots,k$. Then we extend h to $R \cup \{(x,\mu)|x_1 \le x \le \varphi_0(x_1),\ 0 \le \mu \le \varepsilon_1\} = R \cup I$ putting $h_0 = \text{id}$. We assert that h defined as above has the properties we want, is continuous, and that we can extend it to D.

Finally we observe that the conjugacy between two saddle-node fields obtained in [2], gives us a conjugacy between the time-one maps of the saddle-node fields, which has the desired properties. So using this fact and the lemma above, we can prove Theorem 1.

4. Local stability of arcs going through loops.

Let $\{X_\mu\}$ be a one-parameter family of $C^\infty$ vector fields such that $X_{\mu_0} \in \Sigma_1$ has a loop at a saddle p. Suppose $\{X_\mu\}$ unfolds generically at $\mu = \mu_0$ and the other unstable (and stable) separatrices of p do not have the loop as $\omega$-limit set or $\alpha$-limit set. Without loss of generality, we can assume that $\mu_0 = 0$ and that p is a saddle of $X_\mu$ for $\mu$ sufficiently small. Also we assume that trace $(D_p X_0) < 0$. The other case is similar.

Let $\gamma$ be the loop of $X_0$ that is $\gamma$ is the closure of the orbit of some $x \in M$, such that $x \neq p$ and $\alpha(x) = \{p\} = \omega(x)$, and suppose $\gamma$ is two-sided. Let S be a transversal section to the flow at a point of $\gamma$ near p. Let $a, b \in S$ be as in Figure 1(B). That is, $b \in S$ and a is the first intersection of the positive orbit of b with S.

The interval $[a, b] \subset S$ will be called a fundamental domain for the loop. We can assume, that for $\mu$ small enough, say $|\mu| < \varepsilon$, p is a saddle of $X_\mu$ and the positive orbit of $X_\mu$ through $x = b$ intersects S for the first time at $x = a$. As in [4], we know that the unfolding of this loop is as in Figure 1.

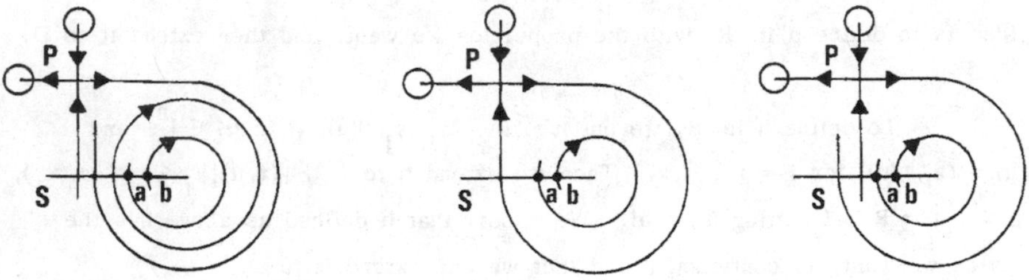


(A) $\mu < 0$ or $\mu > 0$ (B) $\mu = 0$ (C) $\mu > 0$ or $\mu < 0$

Figure 1.

With the assumptions made above we will have for each $|\mu| \leq \varepsilon$ (where $\varepsilon > 0$ is small enough) a Poincaré transformation $\pi_\mu$ defined in the interval $[0, b] \subset S$, where $x = 0$ is the intersection of $W^s(p)$ with S, and $\pi_\mu(0)$ is the first intersection of $W^u(p)$ with S. Notice that $\pi_\mu$ varies differentiably with the parameter $\mu$, so we have a one-parameter family of Poincaré transformations $\varphi: [0, b] \times [-\varepsilon, \varepsilon] \to S \times [-\varepsilon, \varepsilon]$

defined by $\varphi(x,\mu) = (\pi_\mu(x),\mu)$. The fact that $\{X_\mu\}$ unfolds generically at $\mu = 0$ means that $\frac{\partial\varphi}{\partial\mu}(0,0) \neq 0$. Let us suppose that $\frac{\partial\varphi}{\partial\mu}(0,0) < 0$. In this case if b is sufficiently near $x = 0$ and $g_{n,x}$ is the function defined by $g_{n,x}(\mu) = \pi^n_\mu(x)$ for $a \leq x \leq b$, then $g_{n,x}$ is strictly decreasing. The motivation to consider these functions is the following: suppose that q is some other saddle of $X_0$ and $x \in S$, $a \leq x \leq b$, is such that $x \in W^u(q)$. We can suppose that q is a saddle of $X_\mu$, for $|\mu| < \varepsilon$ and $x \in W^u(q)$. If $0 < \mu < \varepsilon$ then $X_\mu$ will have a saddle connection between p and q if and only if $g_{n,x}(\mu) = 0$ for some $n > 0$. So, if $X_0$ has points $x \in S$ as above we can conclude from the properties of the functions $g_{n,x}$, the following facts :

(i) There exists a sequence $\mu_n \to 0$, $\mu_n > 0$ such that $X_{\mu_n}$ has one saddle connection, and $X_\mu \in \Sigma$ if $\mu \neq \mu_n$ for all n.

(ii) There exists $\varepsilon > 0$ such that $\{X_\mu\}_{|\mu|\leq\varepsilon} \in \mathfrak{a}$ .

This is one of the facts that we use to prove that $\mathfrak{a}$ is open in $\mathfrak{B}$.

In this section we assert that if $\{X_\mu\} \in \mathfrak{a}$ then $\{X_\mu\}$ is locally stable at $\mu = 0$, where $X_0$ has a loop, that is, if $\{Y_\mu\}$ is a small perturbation of $\{X_\mu\}$, then, there exists $\varepsilon, \varepsilon' > 0$ such that the arcs $\{X_\mu\}_{|\mu|\leq\varepsilon}$ and $\{Y_\mu\}_{|\mu|\leq\varepsilon'}$ are topologically equivalent. As we observed in the introduction the main point of the proof of this fact is the existence of topological conjugacies between corresponding parametrised Poincaré transformations on the cross-section S. This is sketched in the next theorem.

<u>Theorem</u> : Let $\{X_\mu\} \in \mathfrak{a}$ such that $X_\mu$ has a loop $\gamma$ at $\mu = 0$, and $\{Y_\mu\}$ a small perturbation of $\{X_\mu\}$. Then given a transversal section S through $\gamma$, there exists $\varepsilon, \varepsilon' > 0$, a continuous family of homeomorphisms $\{h_\mu\}$ of S and a reparametrisation $\eta: [\varepsilon, \varepsilon] \to [-\varepsilon', \varepsilon']$ such that $h_\mu$ is a conjugacy between $\pi_\mu$ and $\bar{\pi}_{\eta(\mu)}$, where $\{\pi_\mu\}$ and $\{\bar{\pi}_\mu\}$ are the Poincaré transformations induced on S by $\{X_\mu\}$ and $\{Y_\mu\}$, respectively. Moreover $h_\mu$ preserves saddle separatrices that have the loop as their $\omega$-limit set at $\mu = 0$.

<u>Sketch of the proof.</u>

From the fact that $\{X_\mu\}$ unfolds generically, we know that $Y_\mu$ will have a loop for some $\mu$ and we can suppose that this occurs for $\mu = 0$. We first suppose that $\gamma$ is two sided. From observations made before, we can assume that $\{Y_\mu\} \in \mathfrak{a}$ and so all the properties above are still valid for $\{Y_\mu\}$. Let $[a,b] \subset S$ a fundamental domain for $\gamma$ and $[\bar{a},\bar{b}] \subset S$ a fundamental domain for $\bar{\gamma}$, where $\bar{\gamma}$ is the loop of $Y_0$. Let $a = x_1 < x_2 < \ldots < x_k < b$ the points of $[a,b]$ that are in unstable saddle separatrices of $X_0$, and $\bar{a} = \bar{x}_1 < \bar{x}_2 < \ldots < \bar{x}_k < \bar{b}$ the corresponding points for $Y_0$ (all of these requirements make sense if $\{Y_\mu\}$ is a sufficiently small perturbation of $\{X_\mu\}$). Take $\delta > 0$, small enough, such that all the remarks made above make sense for $X_\mu$ and $Y_\mu$ if $|\mu| \leq \delta$. Let $\varphi$ and $\bar{\varphi}$ be the one-parameter families of Poincaré transformations induced by $\{X_\mu\}$ and $\{Y_\mu\}$ respectively and assume that $\frac{\partial\varphi}{\partial\mu}(0,0) < 0$. Looking at the iterates of the segment $I_b = \{(b,\mu)|0 \leq \mu \leq \delta\}$ by $\varphi$, and the iterates of $I_{\bar{b}} = \{(\bar{b},\mu)|0 \leq \mu \leq \delta\}$ by $\bar{\varphi}$, we see that there are uniquely determined sequences converging to zero, $\mu_n$ and $\bar{\mu}_n$, for $n \geq n_0$, such that $0 < \mu_{n+1} < \mu_n \leq \delta$ (and equivalently $0 < \bar{\mu}_{n+1} < \bar{\mu}_n \leq \delta$), and $\pi^n_{\mu_n}(b) = 0$ (equivalently $\bar{\pi}^n_{\bar{\mu}_n}(\bar{b}) = 0$). For each $n \geq n_0$ let $s_n = (\varphi^n)^{-1}(\{0\} \times (\mu_{n+1},\mu_n])$ and $\bar{s}_n = (\bar{\varphi}^n)^{-1}(\{0\} \times (\bar{\mu}_{n+1},\bar{\mu}_n])$. From the fact that $\varphi$ is continuous we conclude that $s_n$ intersects the horizontal lines $u = c$, for $\mu_{n+1} < c \leq \mu_n$ in one and only one point, and since the functions $g_{n,x}$ defined above are injective, we conclude that $s_n$ intersects vertical lines $x = c$, $a \leq c \leq b$ in a unique point also. Similarly, we have the same properties for $\bar{s}_n$.

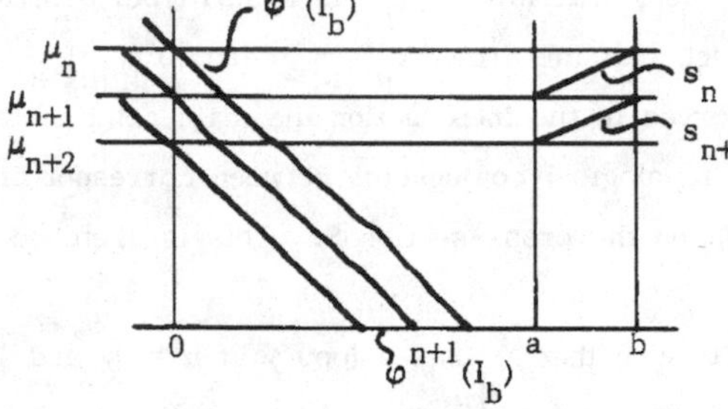


Figure 2.

Let $\varepsilon = \mu_{n_0}$ and $\bar{\varepsilon} = \bar{\mu}_{n_0}$. We first define a homeomorphism sending $R = \{(x,\mu)|a \leq x \leq b,\ |\mu| \leq \varepsilon\}$ to $\bar{R} = \{(x,\mu)|\bar{a} \leq x \leq \bar{b},\ |\mu| \leq \bar{\varepsilon}\}$ and then we will extend it to the iterates of R in such a way it will be a conjugacy between $\varphi$ and $\bar{\varphi}$. We observe that we cannot cover all $S \times [-\varepsilon,\varepsilon]$ with the iterates of R by $\varphi$, but it is easy to see that in this set (that we cannot cover iterating R by $\varphi$) we are free to define the homeomorphism in many ways.

Let $h_0$ be a homeomorphism sending $[a,b]$ to $[\bar{a},\bar{b}]$ such that $h_0$ preserves orientation and $h_0(x_i) = \bar{x}_i$ for $i = 1,\dots,k$, $h_0(b) = \bar{b}$. Now let $0 \leq \mu \leq \varepsilon$. From remarks above there exist unique $n \geq n_0$ and $a \leq y \leq b$, such that the point $(y,\mu)$ belongs to $s_n$. From the same observations, the vertical line $x = h_0(y)$ intersects $\bar{s}_n$ in a unique point $(h_0(y),\bar{\mu})$. Put $\eta(\mu) = \bar{\mu}$. If $\delta \leq \mu \leq 0$ put $\eta(\mu) = \mu$. It is easy to see that $\eta$ is an orientation preserving homeomorphism of $[-\delta,\varepsilon]$ to $[-\delta,\bar{\varepsilon}]$. Define $H(x,\mu) = (h_0(x),\eta(\mu))$ for $(x,\mu) \in R$.

Finally, we observe that the case where $\gamma$ is one-sided is much simpler. In fact, in this case, $\gamma$ is not the $\omega$-limit (or $\alpha$-limit) set of any orbit, and so there is no need for reparametrisations.

## References


1. J. Guckenheimer, "One-parameter families of vector fields on two-manifolds : another non-density theorem", Dynamical Systems, Ed. M. Peixoto, Academic Press, 1973, pp.111-127.

2. S. Newhouse, J. Palis & F. Takens, "Stable families of diffeomorphisms", IMPA preprint, 1980.

3. M. Peixoto, "Structural stability on two dimensional manifolds", Topology 1, 1962, pp. 101-120.

4. J. Sotomayor, "Generic one-parameter families of vector fields on two-dimensional manifolds", Publ. IHES 43, 1974, pp. 5-46.

5. J. Sotomayor, "Structural stability and bifurcation theory", Dynamical Systems, Ed. M. Peixoto, Academic Press, 1973, pp. 549-560.

6. F. Takens, "Normal forms for certain singularities of vector fields", Ann. Inst. Fourier, 23 (1973), (2), pp. 163-195.



I.P. Malta : Departamento de Matemática, Pontifícia Universidade Católica, Rua Marques de S. Vicente, 225, 22.453 - Rio de Janeiro, R.J., Brazil.

J. Palis : Instituto de Matemática Pura e Aplicada, Rua Luiz de Camões, 68, 20.060 - Rio de Janeiro, R.J., Brazil.

# On the dimension of the compact invariant sets of certain non-linear maps.

Ricardo Mañé.

## Introduction.

Let E be a Banach space, $U \subset E$ an open set and $f:U \to E$ a $C^1$ map such that at every $x \in U$ its derivative $D_x f$ can be decomposed as the sum of a compact map and a contraction (i.e. a linear map with norm <1). In [4] J. Mallet-Paret proved that if E is a separable Hilbert space then every compact set $\Lambda \subset E$ such that $f(\Lambda) \supset \Lambda$ has finite Hausdorff dimension. The main purpose of this paper is to show that this property holds for any Banach space. In fact we shall show that not only the Hausdorff dimension but also the limit capacity of $\Lambda$ is finite. The limit capacity is defined as the exponential growth rate of the number of balls of radius exp(-t) required to cover $\Lambda$ when $t \to +\infty$. The limit capacity is always larger than or equal to Hausdorff dimension and we shall construct an example of a compact countable subset of $\ell^2$ that has infinite limit capacity and, since it is countable, Hausdorff dimension zero. We shall also give an upper estimate for the limit capacity of $\Lambda$ that will make it possible to give sufficient conditions that grant the existence of a uniform bound for the dimensions of all the compact sets $\Lambda$ such that $f(\Lambda) \supset \Lambda$. As an example of an application of these results we shall prove that for an RFDE :

$$\dot{x}(t) = F(x_t) \qquad (*)$$

where $F:C^0([-T,0],\mathbb{R}^m) \to \mathbb{R}^m$ is class $C^1$ and $\sup\{\|F(\varphi)\|, \|D_\varphi F\| \mid \varphi \in C^0([-T,0],\mathbb{R}^m)\} < +\infty$ the set $\mathfrak{F}$ of functions $y \in C^0([-T,0],\mathbb{R}^m) \to \mathbb{R}$ such that there exists a bounded solution $x:\mathbb{R} \to \mathbb{R}^m$ of (*) with initial condition $x_0 = y$ has the following property : There exists an integer n such that if $F_0 \subset C^0([-T,0], \mathbb{R}^m)$ is a subspace with $n \le \dim F_0 < \infty$ then $\pi/\mathfrak{F}$ is injective for a residual set of the space of continuous projections $\pi:C^0([-T,0],\mathbb{R}^m) \to F_0$. Moreover n depends only on the numbers T and $\sup\{\|F(\varphi)\|, \|D_\varphi F\| \mid \varphi \in C^0([-T,0],\mathbb{R}^m)\}$.

Another interesting application is RFDE's on compact manifolds. Given a compact boundaryless manifold M and a $C^1$ bounded map $F:C^0([-T,0],M) \to TM$ such that $F(\varphi) \in T_{\varphi(0)}M$ for all $\varphi \in C^0([-T,0],M)$ the RFDE (*) generates a semiflow $\Phi_t$, $t \ge 0$ of $C^1$ maps of $C^0([-T,0],M)$ defined by the property that for all $\varphi \in C^0([-T,0],M)$ the solution $x:[-T,+\infty) \mapsto M$ of (*) with initial condition $\varphi$ (i.e. $x(t) = \varphi(t)$ for

$-T \leq t \leq 0$) satisfies $x_t = \Phi_t \varphi$ for all $t \geq 0$. Define $A(F) = \bigcap_{t\geq 0} \Phi_t(C^0([-T,0],M))$. This set coincides with the set of functions $\varphi \in C^0([-T,0],M)$ such that there exists a solution $x:R \to M$ of (*) with $\varphi$ as initial condition (see [1] Chap. 13). Applying the theorem above we shall prove that the Hausdorff dimension of A(F) is finite.

It is interesting to observe that the topological dimension of A(F) is always larger than or equal to that of M. In fact a stronger property can be easily proved : if $\pi:C^0([-T,0],M) \to M$ is defined by $\pi(\varphi) = \varphi(0)$ then $(\pi/A(F))^*:H^*(M) \to H^*(A(F))$ is injective, where $H^*(.)$ denotes Čech cohomology. Then $H^{\dim M}(A(F))$ is non trivial thus implying [2] that the topological dimension of A(F) is $\geq \dim M$. To prove the injectivity of $(\pi/A(F))^*$ define $\psi:M \to C^0([-T,0],M)$ by $\psi(p)(t) = p$ for all $-T \leq t \leq 0$. Then $\pi\psi$ is the identity and $\psi\pi$ homotopic to the identity. Therefore $\pi^*$ is the identity. But if $i:A(F) \to C^0([-T,0],M)$ denotes the inclusion map we have $(\pi/A(F))^* = (\pi i)^* = i^*\pi^*$ so we have reduced the problem to the injectivity of $i^*$ that, by the continuity property of Čech cohomology, is reduced to showing that if $K_n = \Phi_T^n(C^0([-T,0],M)$ and $i_n:K_n \to C^0([-T,0],M)$ is the inclusion map then $i_n^*$ is injective for all n (recall that $\bigcap_{n\geq 0} K_n = A(F)$). But we can write $\Phi_T^n = i_n g_n$, with $g_n:C^0([-T,0],M) \to K_n$, and then $\Phi_T^{*n} = g_n^* i_n^*$. Now observe that if $\Phi_t^{(\lambda)}$, $t \geq 0$ is the semiflow of maps of $C^0([-T,0],M)$ defined by the RFDE $\dot{x}(t) = \lambda F(x_t)$, then $\Phi_T^{(1)} = \Phi_T$, $\Phi_T^{(0)} = \psi\pi$ and the maps $\psi\pi$ and $\Phi_T$ are homotopic. Hence $g_n^* i_n^* = \Phi_T^{*n} = (\psi\pi)^{*n} = I$ and $i_n^*$ is injective.

Another corollary of the injectivity of $(\pi/A(F))^*$ is that $\pi(A(F)) = M$. Hence for all $p \in M$ there exists a solution $x:\mathbb{R} \to M$ of the RFDE $\dot{x}(t) = F(x_t)$ such that $x(0) = p$.

## 1. General background and statement of the theorem.

Let K be a topological space. We say that K is finite dimensional if there exists an integer n such that for every open covering $\mathfrak{U}$ of K there exists another open covering $\mathfrak{U}'$ refining $\mathfrak{U}$ such that every point of K belongs to at most n+1 sets of $\mathfrak{U}'$. In this case the dimension of K is defined as the minimum n satisfying this property. This concept has several remarkable properties. For instance $\dim \mathbb{R}^n = n$ and if K is a compact finite dimensional space then it is homeomorphic to a subset of $\mathbb{R}^n$ with $n = 2\dim K + 1$ and all its Čech cohomology groups $H^q(K)$ are trivial for $q > \dim K$. Proofs of these properties can be found in [2]. If K is a metric space its Hausdorff dimension is defined as follows : for all $\alpha > 0$, $\varepsilon > 0$ let $\mu_\varepsilon^{(\alpha)}(K)$ be defined as :

$$\mu_\varepsilon^{(\alpha)}(K) = \inf \sum_i r_i^\alpha$$

where the inf is taken over all coverings $B_{r_i}(x_i)$, $i = 1, 2, \ldots$ of K with $r_i < \varepsilon$ for all i ($B_r(x)$ is the open ball of radius r and centre x). Define $\mu^{(\alpha)}(K) = \lim_{\varepsilon \to 0} \mu_\varepsilon^{(\alpha)}(K)$. It is easy to see that if $\mu^{(\alpha)}(K) < \infty$ (resp. $>0$) for some $\alpha$ then $\mu^{(\alpha')}(K) = 0$ if $\alpha' > \alpha$ (resp. $\alpha' < \alpha$). Then we define the Hausdorff dimension $\dim_H(K)$ of K as $\infty$ if $\mu^{(\alpha)}(K) \neq 0$ for all $\alpha$ and

$$\dim_H(K) = \inf\{\alpha > 0 \mid \mu^{(\alpha)}(K) = 0\}$$

otherwise. It is known ([2]) that :

$$\dim(K) \leq \dim_H(K)$$

and $\dim(K) = \dim_H(K)$ when K is a submanifold of a Banach space but little else can be said relating these two dimensions. Moreover in every space $\mathbb{R}^n$ there exist compact zero dimensional subsets with Hausdorff dimension n. For a survey of interesting properties of Hausdorff dimension see [3]. In view of the theorem that will be stated in the next section the following property has great interest :

Lemma 1.1. - If E is a Banach space and $\Lambda \subset E$ is a countable union of compact subsets with $\dim_H(\Lambda) < \infty$ then for every subspace $F \subset E$ with $2 \dim_H(\Lambda) + 1 < \dim F < \infty$ the set of projections $\pi : E \to F$ such that $\pi/\Lambda$ is injective is a residual subset of the space P(F) of projections of E onto F endowed with the norm topology.

Probably this elementary but most enlightening result has been already proved somewhere in the literature on dimension theory. However, as its proof is so simple, it is worthwhile to include it here. Suppose $\Lambda = \bigcup_1^\infty \Lambda_n$ where each $\Lambda_n$ is compact and denote by $P_{n,r}$ the set of projections $\pi \in P(F)$ such that $\operatorname{diam}(\pi^{-1}(p) \cap \Lambda_n) < r$ for all $p \in F$. Clearly $P_{n,r}$ is open and $\bigcap_{n=1}^\infty \bigcap_{m=1}^\infty P_{n,1/m}$ is the set of projections onto F that are injective in $\Lambda$. Hence it is sufficient to prove that every $P_{n,r}$ is dense. Let $A_{n,r} = \{v - w \mid v \in \Lambda_n,\ w \in \Lambda_n,\ \|v-w\| \geq r\}$. Then $\pi \in P_{n,r}$ if and only if

$$\pi^{-1}(0) \cap A_{n,r} = \emptyset .$$

Let p be the canonical homomorphism of E onto E/F. Then $p(A_{n,r}) - \{0\}$ is a countable union of compact sets (namely the sets $\{p(v) | v \in A_{n,r}, \|p(v)\| \geq 1/m\}$). Therefore there exists a sequence $\varphi_i : E/F \to R$ of continuous linear maps such that $\varphi_i(x) = 0$ for all i implies that $x \notin p(A_{n,r}) - \{0\}$. Let

$$A_{n,r,i,j} = \{v \in A_{n,r} \mid |\varphi_i(p(v))| \geq 1/j\}$$

and

$$P_{n,r,i,j} = \{\pi \in P(F) \mid \pi^{-1}(0) \cap A_{n,r,i,j} = \emptyset\} .$$

Then $P_{n,r,i,j}$ is open and $P_{n,r} = \bigcap_i \bigcap_j P_{n,r,i,j}$. Therefore everything is reduced to showing that every $P_{n,r,i,j}$ is dense in P(F). Let $\pi_0 \in P(F)$ and define $\varphi : F - \{0\} \to S = \{v \in F \mid \|v\| = 1\}$ by $\varphi(v) = v/\|v\|$. Then :

$$\dim_H \varphi(\pi_0(A_{n,r})) \leq \sup_{\varepsilon>0} \dim_H \varphi(\pi_0(A_{n,r}) \cap (F - B_\varepsilon(0))) .$$

But $\varphi/(\pi_0(A_{n,r}) \cap (F - B_\varepsilon(0)))$ is Lipschitz. Therefore

$$\dim_H \varphi(\pi_0(A_{n,r}) \cap (F - B_\varepsilon(0))) \leq \dim_H \pi_0(A_{n,r})$$

and then $\dim_H \varphi(\pi_0(A_{n,r})) \leq \dim_H A_{n,r} \leq 2 \dim_H A_n$. Since $\dim_H S = \dim F - 1 > 2 \dim_H(\Lambda)$ there exists $u \in S$ such that $u \notin \varphi(\pi_0(A_{n,r}))$. Given any $\varepsilon > 0$, i and j we define $\pi \in P(F)$ as :

$$\pi = \pi_0 + \varepsilon\, u\, \varphi_i \circ p .$$

Then $\pi \in P_{n,r,i,j}$ because if $\pi(x) = 0$ we have :

$$\pi_0(x) = -\varepsilon\, \varphi_i(p(x))u$$

and if moreover $x \in A_{n,r,i,j}$ we have $\varphi_i(p(x)) \neq 0$ and then $\pi_0(x) \neq 0$. Hence

$$u = -(\varepsilon\, \varphi_i(p(x)))^{-1} \pi_0(x)$$

and then $u = \varphi(u) = \varphi(\pi_0(x)) \in \varphi(\pi_0(A_{n,r}))$ contradicting the choice of u.

Finally let us introduce the concept of limit capacity. Let K be a compact metric space. Define $N(r,K)$ as the minimum number of open balls of radius r needed to cover K. Then define the limit capacity $c(K)$ of K by :

$$c(K) = \limsup_{r\to 0} \frac{\log N(r,K)}{\log (1/r)} .$$

In other words $c(\Lambda)$ is the minimum real number such that for every $\varepsilon > 0$ there exists $\delta > 0$ such that

$$N(r,K) \le \left(\frac{1}{r}\right)^{c(K)+\varepsilon}$$

if $0 < r < \delta$. It is easy to check that

$$\dim_H(K) \le c(K) \quad .$$

Moreover $c(K)$ and $\dim_H(K)$ can be different. Even more we can construct a countable compact subset of $\ell^2$ with infinite limit capacity. To prove this we shall first show that in every Euclidean space $R^m$ endowed with the norm $\|(x_1,\ldots,x_m)\| = (\sum_i x_i^2)^{1/2}$ there exists a countable compact subset $K_m$ with $c(K_m) = m$. To construct this set we start by taking a sequence $a_1 > a_2 > \ldots$ such that $\sum_1^\infty a_n < \infty$. Then take a sequence of cubes $Q_n$ with sides of length $a_n$ such that

$$\|x-y\| \ge a_n \tag{1}$$

for all $x \in Q_n$, $y \in Q_m$, $m \ne n$, $n \ge 1$, and

$$\lim_{n\to+\infty} (\sup_{x\in Q_n} \|x\|) = 0 . \tag{2}$$

In each $Q_n$ take points $x_i^{(n)}$, $1 \le i \le [a_n^{1-n}]^m$ such that its coordinates are separated by intervals of length $a_n^n$. Let $K_m$ be the set $\{0\} \cup \{x_i^{(n)} \mid n \ge 1,\ 1 \le i \le [a_n^{1-n}]^m\}$. By (2) $K_m$ is compact and if we want to cover it by balls of radius $a_n^n/2$ centered at points of $K_m$ we shall need at least one ball centered at each point $x_i^{(n)}$. Hence

$$N(a_n^n/2, K_m) \ge [a_n^{1-n}]^m$$

and then :

$$c(K_m) \geq \limsup_{n\to+\infty} \frac{\log N(a_n^n/2, K_m)}{\log a_n^{-n}/2} \geq \lim_{n\to+\infty} \frac{m \log[a_n^{(1-n)}]}{-\log a_n^n - \log 2} = m .$$

Now take in $\ell^2$ a sequence of countable compact sets $K_m \subset \{x \,|\, 1/m+1 < \|x\| < 1/m\}$ isometric to those constructed above and put $K = \{0\} \cup (\bigcup_1^\infty K_m)$. Then K is countable, compact and $c(K) \geq \sup_m c(K_m) = \infty$.

Now let $E_1, E_2$ be Banach spaces and let $\mathcal{L}(E_1, E_2)$ be the space of bounded linear maps from $E_1$ into $E_2$ endowed with the norm topology. Let $\mathcal{L}_\lambda(E_1, E_2)$ be the space of maps $T \in \mathcal{L}(E_1, E_2)$ that can be decomposed as $T = T_1 + T_2$ with $T_1$ compact and $\|T_2\| < \lambda$. For our next definition we shall need the following lemma :

Lemma 1.2 - For every $L \in \mathcal{L}_\lambda(E_1, E_2)$ there exists a finite dimensional subspace $F \subset E_1$ such that if $L_F : E_1/F \to E_2/L(F)$ is the linear map induced by L then $\|L_F\| < 2\lambda$.

The proof of the lemma is by contradiction. If it is false there exists a sequence $x_1, x_2, \ldots,$ in $E_1$ such that

$$\|x_n\| = 1 \qquad n \geq 1$$
$$\|Lx_n - w\| \geq 2\lambda \qquad n \geq 2$$

for all w in the space spanned by $Lx_1, \ldots, Lx_{n-1}$. Write $L = T + C$ with $\|T\| < \bar{\lambda} < \lambda$ and C compact. Then for all $n > m$ :

$$2\lambda \leq \|Lx_n - Lx_m\| \leq \|Tx_n - Tx_m\| + \|Cx_n - Cx_m\| \leq 2\bar{\lambda} + \|Cx_n - Cx_m\| .$$

Hence $\|Cx_n - Cx_m\| \geq 2(\lambda - \bar{\lambda}) > 0$ for all $n > m$ thus contradicting the existence of a Cauchy subsequence of $Cx_1, Cx_2, \ldots$ .

Definition - If $L \in \mathcal{L}_{\lambda/2}(E_1, E_2)$ we define $\nu_\lambda(L)$ as the minimum integer n such that there exists a subspace $F \subset E$ satisfying $\dim F = n$ and $\|L_F\| < \lambda$.

Theorem 1 - Let E be a Banach space, $U \subset E$ an open set and $f: U \to E$ a $C^1$ map.

If $\Lambda \subset E$ is a compact set such that $f(\Lambda) \supset \Lambda$ and $D_x f \in \mathcal{L}_1(E,E)$ for all $x \in \Lambda$ then

$$c(\Lambda) < \infty .$$

If moreover $D_x f \in \mathcal{L}_{1/4}(E,E)$ for all $x \in \Lambda$ we have

$$c(\Lambda) \le \frac{\log(\nu((2(1+\varepsilon)\lambda + 2K/\lambda\varepsilon)^{\nu})}{\log(1/2\lambda(1+\varepsilon))}$$

where :

$$K = \sup_{x\in\Lambda} \|D_x f\|$$

$$0 < \lambda < 1/2$$

$$0 < \varepsilon < (1/2\lambda) - 1$$

$$\nu = \sup_{x\in\Lambda} \nu_\lambda(D_x f)$$

Observe that the hypothesis $D_x f \in \mathcal{L}_{1/4}(E,E)$ for all $x \in \Lambda$ implies that for some $0 < \lambda < 1/2$ (and then also for all $\lambda < \lambda' < 1/2$) we have that $\nu < \infty$. To see this take for each x a number $0 < \lambda_x < 1/4$ such that $D_x f \in \mathcal{L}_{\lambda_x}(E,E)$. Then there exists a subspace $F(x) \subset E$ with $\dim F(x) < \infty$ such that $\|(D_x f)_{F(x)}\| < 2\lambda_x$. Then for every y in a neighbourhood $U_x$ of x we have $D_y f \in \mathcal{L}_{\lambda_x}(E,E)$ and $\|(D_y f)_{F(x)}\| < 2\lambda_x$, in particular $\nu_{2\lambda_x}(D_y f) \le \dim F(x)$. Then if we take a finite set $x_1, \dots, x_m$ such that $\bigcup_i U_{x_i} \supset \Lambda$ and define $\lambda = 2 \sup_i \lambda_{x_i}$ we obtain

$$\sup_{x\in\Lambda} \nu_\lambda(D_x f) \le \sup_i \dim F(x_i) .$$

Moreover observe that if $D_x f \in \mathcal{L}_1(E,E)$ for all $x \in \Lambda$ then for some power $g = f^n$, $n \ge 1$, of f the property $D_x g \in \mathcal{L}_{1/4}(E,E)$ holds for all $x \in \Lambda_n = \bigcap_0^n f^{-j}(\Lambda)$. To prove this take $0 < \lambda < 1$ such that $D_x f \in \mathcal{L}_\lambda(E,E)$ for all $x \in \Lambda$. This $\lambda$ is obtained by reasoning as in the previous remark. Then $D_x f^n \in \mathcal{L}_{\lambda^n}(E,E)$ for all $x \in \Lambda_n$. Taking n such that $\lambda^n < 1/4$ the property is proved. Therefore Theorem 1 gives an estimate for $c(\Lambda_n)$. But in fact $c(\Lambda) = c(\Lambda_n)$ because $\Lambda_n \subset \Lambda \subset f^n(\Lambda_n)$ thus implying $c(\Lambda_n) \le c(\Lambda) \le c(f^n(\Lambda_n))$. But since $f^n$ is a $C^1$ map, it doesn't increase the capacity of compact sets. Therefore $c(f^n(\Lambda_n)) \le c(\Lambda_n)$ and then $c(\Lambda_n) = c(\Lambda)$.

In the case of maps generated by RFDE's or semilinear parabolic equations their derivatives are compact at every point and the compactness follows from the fact that these maps can be written as a composition of a regularising map and a compact inclusion. More precisely the situation is as follows : E is a Banach space, $U \subset E$ is an open set and $f: U \to E$ is a map such that there exists a subspace $E_0 \subset E$ that when endowed with a norm $|\cdot|$ the inclusion $i:(E_0, |\cdot|) \to (E, \|\cdot\|)$ is compact and $f = i \circ g$ where $g: U \to (E_0, |\cdot|)$ is $C^1$. Then if $\Lambda \subset E$ is compact and $f(\Lambda) \supset \Lambda$, putting

$$k = \sup_{x\in\Lambda} \|D_x g\|$$

we can apply the theorem taking $\varepsilon = 1$ and $\lambda$ of the form $\lambda = k\mu$ where $0 < \mu < \min(1/4, 1/4k)$ and we obtain :

$$c(\Lambda) \le \frac{\log(\nu_{k\mu}(i)(2(2\mu+\|i\|)/\mu)^{\nu_{k\mu}(i)}}{\log(1/4k\mu)} .$$

In particular, if $\sup_{x\in U} \|(D_x g)\| < +\infty$ it follows that there exists m such that $c(\Lambda) \le m$ for every compact $\Lambda \subset U$ that satisfies $f(\Lambda) \supset \Lambda$. Moreover, when $E = U$ and $\sup_x \|(D_x g)\| < \infty$ the map $g:(E, \|\cdot\|) \to (E_0, |\cdot|)$ is Lipschitz and then maps bounded sets to bounded sets. Then if $\Lambda \subset E$ is bounded and $f(\Lambda) \supset \Lambda$ its closure $\bar{\Lambda}$ in E is compact because $\bar{\Lambda} \subset g^{-1}\overline{(i(g(\Lambda)))}$ and $i(g(\Lambda))$ is relatively compact since $g(\Lambda)$ is bounded. Moreover $f(\bar{\Lambda}) \supset \bar{\Lambda}$. Then $c(\bar{\Lambda}) < m$. Now let $\mathfrak{J}$ be the set of points x in E such that there exist $x = x_1, x_2, \ldots \in E$ with $\sup\|x_j\| < \infty$ and $f(x_j) = x_{j-1}$ for all $j \ge 2$. Then $\mathfrak{J} = \cup_n \mathfrak{J}_n$ where $\mathfrak{J}_n$ is the set of points $x \in \mathfrak{J}$ where the sequence $x_1, x_2, \ldots$ can be found satisfying the condition $\sup\|x_j\| \le n$. Then $\mathfrak{J}_n$ is bounded and $f(\mathfrak{J}_n) \supset \mathfrak{J}_n$. Therefore $\dim_H \mathfrak{J} \le \sup_n c(\bar{\mathfrak{J}}_n) \le m$. Combining this result with Lemma 1.1 and observing that m depends only on $\sup_{x\in E} \|D_x g\|$ we obtain the following result :

<u>Corollary</u> - Let E be a Banach space, $E_0 \subset E$ a subspace and $|\cdot|$ a norm on E such that the inclusion $i:(E_0, |\cdot|) \to (E, \|\cdot\|)$ is compact. Then there exists a function $m: \mathbb{R}^+ \to \mathbb{Z}^+$ such that if $g:(E, \|\cdot\|) \to (E_0, |\cdot|)$ is a $C^1$ map with $\sup_{x\in E} \|D_x g\| < \infty$ and $\mathfrak{J}(g)$ is the set of points $x \in E$ such that there exist $x = x_1, x_2, x_3, \ldots \in E$ satisfying $ig(x_j) = x_{j-1}$ for all $j \ge 2$ and $\sup\|x_j\| < \infty$ then for every subspace $F \subset E$ with $m(\sup_x\|D_x g\|) \le \dim F < \infty$ the set of continuous projections $\pi: E \to F$ such that $\pi/F$ is

injective is a residual subset of the space of continuous projections of E onto F endowed with the norm topology.

The result stated in the introduction about RFDE's $\dot{x}(t) = F(x_t)$ where F and its derivative are bounded functions is a straightforward combination of the Corollary and the basic theory of RFDE's. The property $\dim_H A(F) < \infty$ for an RFDE $\dot{x}(t) = F(x_t)$ on a compact boundaryless manifold M can be obtained embedding M in a Euclidean space $\mathbb{R}^k$ and taking a $C^1$ map $\Phi: C^0([-T,0],\mathbb{R}^k) \to \mathbb{R}^k$ such that $\Phi(\varphi) = F(\varphi)$ if $\varphi([-T,0]) \subset M$ and $\Phi(\varphi) = 0$ if $\varphi$ is outside a certain neighbourhood V of the set S of maps $\varphi$ with $\varphi([-T,0]) \subset M$. Moreover let us require that both $\Phi$ and its derivative are uniformly bounded. Applying the previous result to the equation $\dot{x}(t) = \Phi(x_t)$ it follows that $\dim_H A(F) < \infty$. It remains to prove that the extension $\Phi$ exists. Take a bounded neighbourhood U of M in $\mathbb{R}^k$ and a $C^\infty$ retraction $\rho: U \to M$. Let $W = \{\varphi \in C^0([-T,0],\mathbb{R}^k) \mid \varphi([-T,0] \subset U\}$. Define $F_1: W \to \mathbb{R}$ by

$$F_1(\varphi) = 1 - \int_{-T}^{0} \|\rho(\varphi(s)) - \varphi(s)\|^2 ds \ .$$

Then $F_1(\varphi) = 1 \Leftrightarrow \varphi \in S$ and $F_1(\varphi) < 1$ if $\varphi \notin S$. For every $0 < \varepsilon < 1$ let $W_\varepsilon = F_1^{-1}([1-\varepsilon,\infty))$. Fix some $0 < \varepsilon < 1$ and take $\psi: \mathbb{R} \to \mathbb{R}$ class $C^\infty$ satisfying $\psi(t) = 1 \Leftrightarrow t \geq 1$ and $\psi(t) = 0 \Leftrightarrow t \leq 1-\varepsilon/2$. Define $F_2: C^0([-T,0],\mathbb{R}^k) \to \mathbb{R}$ as

$$F_2(\varphi) = \psi(F_1(\varphi))$$

if $\varphi \in W_\varepsilon$ and $F_2(\varphi) = 0$ if $\varphi \notin W_\varepsilon$. Then $F_2$ is $C^\infty$ and satisfies $F_2(\varphi) \leq 1$ for all $\varphi$ and $F_2(\varphi) = 1 \Leftrightarrow \varphi \in S$. Finally define $\Phi$ as $\Phi(\varphi) = 0$ when $\varphi \notin W$ and

$$\Phi(\varphi) = F_2(\varphi) F(\rho \circ \varphi)$$

when $\varphi \in W$.

## 2. Proof of the theorem.

For the proof of the theorem we need the following lemma :

Lemma 2.1 - For every finite dimensional subspace $F \subset E$ we have

$$N(r_1, B^F_{r_2}(0)) \leq m 2^m \left(1 + \frac{r_1}{r_2}\right)^m$$

for all $r_1 > 0$, $r_2 > 0$, where $m = \dim F$, $B^F_{r_2}(0) = \{v \in F \mid \|v\| \le r_2\}$ and $N(r_1, B^F_{r_2}(0))$ is calculated in the metric space F.

Proof. Let $E_m$ be the space $\mathbb{R}^m$ endowed with the norm $|(x_1,\ldots,x_m)| = \sup_i |x_i|$. If $\tilde{B}_r(0) = \{x \in E_m \mid |x| \le r\}$ we have

$$N(r_1, \tilde{B}_{r_2}(0)) \le (1 + \frac{r_1}{r_2})^m .$$

Hence the proof of the lemma is reduced to showing the existence of a linear map $L: E_m \to F$ such that

$$2^{-m}|x| \le \|Tx\| \le m|x|$$

for all $x \in E_m$. To define T take $v_1,\ldots,v_m \in F$ with $\|v_i\| = 1$ for all i and

$$\|v_i + w\| \ge 1 \qquad (*)$$

for all $1 < i \le m$ and w in the space spanned by $v_1,\ldots,v_{i-1}$. Then define T as $T(x_1,\ldots,x_m) = \sum_i x_i v_i$. Obviously $\|Tx\| \le m|x|$ for all x. To prove the other inequality we shall show by induction that :

$$2^{m-j}\|Tx\| \ge |x_j|$$

for all $1 \le j \le m$. It is clear that these inequalities imply $2^m\|Tx\| \ge |x|$. By (*) we have :

$$\|Tx\| = \|x_m v_m + \sum_{i=1}^{m-1} x_i v_i\| \ge |x_m|$$

proving the inequality when $j = m$. Suppose that the inequality holds for $j = m, m-1, \ldots, k+1$. Then by (*) and the induction hypothesis :

$$|x_k| \le \|x_k v_k + \sum_{i=1}^{k-1} x_i v_i\| \le \|\sum_{i=1}^{m} x_i v_i\| + \|\sum_{i=k+1}^{m} x_i v_i\| \le \|Tx\| + \sum_{i=k+1}^{m} |x_i| \le \|Tx\|(1 + \sum_{i=k+1}^{m} 2^{m-i})$$

$$= \|Tx\|(1 + \sum_{i=0}^{m-k-1} 2^i) = 2^{m-k}\|Tx\| .$$

Lemma 2.2 - If $L \in \mathcal{L}(E,E)$ and $F \subset E$ is a subspace with $\dim F = m$ then

$$N((1+\gamma)\lambda r,\ L(B_r(0))) \leq m2^m\left(1 + \frac{\|L\|+\lambda}{\lambda\gamma}\right)^m$$

for all $r > 0$, $\lambda > \|L_F\|$, $\gamma > 0$.

Proof. By the linearity of L it is sufficient to prove the theorem for $r = 1$. Let $\bar{r} = \|L\| + \lambda$. Cover the ball $B_{\bar{r}}(0) \cap L(F)$ by balls $B_{\gamma\lambda}(x_i)$, $i = 1,\dots,k$ with $x_i \in B_{\bar{r}}(0)$ for all i. By Lemma 2.1 we can take :

$$k \leq m2^m\left(1 + \frac{\bar{r}}{\lambda\gamma}\right)^m .$$

The proof will be completed by showing that

$$\bigcup_i B_{(1+\gamma)\lambda}(x_i) \supset L(B_1(0)) .$$

If $\|v\| \leq 1$ we can write Lv as $Lv = v_1 + v_2$ with $\|v_1\| < \lambda$ and $v_2 \in L(F)$. Hence $\|v_2\| \leq \|Lv\| + \|v_1\| \leq \|L\|\,\|v\| + \lambda \leq \|L\| + \lambda = \bar{r}$. Then $\|v_2 - x_i\| \leq \gamma\lambda$ for some i and

$$\|Lv - x_i\| \leq \|v_1\| + \|v_2 - x_i\| \leq \lambda + \gamma\lambda = (1+\gamma)\lambda.$$

Now define $\tilde{N}(r,\Lambda)$ as the minimum number of balls $B_r(x)$, $x \in E$ (and not $x \in \Lambda$ as would be for the definition of $N(r,\Lambda)$) required to cover $\Lambda$. Then

$$N(2r,\Lambda) \leq \tilde{N}(r,\Lambda) \leq N(r,\Lambda) .$$

Hence

$$\limsup_{r\to 0} \frac{\log \tilde{N}(r,\Lambda)}{\log(1/r)} = c(\Lambda) .$$

Now we are ready to prove the theorem. First take $g = f^n$ such that $D_x g \in \mathcal{L}_{\lambda/2}(E,E)$ for all $x \in \Lambda$ and some $0 < \lambda < 1/2$ and

$$\nu = \sup_{x\in\Lambda} \nu_\lambda(D_x f) < \infty .$$

The existence of n and $\lambda$ follows from the remarks after the statement of the theorem. Take $c > 1$ and $\varepsilon > 0$ satisfying $(1+\varepsilon)\lambda c < 1/2$ and $r_0 > 0$ such that

$$g(B_r(x)) \subset g(x) + (D_x g)B_{cr}(x)$$

for all $x \in \Lambda$, $0 < r < r_0$. By Lemma 2.2 :

$$\tilde{N}((1+\varepsilon)\lambda cr,\ g(B_r(x))) \leq \tilde{N}((1+\varepsilon)\lambda cr,\ (D_x g)B_{cr}(x)) \leq N((1+\varepsilon)\lambda cr,\ (D_x g)B_{cr}(x)) \leq$$

$$\leq \nu 2^{\nu}(1+\frac{\|(D_x g)\|+\lambda}{\lambda\varepsilon})^{\nu} .$$

To simplify the notation put

$$\lambda_0 = (1+\varepsilon)c\lambda$$
$$K = \sup_{x\in\Lambda} \|(D_x g)\|$$
$$\lambda_1 = \nu 2^{\nu}(1 + \frac{K+\lambda}{\lambda\varepsilon})^{\nu} .$$

Then if $\Lambda$ can be covered by balls $B_r(x_1),\ldots,B_r(x_n)$, with $x_i \in \Lambda$, $1 \leq i \leq n$, it follows that $\Lambda \subset g(\Lambda) \subset \bigcup_i g(B_r(x_i))$. But by the inequality above $g(B_r(x_i))$ can be covered by less than $\lambda_1$ balls of radius $\lambda_0 r$. Hence $\Lambda$ can be covered by less than $\lambda_1 n$ balls of radius $\lambda_0 r$. In other words :

$$\tilde{N}(\lambda_0 r, \Lambda) \leq \lambda_1 N(r,\Lambda) \leq \lambda_1 \tilde{N}(r/2,\Lambda)$$

for all $0 < r < r_0$. Then :

$$\tilde{N}(2\lambda_0 r, \Lambda) \leq \lambda_1 \tilde{N}(r,\Lambda)$$

for all $0 < r < r_0/2$. Therefore if $0 < r < \lambda_0 r_0$ we can write $r = (2\lambda_0)^n \bar{r}$ with $\lambda_0 r_0 < \bar{r} < r_0/2$, $n \geq 1$, and apply the last inequality :

$$\tilde{N}(r,\Lambda) = \tilde{N}((2\lambda_0)^n \bar{r},\Lambda) \leq \lambda_1^n \tilde{N}(\bar{r},\Lambda) \leq \lambda_1^n \tilde{N}(\lambda_0 r_0,\Lambda)$$

and then, putting $r_1 = \lambda_0 r_0$ :

$$\frac{\log \tilde{N}(r,\Lambda)}{\log(1/r)} \le \frac{n \log \lambda_1 + \log \tilde{N}(r_1,\Lambda)}{n \log (1/2\lambda_0)} = \frac{\log \lambda_1}{\log(1/2\lambda_0)} + \frac{\log \tilde{N}(r_1,\Lambda)/n}{\log(1/2\lambda_0)} \le$$

$$\le \frac{\log \lambda_1}{\log(1/2\lambda_0)} + \frac{\log \tilde{N}(r_1,\Lambda)/(\log r/\log \lambda_0 r_0)}{\log (1/2\lambda_0)} .$$

Taking lim sup when $r \to 0$ we obtain :

$$c(\Lambda) \le \frac{\log \lambda_1}{\log(1/2\lambda_0)} .$$

Since this inequality holds for all $c > 1$ and $\lim_{c \to 1} \lambda_0 = (1+\varepsilon)\lambda$ we obtain

$$c(\Lambda) \le \frac{\log \lambda_1}{\log(1/2(1+\varepsilon)\lambda)}$$

this is precisely the inequality of the theorem.

## References.


1. J. Hale - Theory of functional differential equations, Springer-Verlag (1977). Applied Mathematical Sciences Vol. 3.

2. W. Hurewicz & H. Wallman - Dimension Theory, Princeton University Press (1948).

3. J.P. Kahane - Mesures et dimensions, Turbulence and the Navier-Stokes equation, Lecture Notes in Mathematics 565 (1976), Springer-Verlag.

4. J. Mallet-Paret - Negatively invariant sets of compact maps and an extension of a theorem of Cartwright, Journal Diff. Eqns. 22 (1976).


R. Mañé : I.M.P.A., Rua Luiz de Camões 68, Rio de Janeiro, R.J., Brazil.

# More Topological Entropy for Geodesic Flows.

Anthony Manning.

In [11] we showed that the topological entropy of the geodesic flow on a compact Riemannian manifold is at least the exponential rate at which volume grows in the universal cover and is precisely this number in case the manifold has non-positive curvature. In the present paper we show how to increase the entropy by large local perturbations of the Riemannian metric. On the other hand such a perturbation must increase the curvature drastically, since we find an upper bound for the entropy in terms of the maximum sectional curvature. Nilmanifolds are a good example of where [11] yields no information because the fundamental group is not of exponential growth. We find that the entropy is zero when a nilmanifold is given an invariant metric.

## 1. Increasing entropy by changing the metric.

The sphere with three open discs removed can be given a metric of constant negative curvature as follows. In the Poincaré disc take two copies of a regular hexagon with all corners right angles.

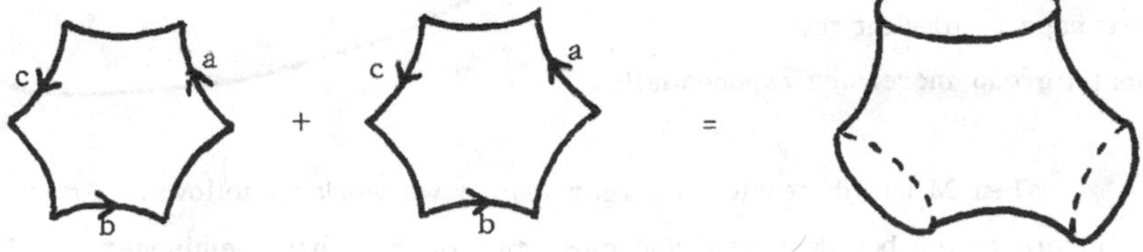


Now join the hexagon along three non-adjacent sides as indicated and we obtain the so-called "pair of pants" P [16] with three geodesic circles as boundary.

Consider these geodesics that remain for all time in P. They cross from the front hexagon to the back one cutting one of the edges a, b and c and then, after time at most the diameter, p say, of P, they must cross back again. After crossing one of a, b, c either of the others is possible for the next crossing. That all such sequences of crossings are realised by a geodesic can be seen by looking at the universal cover of P in the Poincaré disc. Thus our geodesics correspond to points of the subshift of finite type [21] on the three symbols a, b, c given by the matrix $\begin{pmatrix} 0 & 1 & 1 \\ 1 & 0 & 1 \\ 1 & 1 & 0 \end{pmatrix}$. The largest eigenvalue of this matrix is 2 so the subshift has entropy log 2, see [2]. As the time p map of the geodesic flow $\varphi$ has at least all these separated orbits the topological entropy $h(\varphi)$ is at least $p^{-1}\log 2$. This can be made arbitrarily large by taking the diameter p small and the curvature correspondingly large and negative.

Now suppose that M is a two-dimensional Riemannian manifold whose metric we wish to perturb to increase the entropy. Choose a pair of pants P as above small enough to carry the desired entropy. Cut out a small disc somewhere in M and attach a short cylinder with one end sewn to a boundary circle of P. To the other two boundary circles of P attach hemispheres of appropriate constant positive curvature by short connecting cylinders. Finally smooth off the Riemannian metric along the cylinders. Effectively, we have blown a double bubble at one place in M as shown in the diagram. The entropy of the new geodesic flow on M is by [1] at least that on the closed invariant subset of geodesics remaining in P, as required. We have used negative curvature to build up the entropy and positive curvature caps to prevent the fundamental group increasing exponentially.

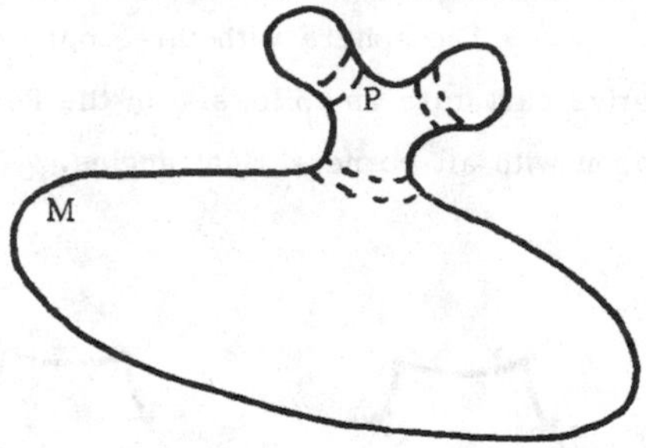


When M has dimension n larger than 2 we work as follows. Attach the three cylinders to the boundary of P and close two of them with hemispheres. Take the product of this with the flat (n-2)-disc $D^{n-2}$. Now we have, $D^2 \times D^{n-2}$ up to diffeomorphism and we can obtain an n-disc by removing a small neighbourhood of the boundary. Glue this in place of a small n-disc cut out of M and smooth off the metric near the join. For the copy of $D^2$ corresponding to the origin in $D^{n-2}$ the geodesics are as before so the entropy is at least its value $p^{-1}\log 2$ found in $P \times \{0\}$.

Since the entropy of the geodesic flow can be multiplied by any constant simply by multiplying the metric by a constant and consequently speeding up the flow along the same orbits, it is natural to normalise a Riemannian metric on M by requiring the total volume to be 1. The procedure above for glueing P into M has changed the total volume of M only slightly so the factor by which we must multiply the new metric to restore the volume to 1 will only change the entropy slightly. We have thus proved

Theorem 1. Any compact differentiable manifold of dimension at least 2 admits Riemannian metrics of volume 1 with arbitrarily high values of the topological entropy of the geodesic flow.

Remark 1. This siutation resembles the case of the topological entropy of diffeomorphisms of a manifold $M^n$. Here Shub's Entropy Conjecture [19] proposed a lower bound for the entropy of a diffeomorphism f in terms of the eigenvalues of the map it induces in homology. There was no upper bound in terms of homology because a $C^0$ small perturbation near a fixed point can introduce a horseshoe [21] with entropy log 2 or indeed log n if there are n branches of the horseshoe and we are willing to increase $\|Df\|$ considerably. In fact Kushnirenko [8] found that an upper bound of the entropy of f with respect to an invariant Riemannian volume is $n \log \sup_{x\in M}\|Df_x\|$. In our perturbations of Riemannian metrics we have had to increase the curvature both positive and negative to increase the entropy of the geodesic flow. In the next section we shall find an upper bound for entropy in terms of the maximum curvature.

Remark 2. The 2-sphere $S^2$ with constant positive curvature has zero entropy for its geodesic flow. However the usual situation near an elliptic closed geodesic is to have flow-invariant tori separated by regions containing hyperbolic periodic orbits with homoclinic orbits [7,12,14,17]. These homoclinic orbits will give rise to suspended horseshoes [21] having positive topological entropy. Thus the geodesic flow on a manifold with everywhere positive curvature can have positive topological entropy.

## 2. An upper bound for the entropy.

Theorem 2. Let M be a compact Riemannian manifold of dimension n and volume 1 and suppose for some positive L that $L^2$ is an upper bound for the modulus of the sectional curvature on M. Then the topological entropy $h(\varphi)$ of the geodesic flow $\varphi$ satisfies

$$h(\varphi) \le (n-1)L .$$

Proof. First consider a Jacobi field Y(t) satisfying the Jacobi equation $Y'' = -RY$ along some geodesic. Define the positive function y(t) by

$$y(t) = L^2\|Y(t)\|^2 + \|Y'(t)\|^2 .$$

Then

$$\begin{aligned} |y'| &= |2L^2\langle Y, Y'\rangle + 2\langle Y', -RY\rangle| \\ &\le 2L^2\|Y\|.\|Y'\| + 2L^2\|Y'\|.\|Y\| = 2L(2L\|Y\|\ \|Y'\|) \\ &\le 2L(L^2\|Y\|^2 + \|Y'\|^2) = 2Ly \end{aligned}$$

so that

$$y(t) \le y(0)e^{2Lt} \qquad \text{for } t \ge 0 .$$

See [9] for this type of growth argument. Eberlein [4] showed that each tangent vector v to the (2n-1)-dimensional unit tangent bundle $T_1M$ is moved by $D\varphi_t$ according to the value and first derivative of a corresponding Jacobi field, Y. In fact

$$\|D\varphi_t v\| = \surd(\|Y(t)\|^2 + \|Y'(t)\|^2).$$

Thus $\|D\varphi_t v\|$ grows no faster than $e^{Lt}$ and hence no Lyapunov exponent for $\varphi$ can be greater than L. See [15] for the definition and properties of these exponents. The argument above applied to negative time shows that exponents must be at least -L. Of the 2n-1 exponents along an orbit of $\varphi$ for which they exist one is zero corresponding to the flow direction and the others sum to zero since the Liouville measure is flow-invariant. Hence the sum of the positive exponents must be at most (n-1)L. Now, according to Dinaburg [3], the topological entropy $h(\varphi)$ is the supremum of the entropy $h_\mu$ of $\varphi$ with respect to any regular invariant Borel probability measure $\mu$ and a result of Margulis [15, 18] says that $h_\mu$ is at most the integral with respect to $\mu$ of the sum of the postive exponents. Hence

$$h(\varphi) \le (n-1)L$$

as claimed.

Remark 3. For the case of negative curvature this result is due to Sinai [20].

Remark 4. It follows immediately from Theorem 2 that the real valued function on the space of all Riemannian metrics on a compact manifold M that attributes to a metric the topological entropy of its geodesic flow is continuous at flat metrics. We observed in [11] that this function is continuous when restricted to metrics of non-positive curvature (and indeed the argument there applies to metrics with no focal points). With the $C^r$ topology on metrics it does not seem likely that this function would otherwise be continuous at general metrics for finite r, see [13].

## 3. Nilmanifolds.

Gromov has shown in [5] that every almost flat manifold (one in which the sectional curvature times the square of the diameter is everywhere sufficiently close to zero) is diffeomorphic to a nilmanifold or else a finite cover of it is. (The basic reference for nilmanifolds is [10].) Moreover every compact nilmanifold admits almost flat metrics. Theorem 2 above says that $h(\varphi)$ is small for such Riemannian manifolds. The lower bound we found in [11] for $h(\varphi)$ in terms of the exponential growth rate of volume yields no information when $\pi_1(M)$ has polynomial growth as in the case of a nilmanifold. (It is apparently not known whether a finitely presented group can have growth larger than polynomial but less than exponential.) Gromov has also shown that a finitely generated group of polynomial growth must have a nilpotent subgroup of finite index, [6]. Here we prove

Theorem 3. Let $N/\Gamma$ be a compact nilmanifold with universal cover the nilpotent Lie group N. Give $N/\Gamma$ a Riemannian metric covered by a left invariant metric on N. Then the topological entropy of the geodesic flow is zero.

Proof. Geodesics in $N/\Gamma$ lift to geodesics in N. If v and w belong to the Lie algebra of N, exp tv and exp t(v+w) are two geodesics through the identity element e. Geodesics through any other point x can be obtained by left translation by x and

$$\begin{aligned} d(x \exp tv, x \exp t(v+w)) &= d(\exp tv, \exp t(v+w)) \\ &= d(e, \exp t(v+w)(\exp tv)^{-1}). \end{aligned}$$

Now $\exp^{-1}(\exp t(v+w).(\exp tv)^{-1})$ can be calculated in the Lie algebra by the

Campbell-Hausdorff formula, which gives it as a finite sum of multiple brackets of tv and t(v+w) since all terms involving very long brackets are zero in a nilpotent Lie algebra. Thus the required Jacobi field and its derivative are polynomial functions of t. All Lyapunov exponents are zero and, by the argument of the last section, the topological entropy is zero.

## References.


1. R. Adler, A. Konheim & M. McAndrew, Topological entropy, Trans. Amer. Math. Soc., 114 (1965) 309-319.

2. R. Bowen, Topological entropy and axiom A, in Global Analysis, Proc. Symp. Pure Math., 14 (1970) 23-42.

3. E.I. Dinaburg, On the relations among various entropy characteristics of dynamical systems, Math. USSR Izv., 5 (1971) 337-378.

4. P. Eberlein, When is a geodesic flow of Anosov type? I, J. Diff. Geom., 8 (1973) 437-463.

5. M. Gromov, Almost flat manifolds, J. Diff. Geom., 13 (1978) 231-241.

6. M. Gromov, Groups of polynomial growth and expanding maps, Preprint, IHES., 1980.

7. W. Klingenberg & F. Takens, Generic propertics of geodesic flows, Math. Ann., 197 (1972) 323-334.

8. A.G. Kushnirenko, An upper bound for the entropy of a classical dynamical system, Sov. Math. Dokl., 6 (1965) 360-362.

9. N. Levinson, The growth of solutions of a differential equation, Duke Math. J., 8 (1941) 1-10.

10. A.I. Mal'cev, On a class of homogeneous spaces, Amer. Math. Soc. Transl., (1) 9 (1962) 276-307.

11. A.K. Manning, Topological entropy for geodesic flows, Ann. Math., 110 (1979) 567-573.

12. R. McGehee & K. Meyer, Homoclinic points of area preserving diffeomorphisms, Amer. J. Math., 96 (1974) 409-421.

13. M. Misiurewicz, Diffeomorphism without any measure with maximal entropy, Bull. Acad. Polon. Sci. Sér. Sci. Math. Astronom. Phys., 21 (1973) 903-910.

14. J. Moser, Stable and random motions in dynamical systems, Princeton University Press, Princeton, 1973.

15. Ya. B. Pesin, Characteristic Lyapunov exponents and smooth ergodic theory, Russ. Math. Surveys, 32 no. 4 (1977) 55-114.

16. V. Poenaru, Travaux de Thurston sur les surfaces, Astérisque, 66-67 (1979) 33-55.

17. R.C. Robinson, Generic properties of conservative systems II, Amer. J. Math., 92 (1970), 897-906.

18. D. Ruelle, An inequality for the entropy of differentiable maps, Bol. Soc. Bras. Mat., 9 (1978), 83-87.

19. M. Shub, Dynamical systems, filtrations and entropy, Bull. Amer. Math. Soc., 80 (1974) 27-41.

20. Ya. G. Sinai, The asymptotic behaviour of the number of closed geodesics on a compact manifold of negative curvature, Amer. Math. Soc. Transl., (2) 73 (1968) 227-250.

21. S. Smale, Differentiable dynamical systems, Bull. Amer. Math. Soc., 73 (1967) 747-817.

A.K. Manning, Mathematics Institute, University of Warwick, Coventry, England.

## Controllability of Multi-Trajectories on Lie Groups.

L. Markus.

### 1. Control Dynamics and Chaotic Flows.

In turbulent or chaotic flows an individual trajectory $x(t)$ may not be effectively determined by its initial state $x_0$. This uncertainty can arise in a strictly deterministic system when the behaviour of the trajectories is extraordinarily sensitive to unnoticeably slight errors in the specification of the initial data (for instance, in ergodic flows with high entropy); or the uncertainties can arise in non-deterministic dynamics when the evolution of the trajectories is affected by interference from external influences. In this latter case the external influences $u(t)$ might be stochastic perturbations introduced by natural causes, or $u(t)$ might be control inputs chosen according to deliberate specifications - hence these two non-deterministic theories have much in common. In any of these mathematical models of turbulence chaotic effects are caused by the evolution of many trajectories - that is, a multi-trajectory - reaching out through the state space $M$ from each prescribed initial state $x_0$ of $M$.

In this paper we shall be concerned only with control dynamics as described by ordinary differential systems

$$1) \qquad \frac{dx}{dt} = f(x, u)$$

in some suitable state space $M$. In particular we take $M$ to be a differentiable n-manifold (connected, separable $C^\infty$-manifold without boundary); and the admissible controllers $u(t)$ will be m-vector functions in some suitable function space, say, $u(t) \in L_1([0,T], \mathbb{R}^m)$ (or $u(t)$ in some prescribed dense subspace thereof - for instance, piecewise constant functions) on various finite durations $0 \le t \le T$.

More technically, we define the control dynamical systems on $M$ as a $\mathbb{R}^m$-parametrised tangent vector field on $M$; this is,

$$f : M \times \mathbb{R}^m \to TM$$

is a $C^\infty$-cross section map into the tangent bundle $TM$ of the manifold $M$. Then, for each initial state $x_0 \in M$ and each choice of control function $u(t)$ on $0 \le t \le T$, the differential system on M

$$\dot{x} = f(x, u(t)) \qquad \text{from} \qquad x(0) = x_0$$

has a response or (local) solution trajectory x(t) in M for some future time duration in $[0, T]$.

Because there are many possible choices for the controller u(t), there are correspondingly many responses or trajectories x(t) from $x(0) = x_0$. Thus the set of all such responses from $x_0$ can be called a multi-trajectory from $x_0$, satisfying the given control dynamical system. The first problem to be met in control dynamics concerns the controllability of such control systems - namely, does the multi-trajectory from $x_0$ reach out to every point of the entire state space M, that is, is all M attainable from $x_0$ along the multi-trajectory?

<u>Definition.</u> Consider the control dynamical system

1) $\dot{x} = f(x, u)$ in $C^\infty$ on $M \times \mathbb{R}^m$

with trajectory x(t) envolving from a given initial state $x_0 \in M$ for controllers $u(t) \in L_1([0, T], \mathbb{R}^m)$. The set of all endpoints x(T), of all responses x(t) initiating at $x_0$, is called the <u>attainable set</u> from $x_0$ in duration $T > 0$ :

$$\mathcal{a}_{x_0}(T) = \{x(T) \mid \text{all responses } x(t) \text{ on } [0, T] \text{ from } x(0) = x_0\} .$$

The eventually attainable set from $x_0$ is :

$$\mathcal{a}_{x_0} = \cup_{T \geq 0} \mathcal{a}_{x_0}(T) .$$

The control dynamical system is called (completely) controllable from $x_0 \in M$ at time T in case :

$$\mathcal{a}_{x_0}(T) = M .$$

The control dynamical systems is called eventually controllable from $x_0$ in case :

$$\mathcal{a}_{x_0} = M .$$

Finally, the control dynamical system is called T-controllable on M (controllable in time $T > 0$), or eventually controllable on M just in case the corresponding condition holds for every $x_0 \in M$.

Remark. It is easy to show that a restriction of the admissible controllers to a dense subspace of $L_1([0,T],\mathbb{R}^m)$ yields an attainable set dense in $\mathcal{A}_{x_0}(T)$. Since, in our examples, these attainable sets will all be suitably closed, they all coincide and we usually ignore the exact specification of the class of the controller. In particular, we often use only piecewise constant controllers for simplicity.

## 2. Control Dynamics on Lie Groups : Classical Examples and Results.

The general control dynamical system

1) $\dot{x} = f(x,u)$ from $x(0) = x_0$

on a differentiable manifold M is too general for any general results. That is, while local or perturbation control results might be obtainable, any global control theory would require further geometric or algebraic demands on the state space M. The most basic analysis deals with the case where M is the n-vector space $\mathbb{R}^n$ and f is linear in the control u, that is, the classical linear control system

2) $\dot{x} = Ax + Bu$

where A is a real constant $n \times n$ matrix and $B = [b_1, b_2, \ldots, b_m]$ is a real constant $n \times m$ matrix (so each column $b_j$ belongs in $\mathbb{R}^n$), see example I) later. We shall be interested in generalising the control-linear system 2) to nonlinear state spaces that are the manifolds of Lie Groups, thus generalising the vector group $\mathbb{R}^n$ to diverse Lie Groups $\mathcal{G}$, [1, 2, 5].

Then take the state space of the control dynamical system to be the manifold of a Lie Group $\mathcal{G}$. Each tangent v at the identity e of $\mathcal{G}$ determines a (right) invariant vector field v(x) at points $x \in \mathcal{G}$. Either v or v(x) can stand for an element of the Lie algebra $\mathfrak{g}$ of $\mathcal{G}$, as usual. The dynamical system on $\mathcal{G}$

$$\dot{x} = v(x) \quad , \quad \text{from } x(0) = x_0 \in \mathcal{G}$$

yields integral curves or trajectories that are the (right) cosets of the 1-parameter subgroup exp(tv).

Now let

$$\mathcal{B} : \mathbb{R}^m \to \mathfrak{g} : u \to v_u$$

be a $C^\infty$-map into the real Lie algebra $\mathfrak{g}$, and consider the control dynamical system on $\mathcal{G}$

$$\dot{x} = v_u(x) .$$

If the map $\mathcal{B}$ is linear, so $v_u = u^1 v_1 + \ldots + u^m v_m$ (in terms of the standard basis on $\mathbb{R}^m$), then we obtain the control-linear dynamical system on $\mathcal{G}$

3) $$\dot{x} = u^1 v_1(x) + u^2 v_2(x) + \ldots + u^m v_m(x) .$$

It has sometimes been suggested that the control system 3) serve as an appropriate generalisation of the linear control system 2), see example II later, especially for the case where A = 0. Possible further generalisations lead to the control-linear systems on Lie Groups $\mathcal{G}$ of the form :

$\tilde{3})$ $$\dot{x} = v_0(x) + u^1 v_1(x) + u^2 v_2(x) + \ldots + u^m v_m(x) .$$

While this introduction of the group-invariant vector field $v_0(x)$ might yield a superficial resemblance to the linear system on $\mathbb{R}^n$

2) $$\dot{x} = Ax + Bu \quad \text{or} \quad \dot{x} = Ax + u^1 b_1 + u^2 b_2 + \ldots + u^m b_m ,$$

the appearances are misleading since the vector field Ax on $\mathbb{R}^n$ is not translation-invariant (not a constant vector on $\mathbb{R}^n$), but is instead the infinitesimal generator of a 1-parameter group of automorphisms $x_0 \to e^{tA} x_0$ of the vector group $\mathbb{R}^n$.

Accordingly we propose as an appropriate extension of linear control systems on $\mathbb{R}^n$, the <u>control-linear systems</u> on Lie Groups $\mathcal{G}$ described by :

4) $$\dot{x} = a(x) + u^1 v_1(x) + u^2 v_2(x) + \ldots + u^m v_m(x) .$$

Here $v_j(x)$ are (right) translation-invariant vector fields on $\mathcal{G}$ , as earlier, and $a(x)$ is a vector field on $\mathcal{G}$ which serves as the infinitesimal generator of a 1-parameter group of automorphisms of $\mathcal{G}$ .

In order to simplify our considerations of the control-linear systems 4) we shall assume that the Lie Group $\mathcal{G}$ is given as a Lie subgroup of the group $GL(q, \mathbb{R})$, for some general linear group of real nonsingular $q \times q$ matrices. Then the state X is a real nonsingular $q \times q$ matrix in $\mathcal{G}$ at each time t.

A tangent vector B to $\mathcal{G}$ at the identity I is a (possibly singular or even zero) real $q \times q$ matrix, and the corresponding right-invariant vector field on $\mathcal{G}$ is given by BX. In this notation the control-linear systems of the type 3) on $\mathcal{G}$ have this expression

$$\dot{X} = B_0 X + u^1 B_1 X + \ldots + u^m B_m X \quad \text{from} \quad X(0) = X_0 \in \mathcal{G},$$

for given matrices $B_0, B_1, B_2, \ldots, B_m$ in the Lie algebra $\mathfrak{g}$. In particular, the strictly linear (or homogeneous) systems on $\mathcal{G}$ of the form demanded by 3) correspond to $B_0 = 0$. Before realising the appropriate control-linear system 4) on a matrix Lie group $\mathcal{G}$ , we examine the motivation and present some examples to clarify this conceptual and notational development.

The reasons for investigating control dynamics on Lie groups are both mathematical and physical. From the mathematical viewpoint there is a long-standing superstitious sentiment that every piece of global analysis on linear spaces must also be duplicated in some generalisation on arbitrary Lie groups - since Lie groups are the appropriate geometric and algebraic spaces that are the natural nonlinear generalisations of the vector group $\mathbb{R}^n$. From the physical viewpoint we can recognise several engineering control systems whose state spaces are nonlinear matrix groups. For instance, the attitudes of a rigid rotor spinning about its centroid (say, a tumbling space vehicle) are described as states in the rotation group $SO(3, \mathbb{R})$. In such an example we have the state space $\mathcal{G} = SO(3, \mathbb{R})$ with the Lie algebra $\mathfrak{g}$ of all real $3 \times 3$ skew symmetric matrices.

Next let us turn to examples from the classical literature concerning the controllability of linear systems in $\mathbb{R}^n$, and control systems on matrix Lie groups.

<u>Example I.</u> The theory of linear control systems in $\mathbb{R}^n$ has been developed extensively over the past two decades [7]. Consider the state $x \in \mathbb{R}^n$ satisfying the control dynamics

2) $\dot{x} = Ax + Bu$ or $\dot{x} = Ax + u^1 b_1 + \ldots + u^m b_m$ ,

for constant matrices A and $B = [b_1, b_2, \ldots, b_m]$, as earlier. Let $x_0 \in \mathbb{R}^n$ be the fixed initial state, and for any controller u(t) on $0 \leq t \leq T$ we compute the response

$$x(t) = e^{At}x_0 + e^{At}\int_0^t e^{-As}[u^1(s)b_1 + \ldots + u^m(s)b_m]ds .$$

Then, temporarily ignoring the fixed translation vector $e^{AT}x_0$, and the invertible linear map in $\mathbb{R}^n$ given by the matrix $e^{AT}$, we note the significance of the linear space

$$\{\int_0^T e^{-As}[u^1(s)b_1 + \ldots + u^m(s)b_m]ds\} \subset \mathbb{R}^n .$$

It is easy to see that a spanning set for this space is

$$\{b_1, b_2, \ldots, n_m, Ab_1, \ldots, Ab_m, A^2 b_j, A^3 b_j, \ldots\} , \; j = 1, \ldots, m .$$

In fact, by the Cayley-Hamilton theorem $A^n$ depends linearly on the preceding powers of A, so the spanning set can be terminated with $\{A^{n-1}b_1, \ldots, A^{n-1}b_m\}$. Thus we obtain the classical fundamental theorem :

<u>Theorem.</u>

$\mathcal{Q}_{x_0}(T) = \mathbb{R}^n$ <u>if and only if</u>

dim linear span $\{b_1, \ldots, b_m, Ab_1, \ldots, Ab_m, \ldots, A^{n-1}b_1, \ldots, A^{n-1}b_m\} = \dim \mathbb{R}^n$ .

We note that the algebraic nature of the controllability condition indicates that it does not depend on the choice of $x_0 \in \mathbb{R}^n$, or $T > 0$, or the class of admissible controllers u(t). It is this global algebraic and geometric analysis we seek to generalise to the case of Lie Groups $\mathcal{G}$ other than the vector group $\mathbb{R}^n$.

In order to cast the linear dynamics 2) into the framework of matrix Lie groups we first take a standard isomorphic embedding of $\mathbb{R}^n$ as a Lie subgroup $\mathcal{G}$ in GL(n+1), $\mathbb{R}$), namely

$$\mathbb{R}^n \to \mathcal{G} \subset GL(n+1, \mathbb{R}) : x \to X = \begin{pmatrix} I & x \\ 0 & 1 \end{pmatrix} .$$

In this case the state $x \in \mathbb{R}^n$ is equally well represented by the $(n+1)\times(n+1)$ matrix X.

Define the matrices

$$B_1 = \begin{pmatrix} 0 & b_1 \\ 0 & 1 \end{pmatrix} , \dots , B_m = \begin{pmatrix} 0 & b_m \\ 0 & 1 \end{pmatrix}$$

in the Lie algebra $\mathcal{G}$ of $\mathcal{G}$ , and consider the control-linear system on $\mathcal{G}$ given by

$$\tilde{2}) \qquad \dot{X} = (\tilde{A}X - X\tilde{A}) + u^1 B_1 X + \dots + u^m B_m X \quad \text{from} \quad X(0) = \begin{pmatrix} I & x_0 \\ 0 & 1 \end{pmatrix} .$$

Here $\tilde{A} = \begin{pmatrix} A & 0 \\ 0 & 0 \end{pmatrix}$ does not belong to $\mathcal{G}$ (unless $A = 0$), but the vector field $(\tilde{A}X - X\tilde{A})$ if tangent to $\mathcal{G}$ since it generates the 1-parameter group of automorphisms of $\mathcal{G}$ according to :

$$X_0 \to \Phi_t(X_0) = e^{\tilde{A}t} X_0 e^{-\tilde{A}t}$$

so

$$\left.\frac{d\Phi_t}{dt}\right|_{t=0} = \tilde{A}X_0 - X_0\tilde{A} = \begin{pmatrix} 0 & Ax_0 \\ 0 & 0 \end{pmatrix} .$$

Thus the matrix system $\tilde{2}$) is precisely the image of the vector system 2) under the diffeomorphic map $\mathbb{R}^n$ onto $\mathcal{G}$ .

This completes our comments on the linear control systems of Example I and next we turn to Example II of control-linear systems on Lie groups.

Example II. Consider the control dynamical system on the matrix Lie group $\mathcal{G} \subset GL(q, \mathbb{R})$ :

$$\dot{X} = B_0 X + u^1 B_1 X + \dots + u^m B_m X \quad \text{from } X(0) = X_0 \in \mathcal{G} .$$

Such control systems on Lie groups (although they do not correspond with our class of control-linear systems 4)) have been studied extensively for the past decade [1,5]. The special case $B_0 = 0$ is especially easy to analyse.

Lemma. (Lightning-Fast). Consider the control system.

$$\dot{X} = (u^1(t)B_1 + \dots + u^m(t)B_m)X \quad \text{from } X(0) = X_0 \in \mathcal{G} ,$$

as above (for $B_0 = 0$). Then the attainable set $\mathcal{A}_{X_0}(T)$ does not depend on $T > 0$.

In more detail

$$\mathcal{A}_{X_0}(T) = \mathcal{A}_I(1)X_0 ,$$

where $\mathcal{A}_I(1)$ is the Lie subgroup $\mathcal{G}_0$ of $\mathcal{G}$ whose Lie algebra $g_0$ is generated by $\{B_1, \ldots, B_m\}$.

Proof.

By a linear change of scale on the time axis, and a corresponding magnification of the controllers $u^j(t)$, we find that each point in $\mathcal{A}_I(T)$ is also in $\mathcal{A}_I(1)$, and vice versa. Thus each target point in $\mathcal{A}_I(1)$ can be reached by a "lightning-fast" controlled trajectory in an arbitrarily short time $T > 0$.

The fact that $\mathcal{A}_I(1)$ is a Lie subgroup $\mathcal{G}_0$ of $\mathcal{G}$, which is generated by the exponentials $e^{B_1 t_1}, \ldots, e^{B_m t_m}$ (for constants $t_j \in \mathbb{R}$) and which has the corresponding Lie algebra $g_0$ generated by $\{B_1, \ldots, B_m\}$, is essentially the fundamental discovery of Sophus Lie.

Moreover $\mathcal{A}_{X_0}(1) = \mathcal{A}_I(1)X_0$ is the right coset of the group $\mathcal{G}_0$ through the given point $X_0$, since all the vector fields of the control dynamics are right-invariant on $\mathcal{G}_0$. □

Remarks. To clarify the lemma let us assume, for the moment, that piecewise constant controllers $u^j(t)$ are to be used. Then, taking $u^2(t) = u^3(t) = \ldots = u^m(t) \equiv 0$ and $u^1(t) = \pm 1$ we find the responses (from I) to be the exponential 1-parameter subgroup $X(t) = e^{B_1 t}$ for all $t \in \mathbb{R}$. Then, taking $u^2(t) = \pm 1$, and the other controllers zero, we obtain the trajectory (from $e^{B_1 t_1}$) to be $e^{B_2 t}(e^{B_1 t_1})$. In this way we compute that $\mathcal{A}_I(1)$ must contain all such products of exponentials. Clearly $\mathcal{A}_I(1)$ is a subgroup of $\mathcal{G}$, and by the theory of S. Lie (or the H. Yamabe theorem on arc-connected subgroups of $\mathcal{G}$) we conclude that $\mathcal{A}_I(1)$ is a Lie subgroup $\mathcal{G}_0$ in $\mathcal{G}$, see [5].

The fact that the tangent vector to $e^{B_2 \alpha_2 t} e^{B_1 \alpha_1 t}$ at $t = 0$ is the linear conbination $\alpha_1 B_1 + \alpha_2 B_2$, for real constants $\alpha_j$, shows that the linear span $\{B_1, \ldots, B_m\}$ lies in the tangent space to $\mathcal{G}_0$ at the identity. But the familiar calculation using

$e^{-B_2t} e^{-B_1t} e^{B_2t} e^{B_1t}$ with the geometric "almost parallelogram" construction, shows that the commutator or Lie bracket $[B_1B_2] = B_1B_2 - B_2B_1$ also defines a tangent vector to $\mathcal{G}_0$ at the identity. The Lie theory verifies that the Lie algebra $g_0$, namely Lie span $\{B_1,\ldots,B_m\}$ (that is, smallest algebra containing $B_1,\ldots,B_m$ and closed under the operations of taking linear combinations and Lie bracket products), is precisely the tangent space to $\mathcal{G}_0$ at I.

Finally allow general controllers that are in $L_1([0,T],\mathbb{R}^m)$. In any case, all the tangents to control trajectories at $X_0 \in \mathcal{G}_0$ are just right-translations of vectors in $g_0$ and hence such trajectories X(t) are tangent to the manifold $\mathcal{G}_0$. Thus no control trajectory that meets $\mathcal{G}_0$ can ever leave $\mathcal{G}_0$, and so $\mathcal{Q}_I(1) = \mathcal{G}_0$, even allowing arbitrary integrable controllers. These arguments explain the known result [5] :

<u>Theorem.</u> <u>The control-linear system on the matrix Lie group</u> $\mathcal{G} \subset GL(q,\mathbb{R})$

$$\dot{X} = u^1B_1X + \ldots + u^mB_mX \quad \text{from} \quad X(0) = X_0 \in \mathcal{G}$$

<u>if T-controllable (arbitrary T > 0) in</u> $\mathcal{G}$ <u>if and only if</u>

$$\dim \text{Lie span } \{B_1,\ldots,B_m\} = \dim \mathcal{G} .$$

If we study the nonhomogeneous right-invariant control system on $\mathcal{G}$

$$\dot{X} = B_0X + u^1B_1X + \ldots + u^mB_mX \quad ,$$

then the analysis is greatly complicated by the requirement that (for all $u^j(t) \equiv 0$) the dynamical system $\dot{X} = B_0X$ can be solved only for future times, and hence the trajectory (in the sense of control dymanics) is only a semi-group. It is clear that a sufficient condition for T-controllability, $\mathcal{Q}_I(T) = \mathcal{G}$ , is that

$$\dim \text{Lie span } \{B_1,\ldots,B_m\} = \dim \mathcal{G} .$$

This follows by easy modifications of the "lightning-fast" lemma. Moreover the algebraic condition

$$\dim \text{Lie span } \{B_0, B_1,\ldots,B_m\} = \dim \mathcal{G}$$

is necessary for eventual controllability in $\mathcal{G}$ , and the condition is also sufficient when the group $\mathcal{G}$ is assumed compact [5]. If $\mathcal{G}$ is not compact, then the search for algebraic necessary and sufficient conditions for controllability is very complicated, intricate, and only partially satisfactorily resolved [6].

This completes our resumé of the known theory for the control systems for Example II.

As distinct from the material in sections 1) and 2) that review known results, section 3) will present some new theorems on the controllability of control-linear systems on Lie groups.

## 3. Control-linear Systems on Lie Groups.

We shall consider control-linear systems on Lie group state spaces $\mathcal{G}$ :

$$4) \qquad \dot{x} = a(x) + u^1 v_1(x) + \dots + u^m v_m(x) .$$

For simplicity of exposition we take $\mathcal{G} \subset GL(q, \mathbb{R})$ to be a matrix Lie group and the right-invariant vector fields $v_j(x) = B_j X$ for matrices $B_j$ in the Lie algebra $\mathfrak{g}$ of $\mathcal{G}$ . The vector field $a(x)$ will be assumed to be the infinitesimal generator of a 1-parameter group $\Phi_t$ of automorphisms of $GL(q, \mathbb{R})$ having the special form (on the state x or $X \in \mathcal{G}$ )

$$\Phi_t(X) = e^{At} X e^{-At}$$

(some fixed matrix A in the general Lie algebra $g\ell(q, \mathbb{R})$), so

$$a(x) = \left.\frac{d\Phi_t}{dt}(X)\right]_{t=0} = AX - XA \quad .$$

The requirement that (AX-XA) is tangent to $\mathcal{G}$ (that is, $\Phi_t$ defines automorphisms of $\mathcal{G}$ onto itself), is guaranteed by the condition that (adA) maps $\mathfrak{g}$ into itself (recall that $(\mathrm{ad}A)B = [AB] = AB - BA$ is the infinitesimal generator of the corresponding 1-parameter group of automorphisms of $g\ell(q, \mathbb{R})$) - and we henceforth assume this condition.

Then consider the control-linear system on $\mathcal{G}$

5) $$\dot{X} = (AX - XA) + u^1 B_1 X + \dots + u^m B_m X$$

for initial states $X(0) = X_0 \in \mathcal{G}$. First consider the free dynamics where all $u^j(t) = 0$. Then clearly

$$X(t) = e^{At} X_0 e^{-At} \qquad \text{(at least for } t \geq 0\text{)}.$$

In order to study the control dynamics 5) with arbitrary (say, piecewise constant) controllers, use the method of "variation of constants" and define $C(t)$ by

$$X(t) = e^{At} C(t) e^{-At} \quad .$$

Direct calculation shows that $C(t)$ must satisfy the differential system

$$\dot{C}(t) = (u^1(t) B_1^{(t)} + \dots + u^m(t) B_m^{(t)})\, C \; ,$$

where we define

$$B_j^{(t)} = e^{-At} B_j e^{At} \qquad \text{for each } t \geq 0.$$

Hence $C(t)$, from $C(0) = X_0 \in \mathcal{G}$, satisfies a control system that resembles a homogeneous control-linear system except that the constant vectors $B_j \in \mathfrak{g}$ are replaced by non-constant vectors $B_j^{(t)}$ that vary within $\mathfrak{g}$ - since familiar calculations show

$$e^{-At} B e^{At} = B - t(\operatorname{ad} A)B + \frac{t^2}{2!}(\operatorname{ad}^2 A)B - \dots = e^{-t \operatorname{ad} A} B,$$

using the notation $(\operatorname{ad}^{k+1} A)B = (\operatorname{ad} A)(\operatorname{ad}^k A)B$ .

<u>Lemma 1.</u> <u>Consider all solutions</u> $C(t)$ <u>on</u> $0 \leq t \leq T$ <u>from</u> $C(0) = I$, <u>for the control dynamics in</u> $\mathcal{G}$

$$\dot{C} = (u^1(t) B_1^{(t)} + \dots + u^m(t) B_m^{(t)}) C \quad .$$

<u>Then the attainable set</u> $\{C(T)\}$ <u>has a closure in</u> $\mathcal{G}$

$$\overline{\{C(T)\}} \supset \bigcup_{s \geq 0}^{T} \mathcal{G}_{-s} \quad .$$

<u>Here</u> $$\mathcal{G}_{-s} = e^{-As} \mathcal{G}_0 e^{As}$$
<u>is a Lie group (with Lie algebra</u> $\mathfrak{g}_{-s}$<u>),</u> and $\mathcal{G}_{-s}$ <u>is generated by</u>

$\mathcal{g}_{-s}$ = Lie span $\{B_1^{(s)},\ldots,B_m^{(s)}\}$.

Proof.

If we take $u^j(t) \equiv 0$ for $\varepsilon \le t \le T$, for small positive $\varepsilon$, then the control system approximates

$$\dot{\tilde{C}} = (u^1(t)B_1 + \ldots + u^m(t)B_m)\tilde{C} \quad , \quad \tilde{C}(0) = I \quad ,$$

for the sort duration $0 \le t \le \varepsilon$. As seen earlier, each point $C_1 \in \mathcal{G}_0$ can be reached using a "lightning-fast" control trajectory in this arbitrarily short duration. By standard approximation arguments $\tilde{C}(t)$ is very near to the corresponding control trajectory $C(t)$ on $0 \le t \le \varepsilon$, and hence we conclude that $\mathcal{G}_0$ is contained in the closure $\overline{\{C(\varepsilon)\}}$. But since $\dot{C}(t) = 0$ for $\varepsilon < t \le T$, when the controllers $u^j(t)$ vanish, we find that $C(\varepsilon) = C(T)$ and thus we conclude that $\mathcal{G}_0$ lies in the closure of $\{C(T)\}$.

But the same argument can be employed for any starting time $s \in [0,T)$ to prove that $\mathcal{G}_{-s}$, the Lie group whose Lie algebra is $\mathcal{g}_{-s}$ = Lie span $\{B_1^{(s)},\ldots,B_m^{(s)}\}$, lies in the closure of $\{C(T)\}$. Therefore $\cup_{s\ge 0}^{T-\delta} \mathcal{G}_{-s}$, for each $\delta > 0$, belongs to the closure of $\{C(T)\}$, and by continuity arguments the same holds for $\cup_{s\ge 0}^{T} \mathcal{G}_{-s}$. □

Lemma 2. Consider the control-linear system on $\mathcal{G}$

5) $$\dot{X} = (AX - XA) + u^1 B_1 X + \ldots + u^m B_m X$$

for $X(0) = I$. Then the attainable set on $0 \le t \le T$ has a closure in $\mathcal{G}$

$$\overline{\{X(T)\}} \supset \cup_{s\ge 0}^{T} \mathcal{G}_s \quad .$$

Proof.

Compute with $X(T) = e^{AT}\, C(T)\, e^{-AT}$ so

$$\overline{\{X(T)\}} \supset e^{AT}(\cup_{s\ge 0}^{T} \mathcal{G}_{-s})\, e^{-AT} = \cup_{s\ge 0}^{T} e^{A(T-s)}\, \mathcal{G}_0\, e^{-A(T-s)} \quad .$$

Let $\sigma = T - s$ in the union holding for $T \ge \sigma \ge 0$ to yield

$$\overline{\{X(T)\}} \supset \cup_{\sigma\ge 0}^{T} e^{A\sigma}\, \mathcal{G}_0\, e^{-A\sigma} = \cup_{\sigma\ge 0}^{T} \mathcal{G}_\sigma \quad .$$

as required. □

Remark 1. Since the piecewise constant controllers u(t) on [0,T] are dense in $L_1([0,T], \mathbb{R}^m)$, the closure of $\{X(T)\}$ is the same set regardless of the space of admissible controllers.

2. At first glance it might appear that $\dot{C}(t)$ is tangent to $\mathcal{G}_{-t}$ at each instant $t > 0$ so that the set inclusion should be replaced by set equality. However this is not a correct conclusion since the points of $\mathcal{G}_0$ should serve to lead into (approximately) the right cosets of the other groups $\mathcal{G}_{-t}$ and so the closure of $\{X(T)\}$ properly includes the set $\cup_{\sigma\geq 0}^{T} \mathcal{G}_\sigma$.

But in the important special case where $\mathcal{G}$ is commutative, so its simply-connected covering group in $\mathbb{R}^n$, then $\{X(T)\} = \mathcal{Q}_I(1)$ is a Lie subgroup of $\mathcal{G}$. The same conclusion holds if we assume only that the groups $\mathcal{G}_s$ all are subgroups of some commutative Lie group $\mathcal{G}_\infty$. But these cases then reduce to the linear control system in $\mathbb{R}^n$ (or its projection onto some homomorphic image, like a torus) as described in the prior Example I.

We summarise the results of the two lemmata in the next theorem.

Theorem 1. Consider the control linear system in a matrix Lie group $\mathcal{G} \subset GL(q, \mathbb{R})$ :

$$5) \qquad \dot{X} = (AX - XA) + (u^1 B_1 + \dots + u^m B_m) X ,$$

from $X(0) = I$, as described above. Then, for each $T > 0$, the attainable sets have the closures in $\mathcal{G}$:

$$\bar{\mathcal{Q}}_I(T) \supset \cup_{s\geq 0}^{T} \mathcal{G}_s \quad \text{and} \quad \bar{\mathcal{Q}}_I \supset \cup_{s\geq 0} \mathcal{G}_s .$$

Finally we present some algebraic criteria for the controllability of the control-linear system 5) on $\mathcal{G}$.

Theorem 2. Consider the control-linear system in a matrix Lie group $\mathcal{G} \subset GL(q, \mathbb{R})$ :

$$5) \qquad \dot{X} = (AX - XA) + (u^1 B_1 + \dots + u^m B_m) X$$

from $X(0) = I$, as described above.

<u>If</u>

$$\dim \text{ Lie span } \{B_1, \ldots, B_m\} = \dim \mathcal{G} \quad ,$$

<u>then</u> $\mathcal{A}_I(T) = \mathcal{G}$ <u>for each</u> $T > 0$ <u>so</u> 5) <u>is then T-controllable.</u>

<u>On the other hand if</u> 5) <u>is eventually controllable</u> $\mathcal{A}_I = \mathcal{G}$ ,

<u>then</u>

$$\dim \text{ Lie span } \{B_1, \ldots, B_m (\text{ad}A)B_1, \ldots, (\text{ad}A)B_m, (\text{ad}^2 A)B_j, \ldots, (\text{ad}^{q^2-1}A)B_j\} = \dim \mathcal{G} .$$

**Proof.**

Assume $\{B_1, \ldots, B_m\}$ generate the Lie algebra $\mathfrak{g}$ of the Lie group $\mathcal{G}$. Then by "lightning-fast control" we note that the attainable set from I is dense in the full group $\mathcal{G}$; that is, $\overline{\mathcal{A}}_I(T) = \mathcal{G}$, for each $T > 0$. Moreover each free trajectory, say $X = \varphi_0(t)$ (for $u(t) \equiv 0$) is locally controllable. That is, $\mathcal{A}_{\varphi_0(0)}(t)$ contains a full ball neighbourhood $W(t)$ for $\varphi_0(t)$ for each $t > 0$, and moreover the radius of the neighbourhood $W(t)$ depends only on T (assuming all relevant trajectories lie within some prescribed compact subset of $\mathcal{G}$), see [3, 4].

Select any target state $X_1 \in \mathcal{G}$ and examine the free trajectory backwards from $X_1$ for duration $T/2$ to some point $X_{-1}$. Use the lightning-fast controller to steer I to some $\hat{X}_{-1}$ that is very near to $X_{-1}$ and do this in duration $T/2$. Then the free trajectory $\hat{\varphi}_0(t)$ forwards from $\hat{X}_{-1}$ passes very near to $X_1$ in time $T/2$. By means of the local controllability along $\hat{\varphi}_0(t)$, verify that the corresponding neighbourhood $\hat{W}(T/2)$ contains the target point $X_1$. Hence $\mathcal{A}_I(T)$ contains the target point $X_1$ and so $\mathcal{A}_I(T) = \mathcal{G}$ .

Now assume the eventual controllability that $\mathcal{A}_I = \mathcal{G}$ . Suppose that

$$\mathfrak{g}_\infty = \text{Lie span } \{B_j, (\text{ad}A)B_j, \ldots, (\text{ad}^{q^2-1}A)B_j\} \neq \mathfrak{g}, \qquad \text{for } j = 1, \ldots, m.$$

Then, since (adA) is a linear operator on the (at most) $q^2$-dimensional vector space $\mathfrak{g}$, we conclude that

$$\text{Lie span } \{B_j, e^{s \, \text{ad}A} B_j\} = \mathfrak{g}_\infty, \qquad \text{for all } s \in \mathbb{R} .$$

The Lie group $\mathcal{G}_\infty$ generated by the Lie algebra $\mathfrak{g}_\infty$ is thus a proper subgroup of $\mathcal{G}$ but it contains all the Lie groups $\mathcal{G}_s = e^{As} \mathcal{G}_0 e^{-As}$, in the prior notation.

But then the differential system for C(t)

$$\dot{C} = (u^1(t)B_1^{(t)} + \ldots + u^m(t)B_m^{(t)})C \qquad \text{from } C(0) = I$$

shows that C(t) is always tangent to the manifold $\mathcal{G}_\infty$. Thus, provided $X_0 \in \mathcal{G}_\infty$ we conclude that $C(t) \subset \mathcal{G}_\infty$. Hence $X(t) = e^{At}C(t)e^{-At}$ also lies within $\mathcal{G}_\infty$ and therefore the attainable set $\mathcal{A}_I(T) \subset \mathcal{G}_\infty$. In such a case $\mathcal{A}_I \subset \mathcal{G}_\infty$ which is a proper subgroup of $\mathcal{G}$ ; but this contradicts the supposition that $\mathcal{A}_I = \mathcal{G}$ . Therefore we conclude that

$$\text{Lie span } \{B_j, (\text{ad}A)B_j, (\text{ad}^2 A)B_j, \ldots, (\text{ad}^{q^2-1} A)B_j\} = \mathfrak{g} \;,$$

as required. □

<u>Remark.</u> Assume that

$$\mathfrak{g}_\infty = \text{Lie span } \{B_j, (\text{ad}A)B_j, \ldots, (\text{ad}^{q^2-1} A)B_j\} \neq \mathfrak{g} \;.$$

Then the attainable set from any initial state $X_0 \in \mathcal{G}$ lies within the coset $\mathcal{G}_\infty X_0$. Hence, in this case, the system on $\mathcal{G}$ fails to be eventually controllable from each $X_0 \in \mathcal{G}$ .

## References.


1. R. Brockett, System Theory on Group Manifolds and Coset Spaces, SIAM J. Control (1972) pp. 265-284.

2. R. Brockett & A. Willsky, Some structural properties of automata defined on Groups, Lecture Notes in Computer Science, Vol. 25, pp. 112-118, <u>Category Theory Applied to Computation and Control</u>, Springer-Verlag, N.Y., 1974.

3. H. Hermes, On Local and Global Controllability, SIAM J. Control (1974) pp. 252-261.

4. V. Jurdjevic & H. Sussmann, Controllability of Nonlinear Systems, J. Diff. Eqs. (1972) pp. 95-116.

5. V. Jurdjevic & H. Sussmann, Control Systems on Lie Groups, J. Diff. Eqs. (1972) pp. 313-329.

6. I. Kupka & V. Jurdjevic, "Etude de l'assessibilité pour les systèmes de contrôles bilinéares sur les groupes de Lie semi-simples." Thèse d'état I. Kupka, Dijon 1978.

7. E.B. Lee & L. Markus, <u>Foundations of Optimal Control Theory,</u> Wiley, N.Y., 1967.

L. Markus, Department of Mathematics, University of Minnesota, Minneapolis, Minnesota 55455, U.S.A., and Mathematics Institute, University of Warwick, Coventry, England.

## Characterising diffeomorphisms with modulus of stability one.

W. de Melo, J. Palis* & S.J. van Strien

In the theory of dynamical systems one tries to classify diffeomorphisms according to their orbit structure. One of the most useful equivalence relations in this context is topological conjugacy. In [P1] and [P.S.] it was found that the orbit structure of a diffeomorphism with finite limit set is stable under perturbations if it is "Morse-Smale".

Of course, not all diffeomorphisms are stable. For example, in dimension 2, we have that a diffeomorphism $\Phi_0$ with a saddle-connection, as in figure 1(b) is not conjugate to the nearby diffeomorphisms without the saddle-connections (see figure 1(a) and 1(c)).

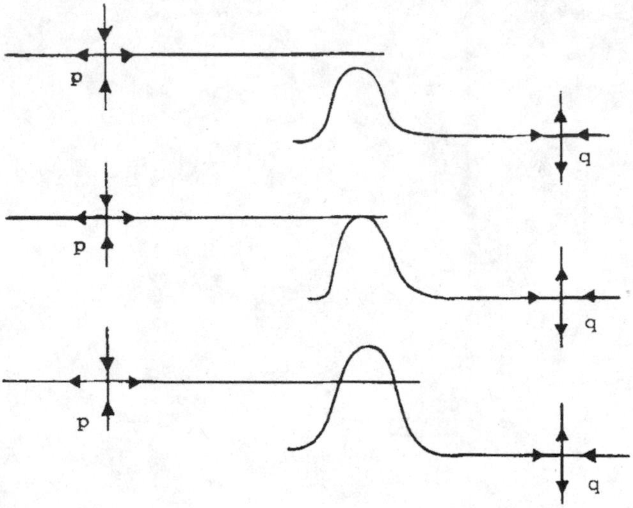


Figure 1(a) $\Phi_\mu$ $\mu < 0$

Figure 1(b) $\Phi_0$

Figure 1(c) $\Phi_\mu$ $\mu > 0$

It is even not true that there are only three equivalence classes of diffeomorphisms near this saddle-connection diffeomorphism $\Phi_0$. If we denote the contracting eigenvalue of $\Phi_0$ at p by a and the expanding eigenvalue at q by b, then for diffeomorphisms as in figure 1(b), $P(\Phi) = \log|a|/\log|b|$ is a topological invariant, i.e., two diffeomorphisms $\Phi$ and $\tilde{\Phi}$ as in figure 1(b) can only be conjugate if $P(\Phi) = P(\tilde{\Phi})$. Such a condition is called a modulus of stability.

In this paper we shall give a characterisation of a large class of diffeomorphisms with modulus of stability one. In a previous paper, [M.P.], it was proved that these

* The second author gratefully acknowledges the financial support of the Stiftung Volkswagenwerk for a visit to the IHES, during which part of this work was developed.

diffeomorphisms, under certain conditions, have only one modulus. The conditions imposed on the diffeomorphisms are quite natural and in fact necessary ones as we show here.

## §1. Statement of Results.

Let M be a compact $C^\infty$ manifold without boundary and Diff(M) the set of $C^\infty$ diffeomorphisms with the $C^\infty$ topology. For $\Phi$ and $\tilde{\Phi}$ in Diff (M) we say that $\Phi$ and $\tilde{\Phi}$ are conjugate if there exists a homeomorphism h on M such that $h \circ \Phi = \tilde{\Phi} \circ h$. If each $\tilde{\Phi}$ sufficiently near $\Phi$ is conjugate to $\Phi$ then $\Phi$ is called stable. If the equivalence classes in small neighbourhoods of $\Phi$ can be parametrised by k real parameters then the modulus of stability is called k, (assume that k is chosen minimal). We also say that the modulus is zero if the number of equivalence classes is finite.

Let p be a hyperbolic periodic point for $\Phi$. Then $\Phi$ has a strictly weakest contracting eigenvalue a at p, if the eigenvalue has multiplicity one and if for any other eigenvalue $\lambda \neq a$, a at p, with $|\lambda| < 1$ one has $|\lambda| < |a|$. In this case there is a unique invariant submanifold $W^{ss}(p)$, called the strong stable manifold, which is tangent to the eigenspace corresponding to eigenvalues with norm smaller that $|a|$, see [H.P.S.]. Moreover there is a uniquely defined foliation $F^{ss}(p)$ contained in $W^s(p)$ with smooth leaves such that $W^{ss}(p)$ is a leaf and such that $\Phi$ maps leaves to leaves (see [H. P.S.]). Similarly for the strictly weakest expanding eigenvalue.

We say that p is s-critical if there is some periodic point z such that $W^u(z)$ intersects some leaf of $F^{ss}(p)$ non-transversally.

When the non-wandering set $\Omega(\Phi)$ is finite, we say that $\Omega(\Phi)$ has no cycles if for any sequence $p_1, \ldots, p_n$ of distinct periodic orbits such that $W^s(p_i) \cap W^u(p_{i+1}) \neq \emptyset$, $1 \leq i \leq n-1$, we have that $W^s(p_n) \cap W^u(p_1) = \emptyset$.

We will consider the rather large class of diffeomorphisms with finite hyperbolic non-wandering set having no-cycles. This class is denoted by $\mathcal{a}$.

Definition (1.1). Let $\mathcal{B} \subset \mathcal{a}$ be the set of diffeomorphisms $\Phi$ such that :

(1) the stable and unstable manifolds of the periodic orbits of $\Phi$ are either transversal or quasi-transversal along their orbits of intersection. (See [N.P.T.] for the definition of quasi-transversality).

(2) for each pair of periodic orbits p,q such that $W^u(p)$ and $W^s(q)$ has an orbit of quasi-transversal intersection, the strictly weakest contracting eigenvalue at p and the strictly weakest expanding eigenvalue at q are defined;

(3) for each such pair of periodic points p and q, there are $C^1$ linearising coordinates near p and q (see [Ste]);

(4) let $W^{cu}(p)$ be an invariant $C^1$ manifold tangent at p to the direct sum of the expanding eigenspace and the weakest contracting eigenspace. These centre-unstable manifolds $W^{cu}(p)$ are not unique, but they are tangent to each other at $W^u(p)$. Therefore we can demand $W^s(q)$ to be transversal to $W^{cu}(p)$. Similarly, we assume that $W^u(p)$ is transversal to $W^{cs}(q)$.

We point out that $\mathcal{A}$-$\mathcal{B}$ is of codimension bigger than one in $\mathcal{A}$ : most (open and dense) smooth one-parameter families of diffeomorphisms in $\mathcal{A}$ miss $\mathcal{A}$-$\mathcal{B}$. Our result really concerns diffeomorphisms in $\mathcal{A}$; we will restrict ourselves to $\mathcal{B}$ for technical reasons.

We can now state a characterisation of diffeomorphisms in $\mathcal{B}$ with modulus of stability one. Let $C \subset \mathcal{B}$ be the subset of $\Phi$ such that :

(1) there is a pair of periodic orbits p and q such that $W^u(p)$ and $W^s(q)$ have a unique orbit of non-transversal intersection; along all other orbits the stable and unstable manifolds meet transversally;

(2) for such a pair of periodic orbits p and q, the strictly weakest contracting eigenvalue at p and the weakest expanding eigenvalue at q are real. Furthermore, p is not s-critical and q is not u-critical.

In [M-P] the following theorem was proved.

<u>Theorem.</u> If $\Phi \in C$ then $\Phi$ has modulus of stability one.

Here we prove the converse.

Main Theorem. If $\Phi \in \mathcal{B}$ has modulus of stability one then $\Phi \in \mathcal{C}$.

We also show that the no-cycle condition is necessary for a diffeomorphism to have modulus of stability one.

The organisation of this paper is as follows. Let $\Phi$ be a diffeomorphism which has modulus of stability one.

In Section 2 we assume that (i) $W^u(p)$ and $W^s(q)$ have a quasi-transversal intersection, (ii) the weakest eigenvalues at p and q exist and are real. We then prove that $W^u(p)$ and $W^s(q)$ have only one orbit of quasi-transversal intersection, and furthermore that p is not s-critical and q is not u-critical. If any of these conditions is not satisfied, we exhibit at least two (real independent) conjugacy invariants and thus $\Phi$ has modulus of stability bigger than one.

In Section 3 we prove that the weakest eigenvalues must be real.

In Section 4 the proof of the main theorem is completed.

In Section 5 we prove that the no-cycle hypothesis is necessary.

Throughout this paper suppose that h is a conjugacy between two diffeomorphisms $\Phi$ and $\tilde{\Phi}$. We denote the analogue of $W^u(p)$ by $W^u(\tilde{p})$, of $W^s(q)$ by $W^s(\tilde{q})$ and so on.

Furthermore we adopt the following notation. Let $c_i, d_i$ be two sequences of numbers. Then $c_i \lesssim d_i$ means that $c_i/d_i$ has a supremum not bigger than one and $c_i \cong d_i$ means that $c_i/d_i$ converges to one.

## §2. Criticallity and number of orbits of non-transversal intersection.

In this section we suppose that

(i) $\Phi$ has periodic points p and q such that $W^u(p)$ and $W^s(q)$ have a quasi-transversal intersection in some orbit O(r).

(ii) the strictly weakest contracting eigenvalue a at p exists and is real. Similarly for the weakest expanding eigenvalue b at q.

(iii) Furthermore suppose that conditions (3) and (4) of Definition (1.1) are satisfied.

§2.a. The strong-stable foliation is preserved.

Theorem (2.1). Let $\Phi$ and $\tilde{\Phi}$ be as above and h a conjugacy between them. Consider the restriction of h to $W^s(p)$; i.e., $h: W^s(p) \to W^s(\tilde{p})$. Then :

(1) The unique strong stable foliations in $W^s(p)$, $W^s(\tilde{p})$ are preserved by h; i.e., h sends leaves of $F^{ss}(p)$ onto leaves of $F^{ss}(\tilde{p})$.

(2) Suppose that $\frac{\log|a|}{\log|b|} \notin \mathbb{Q}$. In this case the induced map $h_*$ on the fibre-space of these foliations is smooth except at $W^{ss}(p)$. It is, in fact, nearly uniquely defined in the following sense. If $a > 0$, take leaves $F_1, F_2$ of $F^{ss}(p)$ in the two components of $W^s(p) - W^{ss}(p)$. Then $h_*$ is completely determined as soon as one chooses the image of two such leaves. For $a < 0$, the image of one leaf determines $h_*$. Similarly for the map induced on the space of leaves of $F^{uu}(q)$.

Remark. Take linearising coordinates $(y_1, \ldots, y_n)$ near p, such that the $y_1$-axis is the eigenspace corresponding to the weakest contracting eigenvalue (i.e. a), and similarly for p. Then for $y \in W^s(p)$

$$\tilde{\pi}_1 \circ h(y) = \alpha \circ (\pi_1(y))^{\delta} ,$$

where $\alpha$ is some constant $\delta = \frac{\log|\tilde{a}|}{\log|a|}$ and $\pi_1(y_1, \ldots, y_n) = y_1$.

For further use assume that $s = \dim W^s(q)$ and $u = \dim W^u(q)$ .

Proof of Theorem (2.1). Assume that $a, b > 0$, otherwise take $\Phi^2$ instead of $\Phi$. We want to restrict our attention to a centre-unstable manifold $W^{cu}(p)$. The problem now is that

$W^{cu}(p)$ is not unique and it has no topological characterisation. Thus $h(W^{cu}(p))$ is a topological manifold which might not coincide with a $C^1$ centre-unstable manifold of p. But $h(W^u(p)) = W^u(\tilde{p})$ and so $h(W^{cu}(p))$ cannot be too far from $W^{cu}(\tilde{p})$. In fact we will show that $h(W^{cu}(p))$ lies in a cone tangent to $W^{cu}(\tilde{p})$ at $W^u(\tilde{p})$. First we will compare distances on M arising from different Riemannian metrics. Let $S \subset W$ be closed submanifolds of M and d be a distance function on M induced by a Riemannian metric. For $c > 0$ and $\tau > 1$ we consider the cone $C_d(S,W;c) = \{x \in M;\ d(x,W) \leq c(d(x,S))^\tau\}$. Since any two Riemannian matrics on M are equivalent it follows that they are cone-wise equivalent : if d' is another distance function there exist $c', c'' > 0$ such that $C_{d'}(S,W;c') \subset C_d(S,W;c) \subset C_{d'}(S,W;c'')$. The set $C_d(S,W;c)$ contains a neighbourhood of W-S and it is "tangent to W at S" in the following sense : if $\gamma:(-\varepsilon,\varepsilon) \to M$ is $C^1$ curve such that $\alpha(0) \in S$ and $\alpha(t) \in C_d(S,W;c)$ for values of t arbitrarily near 0, then $\alpha'(0) \in T_{\alpha(0)}W$.

<u>Lemma</u> 2.2. Let $d_i$, $i = 1,2$, be distance functions on M induced by Riemannian metrics $\langle\ ,\ \rangle^i$. Let $S \subset W$ be closed submanifolds of M such that S has codimension one in W. If $x \in S$ there exists a positive real number $\alpha$ such that for any sequence $x_n \in C_{d_1}(S,W;c)$-S converging to x we have that $\dfrac{d_1(x_n,S)}{d_2(x_n,S)}$ converges to $\alpha$.

<u>Proof.</u> Since it is a local problem we may assume that $M = \mathbb{R}^m$, S and W are subspaces of $\mathbb{R}^m$ and x is the origin. Let $\exp^i_y$ be the exponential mapping induced by the Riemannian metric $\langle\ ,\ \rangle^i$ at y. We have $\exp^i_y(v) = y + v + \varphi_i(y,v)$, where $\dfrac{\varphi_i(y,v)}{\|v\|}$ tends to zero as $\|v\| \to 0$. Here $\|\ \|$ is any norm in $\mathbb{R}^m$. For z near 0 there is a unique $y_i$ near 0 such that $d_i(z,S) = d_i(z,y_i)$. Hence $v_i = (\exp_{y_i})^{-1}(z)$ is orthogonal to S with respect to the inner product $\langle\ ,\ \rangle^i_{y_i}$ and $d_i(z,S) = \|v_i\|^i_{y_i} = (\langle v_i, v_i\rangle^i_{y_i})^{1/2}$. Thus

$$z = y_1 + d_1(z,S)\,\frac{v_1}{\|v_1\|^1_{y_1}} + \varphi_1(y_1,v_2) = y_2 + d_2(z,S)\,\frac{v_2}{\|v_2\|^2_{y_2}} + \varphi_2(y_2,v_2)\ .$$

Let $e_i \in W$, $i = 1,2$, be such that $\langle e_i, e_i\rangle^i_0 = 1$, $e_i$ is orthogonal to S with respect to the inner product $\langle\ ,\ \rangle^i_0$ and $\alpha = \langle e_1, e_2\rangle^1_0$ is positive. Since $\langle e_1, y_j\rangle^1_0 = 0$ for $j = 1,2$ we have

$$d_1(z,S)\,\langle e_1, \frac{v_1}{\|v_1\|^1_{y_1}}\rangle^1_0 + \langle e_1, \varphi_1(y_1,v_1)\rangle^1_0 =$$

$$= d_2(z,S)\,\langle e_1, \frac{v_2}{\|v_2\|^2_{y_2}}\rangle^1_0 + \langle e_1, \varphi_2(y_2,v_2)\rangle^1_0 .$$

Since

$$\frac{v_1}{\|v_1\|^1_{y_1}} \to \pm e_1, \qquad \frac{v_2}{\|v_2\|^2_{y_2}} \to \pm e_2, \qquad \frac{\varphi_i(y_i,v_i)}{\|v_i\|^i_{y_i}} \to 0$$

and $\frac{d_1(z,S)}{d_2(z,S)}$ is bounded it follows that $\frac{d_1(z,S)}{d_2(z,S)}$ converges to $\alpha = \langle e_1, e_2\rangle^1_0$. This finishes the proof of the lemma.

Proposition (2.3). Take a small compact neighbourhood V of r, and some centre-unstable manifolds $W^{cu}(p)$ and $\tilde{W}^{cu}(\tilde{p})$. Let d be a distance function on M induced by a Riemannian metric.

(1) For c sufficiently big and $\tau>1$ sufficiently near 1 the set $h(W^{cu}(p)) \cap V$ is contained in the cone $D^{cu}(\tilde{p}) = \{x \mid d(x, W^{cu}(\tilde{p})) \le c.(d(x, W^{u}(\tilde{p})))^{\tau}\}$.

(2) Let $D^{cu}(p)$ and $D^{cs}(q)$ be cones similar to $D^{cu}(\tilde{p})$. Consider sequences $r^1_n, r^2_n \in D^{cu}(p)$, $r^1_n, r^2_n \to r$ with $\Phi^{-k_n}(r^{(i)}_n) \to s^{(i)}$, $s^{(i)} \in W^s(p) - \{p\}$. We claim that $s^{(i)} \in W^s(p)\backslash W^{ss}(p)$ and $s^1, s^2$ are both contained in the same leaf of $F^{ss}(p)$ if and only if

$$d(r^1_n, W^u(p)) \cong d(r^2_n, W^u(p)) .$$

For such a sequence $r_n$ one has :

$$d(r_n, W^u(p)) \cong \sigma_p . (a)^{k_n} . |\pi_1(s)| ;$$

where $\pi_1(s)$ is the first coordinate of s (assuming that s is contained in the linearising neighbourhood near p as above) and $\sigma_p > 0$ is some "transition constant", depending only on d and r.

Proof. Take a full fundamental neighbourhood S of $W^s(p)$. Since we have that $h(W^{ss}(p)) = W^{ss}(\tilde{p})$, see [N.P.T.], we certainly have that $h(W^{cu}(p)) \cap W^{ss}(\tilde{p}) = \tilde{p}$. Therefore by compactness $h(W^{cu}(p)) \cap S$ is contained in the set $D^{cu}(\tilde{p}) = \{x : d(x, W^{cu}(\tilde{p})) \le C_1 . (d(x, W^u(\tilde{p})))^{\tau}\}$ for $C_1$ big. Since the metrics are cone-wise equivalent, we may assum that the metric defining the distance d is, in a neighbourhood of $\tilde{p}$, induced from the usual metric of $R^n$ by the coordinate system linearising $\tilde{\Phi}$. Since the eigenvalue $\tilde{a}$ is strictly weaker than other contracting eigenvalues at $\tilde{p}$, $D^{cu}(\tilde{p})$ is positively invariant in a small neighbourhood U of $\tilde{p}$, provided $\tau > 1$ is sufficiently near 1. Therefore $h(W^{cu}(p)) \cap U \subset D^{cu}(\tilde{p})$. See figure 2. It is now easy to translate such a relation to any point in the orbit of r (taking a finite number of iterates by $\Phi$ and $\tilde{\Phi}$). This concludes part (1) of the proposition.

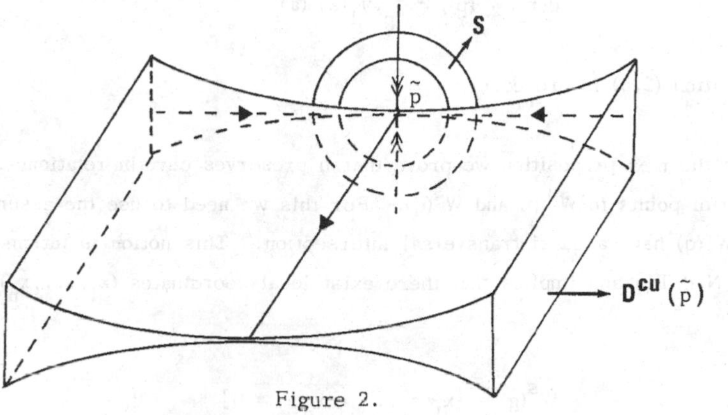


Figure 2.

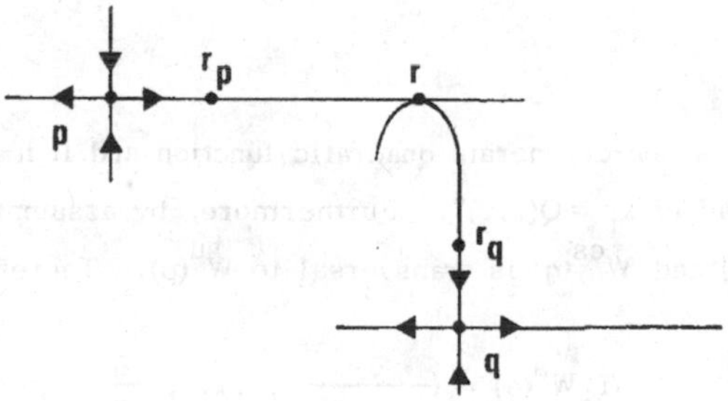


Figure 3.

For some $k, \ell \in \mathbb{N}$ we have that $r_p = \Phi^{-k}(r)$ and $r_q = \Phi^{\ell}(r)$ are contained in the linearising boxes near p, respectively q. In the linearising coordinates $(y_1, \ldots, y_n)$ let $\pi_1(y_1, \ldots, y_n) = y_1$, where the $y_1$-axis is the weakest contracting direction. Since $W^u(p)$ is a codimension one submanifold of $W^{cu}(p)$, it follows from Lemma (2.2) that, for any sequence $r_n \in D^{cu}(p)$, $r_n \to r$,

$$d(r_n, W^u(p)) \cong \sigma'_p |\pi_1(\Phi^{-k}(r_n))|$$

for some constant $\sigma'_p > 0$ independent of the sequence $\{r_n\}$. But since $(y_1,\dots,y_n)$ are linearising coordinates

$$\pi_1(\Phi^{-k_n}(r_n)) = (a)^{k-k_n} \cdot \pi_1(\Phi^{-k}(r_n)) .$$

Since $\Phi^{-k_n}(r_n) \to s$ this implies

$$d(r_n, W^u(p)) \cong \sigma_p \cdot \pi_1(s) \cdot (a)^{k_n} .$$

Thus Proposition (2.3) is proved.

In the next proposition we prove that h preserves certain relations between the distance of points to $W^u(p)$ and $W^s(q)$. For this we need to use the assumption that $W^u(p)$ and $W^s(q)$ have a quasi-transversal intersection. This notion is intrinsically defined in [N.P.T.] and implies that there exist local coordinates $(x_1,\dots,x_n)$ near r such that :

$$W^s(g) = \{x_1 = \dots = x_{n-s} = 0\}$$

$$W^u(p) = \{x_{u+2} = \dots = x_n = 0, \quad x_1 = Q(x_{n-s+1},\dots,x_{u+1})\}$$

and $n-s+1 \leq u+2$ .

Here Q is a non-degenerate quadratic function and if $n-s+1 = u+2$ one should read "$x_1 = 0$" instead of $x_1 = Q(\dots)$". Furthermore, by assumption, $W^{cu}(p)$ is transversal to $W^s(q)$ and $W^{cs}(q)$ is transversal to $W^u(p)$. Therefore we can write

$$T_r W^s(q) = \langle \frac{\partial}{\partial x_{n-s+1}} , \dots , \frac{\partial}{\partial x_n} \rangle ,$$

$$T_r W^u(p) = \langle \frac{\partial}{\partial x_2} , \dots , \frac{\partial}{\partial x_{u+1}} \rangle$$

meaning that the subspaces on the left are generated by the vectors on the right hand side. Since $n-s+1 \leq u+2$ we have that, by the transversality assumptions, there exists a vector v at r such that

$$T_r W^{cs}(q) = \langle v, \frac{\partial}{\partial x_{n-s+1}} , \dots , \frac{\partial}{\partial x_n} \rangle$$

$$T_r W^{cu}(p) = \langle v, \frac{\partial}{\partial x_2}, \ldots, \frac{\partial}{\partial x_{u+1}} \rangle .$$

Using this we will prove the next proposition. This proposition originates from [Str].

Proposition (2.4). (1) If the non-degenerate quadratic function Q has 0 as a saddle-point or if $n-s+1 = u+2$ then there exist distance functions $d, \tilde{d}$ on M and for each small $\alpha > 0$ a point $x^\alpha \in W^{cu}(p) \cap W^{cs}(q)$ such that

$$\alpha = d(x^\alpha, W^u(p)) \cong d(x^\alpha, W^s(q)) ,$$

and $\tilde{d}(h(x^\alpha), W^u(\tilde{p})) \cong \tilde{d}(h(x^\alpha), W^s(\tilde{q}))$ .

(2) If the function Q has a maximum or a minimum (and $n-s+1 \leq u+1$) then there exist a point $x_i^\alpha \in W^{cu}(p) \cap W^{cs}(q)$, for $i = 1,2$, such that

$$\alpha = d(x^\alpha, W^u(p)) \cong d(x^\alpha, W^s(q)) ,$$
$$d(h(x_1^\alpha), W^u(p)) \leq d(h(x_1^\alpha), W^s(q))$$

and

$$d(h(x_2^\alpha), W^u(p)) \geq d(h(x_2^\alpha), W^s(q)) .$$

Proof. As anywhere else in this section we have assumed that $\Phi$ and $\tilde{\Phi}$ are conjugate by the homeomorphism h. Now the index of the quadratic function Q determines the intersection pattern of $W^s(q)$ and $W^u(p)$. Therefore Q and $\tilde{Q}$ have the same index.

(1) First assume that $n-s+1 = u+2$. Let $E = W^{cu}(p) \cap W^{cs}(q)$ and $\tilde{E} = D^{cu}(\tilde{p}) \cap D^{cs}(\tilde{q})$. These sets contain one-dimensional curves, since $T_r W^{cu}(p) \cap T_r W^{cs}(q) = \langle v \rangle$. We may choose a Riemannian metric for which the vector v is orthogonal to $T_r W^u(p)$ and $T_r W^s(q)$. If d denotes the induced distance function, then $d(x_n, W^s(q)) \cong d(x_n, W^u(q))$ for any sequence $x_n \in E$ converging to r. Similarly we can choose a distance function $\tilde{d}$ so that $\tilde{d}(y_n, W^s(\tilde{q})) \cong \tilde{d}(y_n, W^s(\tilde{p}))$ for any sequence $y_n \in \tilde{E}$ converging to $\tilde{r}$. By Proposition (2.3) one has $h(E) \subset \tilde{E}$, which proves the statement in this case.

Now assume that $n-s+1 \leq u+1$ and that Q and $\tilde{Q}$ have saddle points. Consider the set

$$E = (W^{cu}(p) \cap W^{cs}(q)) \cap \{Q(x_{n-s+1}, \ldots, x_{u+1}) = 0\}$$

and choose a distance function d such that $d(x, W^s(q)) \cong d(x, W^u(p))$ for $x \in E$. Let $E_\alpha = E \cap \{d(x, W^u(p)) = \alpha\}$. Since $T_r W^{cu}(p) \cap T_r W^{cs}(q)$ is equal to $\langle v, \frac{\partial}{\partial x_{n-s+1}}, \ldots, \frac{\partial}{\partial x_{u+1}} \rangle$, $E_\alpha$ is not empty for $\alpha$ small and this set is homeomorphic to a cone. Similarly let $\tilde{E} = (D^{cu}(\tilde{p}) \cap D^{cs}(\tilde{q})) \cap \{\tilde{Q}(\tilde{x}_{n-s+1}, \ldots, \tilde{x}_{u+1}) = 0\}$ and $\tilde{d}$ be a distance function such that $\tilde{d}(x, W^s(\tilde{q})) \cong \tilde{d}(x, W^u(\tilde{q}))$ for $x \in \tilde{E}$.

The second step in the proof now is to show that, for $\alpha$ sufficiently small, $h(E_\alpha) \cap \tilde{E} \neq \emptyset$. From Proposition (2.3) it follows that $h(W^{cu}(p)) \subset D^{cu}(\tilde{p})$ so it suffices to show that if $h(E_\alpha)$ is contained in the set $\{\tilde{x} \mid \tilde{Q}(\tilde{x}) > 0\}$ then we get a contradiction. By assumption we had that Q has a saddle-point, i.e. we can write $\tilde{Q}(\tilde{x}_{n-s+1}, \ldots, \tilde{x}_{u+1}) = + \tilde{x}^2_{n-s+1} + \ldots + \tilde{x}^2_\xi - \tilde{x}^2_{\xi+1} - \ldots - \tilde{x}^2_{u+1}$ for some choice of the local coordinates $(\tilde{x}_1, \ldots, \tilde{x}_n)$. Let $V = \{\tilde{x}_1 = \ldots = \tilde{x}_{n-s} = 0, \tilde{x}_{\xi+1} = \ldots = \tilde{x}_{u+1} = \tilde{x}_{u+2} = \ldots = \tilde{x}_n = 0\}$ and $\mathrm{proj}(\tilde{x}_1, \ldots, \tilde{x}_n) = (0, \ldots, 0, \tilde{x}_{n-s+1}, \ldots, \tilde{x}_\xi, 0, \ldots, 0)$. Now $E_0 = W^s(q) \cap W^u(p)$ and, since h is a conjugacy, $h(E_0) = W^s(\tilde{q}) \cap W^u(\tilde{p})$. Hence $\mathrm{proj}(h(E_0))$ is a neighbourhood of 0 in V. This implies that, for $\alpha$ small, $\mathrm{proj}(h(E_\alpha))$ contains a neighbourhood of 0. On the other hand, if $h(E_\alpha)$ is contained in the set $\{\tilde{x}; Q(\tilde{x}) > 0\}$ then $\mathrm{proj}(h(E_\alpha))$ cannot contain $0 \in V$. This contradiction proves Proposition (2.4) (1). To prove the second part of the proposition we let $W = W^{cu}(p) \cap W^{cs}(q)$, $E \subset W$ be a $C^1$ curve through r transversal to $W^s(q)$. Let $W_1$ (resp. $W_2$) be the connected component of $W - W^u(p)$ (resp. $W - W^s(q)$) such that $W_1 \cap W^s(q)$ (resp. $W_2 \cap W^u(p)$) does not intersect a neighbourhood of r as in the figure 4.

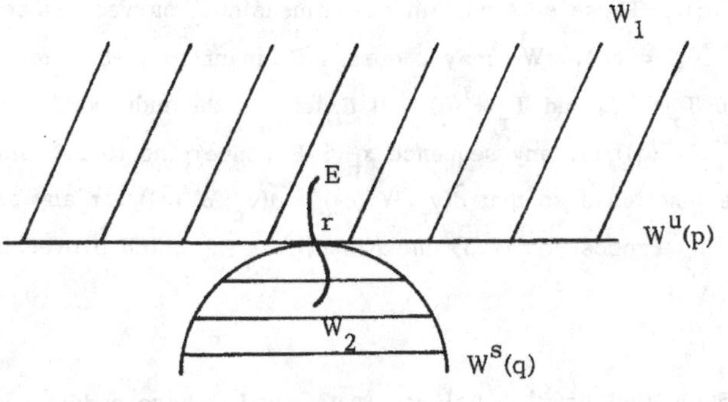


Figure 4.

Similarly we let $\tilde{W} = D^{cu}(\tilde{p}) \cap D^{cs}(\tilde{q})$ and $\tilde{W}_1$ (resp. $\tilde{W}_2$) be the connected component of $\tilde{W}-W^u(\tilde{p})$ (resp. $\tilde{W}-W^s(\tilde{q})$) such that $\tilde{W}_1 \cap W^s(\tilde{q})$ (resp. $\tilde{W}_2 \cap W^u(\tilde{p})$) does not intersect a small neighbourhood of $\tilde{r}$. Clearly $h(W_i) \subset \tilde{W}_i$ and if $\tilde{r}_n^i \to \tilde{r}$, $\tilde{r}_n^i \in \tilde{W}_i$ then

$$d(\tilde{r}_n^1, W^u(\tilde{p})) \leq d(\tilde{r}_n^1, W^s(\tilde{q}))$$
$$d(\tilde{r}_n^2, W^s(\tilde{q})) \leq d(\tilde{r}_n^2, W^u(\tilde{p})) \quad .$$

Since E is transversal to $W^s(q)$ at r we have, for any sequence $r_n \in E$ converging to r,

$$d(r_n, W^s(q)) \cong d(r_n, W^u(p)) \; .$$

Choose $x_i^\alpha \in E \cap W_i$ such that $\alpha = d(x_i^\alpha, W^u(p))$. Clearly $x_i^\alpha$ satisfy condition (2) of Proposition (2.4). This concludes the proof of Proposition (2.4).

Conclusion of the Proof of Theorem (2.1).

We will prove the Theorem for the case that we are in the siutation of Proposition (2.4) (1). The other case is similar. Choose points $s^1, s^2$ in $W^s(p) - W^{ss}(p)$ in the same leaf F of $F^{ss}(p)$ and take two centre-unstable manifolds $W_1^{cu}(p)$, $W_2^{cu}(p)$ such that $s^i \in W_i^{cu}(p)$. Consider a sequence $r_n \to r$, $r_n \in W_1^{cu}(p) \cap W^{cs}(q)$, such that $\Phi^{-n}(r_n) \to s^1$. By Proposition (2.3), we have $\alpha_n = d(r_n, W^u(p)) \cong \sigma_p |a|^n \pi_1(s_1)$. For these $\alpha_n$, we have from Proposition (2.4) that there are sequences $r_n^i \in W_i^{cu}(p) \cap W^{cs}(q)$ such that $\alpha_n = d(r_n^i, W^u(p)) \cong d(r_n^i, W^s(q))$ and $\tilde{d}(h(r_n^i), W^u(\tilde{p})) \cong \tilde{d}(h(r_n^i), W^s(\tilde{q}))$. But then, again by Proposition (2.3), we have $\Phi^{-n}(r_n^i) \to s^i$. Take now a subsequence $k_n$ of intergers such that $\Phi^{k_n}(r_{k_n}^1) \to u \in W^u(q)$. By Proposition (2.3) $d(r_{k_n}^1, W^s(q)) \cong \sigma_q |b|^{-k_n} |\pi_1(u)|$. Since $d(r_{k_n}^1, W^s(q)) \cong d(r_{k_n}^1, W^u(p)) = d(r_{k_n}^2, W^u(p)) \cong d(r_{k_n}^2, W^s(q))$, we have that $d(r_{k_n}^2, W^s(q)) = \sigma_q |b|^{-k_n} |\pi_1(u)|$. But $r_k^2 \in W^{cs}(q)$ and thus by Proposition (2.3) we conclude that $\Phi^{k_n}(r_k^2)$ also converges to $u \in W^u(q)$. Thus $\tilde{\Phi}^{-n}(h(r_n^i)) \to h(s_i) \in W^s(\tilde{p})$ as well as $\tilde{\Phi}^{k_n}(h(r_{k_n}^i)) \to h(u) \in W^u(\tilde{q})$. From this and Proposition (2.3), we have that $\tilde{d}(h(r_{k_n}^1), W^s(\tilde{q})) \cong \tilde{d}(h(r_{k_n}^2), W^s(\tilde{q}))$ and from this, $\tilde{d}(h(r_{k_n}^1), W^u(\tilde{p})) \cong \tilde{d}(h(r_{k_n}^2), W^u(\tilde{p}))$. The last equivalence implies by Proposition (2.3) that $|\tilde{\pi}_1(h(s_1))| = |\tilde{\pi}_1(h(s_2))|$ and by continuity of h, $\tilde{\pi}_1(h(s_1)) = \tilde{\pi}_1(h(s_2))$. This finishes part (1) of Theorem (2.1). Let us prove part (2). Since $\frac{\log|a|}{\log|b|} \notin \mathbb{Q}$, we have as in [M] that for any given $u \in W^{cs}(q) \cap W^u(q)$ there exists a sequence of integers $k_n$ such that $\Phi^{k_n}(r_n^1) \to u$. In fact this follows from the relation $\sigma_p|\pi_1(s_1)||a|^{k_n} \cong \sigma_q|\pi_1(u)||b|^{-k_n}$ obtained above. But then $\tilde{\sigma}_{\tilde{p}}|\tilde{\pi}_1(h(s_1))||\tilde{a}|^{k_n} \simeq \tilde{\sigma}_{\tilde{p}}|\tilde{\pi}_1 h(u)||\tilde{b}|^{-k_n}$. But since $P(\Phi) = P(\tilde{\Phi})$, we can write $\tilde{a} = a^\delta$ and $\tilde{b} = b^\delta$ where $\delta = \frac{\log \tilde{a}}{\log a}$. Therefore, $\tilde{\pi}_1(h(u)) = K(\pi_1(u))^\delta$ for all $u \in W^{cs}(q) \cap W^u(p)$, where

$$K = \left(\frac{\sigma_q}{\sigma_p}\right)^\delta \frac{\tilde{\sigma}_{\tilde{p}}|\tilde{\pi}_1(h(s_1))|}{\tilde{\sigma}_{\tilde{q}}|\pi_1(s_1)|} .$$

This proves part (2) and the Remark following Theorem (2.1) for the induced map on the space of leaves of $F^{uu}(q)$. Similarly for the map induced on the space of leaves of $F^{ss}(p)$.

## §2.b. Moduli due to orbits of non-transversal intersection.

Theorem (2.5). If $W^u(p)$ and $W^s(q)$ have k orbits of tangency (quasi-transversal) then there are at least k moduli.

Proof. According to the last formula of Section 2.a one has that the conjugacy h as a map on the set of leaves of $F^{ss}(p)$ determines the conjugacy on the set of leaves of $F^{uu}(q)$, for each tangency, namely $\tilde{\pi}_1(h(u)) = K_i(\pi_1(u))^\delta$ for $i = 1,2,\dots,k$. Therefore $K_i = K_1$ for all $1 \le i \le k$. This gives rise to k-1 new topological invariants.

## §2.c. Moduli due to criticallity.

Suppose that $\Phi$ is as above and that z is a hyperbolic periodic point such that $W^u(z)$ intersects some leaf of $F^{ss}(p)$ non-transversally and furthermore that

(1) There exists a strictly weakest contracting eigenvalue c at z. In this section we assume c is real. In §3 we will deal with the case that c is complex.

(2) $W^u(z)$ intersects a leaf F of $F^{ss}(p)$ either transversally or quasi-transversally.

(3) $W^u(z)$ is transversal to $W^s(p)$ and $W^{cu}(z)$ is transversal to each leaf F of $F^{ss}(p)$, for which $W^u(z)$ is quasi-transversal to F.

Notice that the conditions (1), (2), (3) are open-dense conditions.

Theorem (2.6). $\Phi$ and $\tilde{\Phi}$ as above can only be conjugate if

(1) $$\frac{\log a}{\log b} = \frac{\log \tilde{a}}{\log \tilde{b}}$$

(2) $$c = \tilde{c}$$

Proof. The equation (1) is proved in [N.P.T.]. To prove equation (2) let us first look at the two-dimensional case :

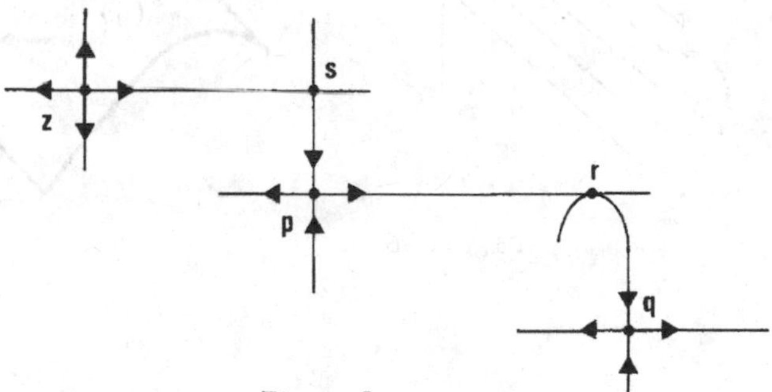


Figure 5.

Since it is proved, in Theorem (2.4) that the conjugacy $h|W^s(p)$ is differentiable near s, we have the following :
Take a sequence $x_n \in W^s(p)$, $x_n \to s$. Then for a proper choice of $x_n$ one has that $\Phi^{-n}(x_n)$ converge to some point $t \in W^s(z)$. Therefore

$$d(x_n, s).|c|^{-n} \simeq \alpha$$

and $$d(h(x_n), h(s)).|\tilde{c}|^{-n} \simeq \beta$$

where $\alpha$ and $\beta$ are positive real numbers.

Since $h/W^s(p)$ is differentiable near s, there exists $\gamma > 0$ such that

$$d(x_n, s) \cong \gamma d(h(x_n), h(s))$$

and therefore $|c| = |\tilde{c}|$. Clearly if $\Phi$ and $\tilde{\Phi}$ are conjugate then c and $\tilde{c}$ also have the same sign.

Now let us deal with the higher dimensional case, see figure 6, (but let $c \in \mathbb{R}$). The conjugacy preserves $W^u(z)$, the foliation $F^{ss}(p)$, the the topological type of the intersection of $W^u(z)$ and a leaf F of $F^{ss}(p)$. Therefore if s is a point where $W^u(z)$

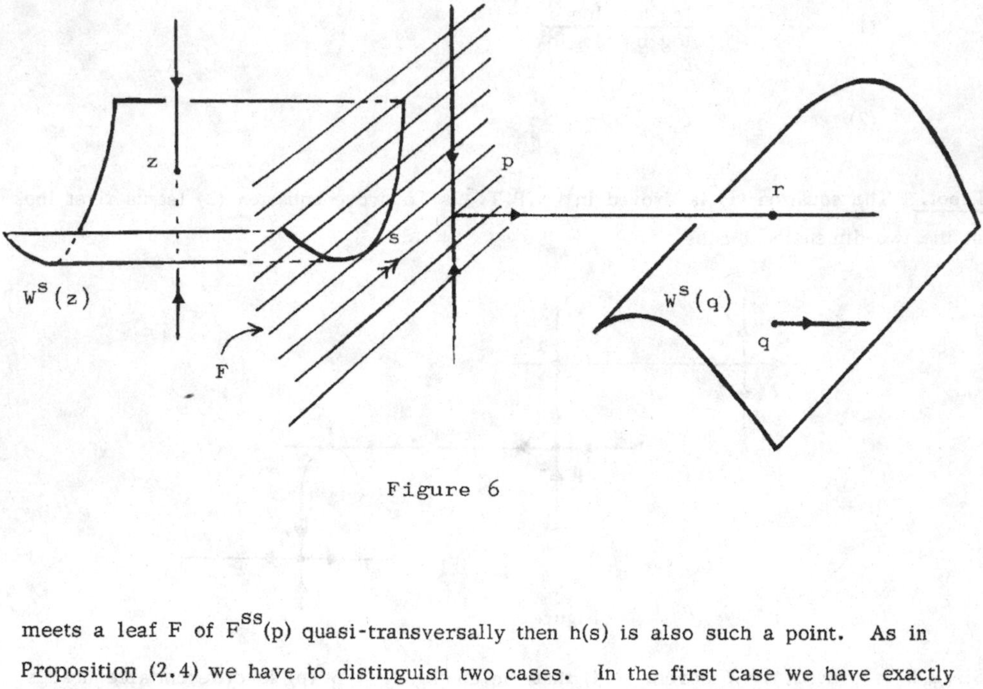


Figure 6

meets a leaf F of $F^{ss}(p)$ quasi-transversally then h(s) is also such a point. As in Proposition (2.4) we have to distinguish two cases. In the first case we have exactly as in Proposition (2.4) (1), a point $x^\alpha \in W^{cu}(z) \cap W^s(p)$ such that $d(x, W^u(z)) \cong d(x, F) = \alpha$ and $d(h(x), W^u(z)) \cong d(h(x), h(F))$, for each $\alpha$ sufficiently small. (We remark here that F has codimension one in $W^{cu}(z)$.) But since h is differentiable as a map on leaves of $F^{ss}$, outside $W^{ss}(p)$ :

$$d(x, F) \cong \gamma d(h(x), h(F)), \quad \gamma > 0 .$$

Then the rest of the proof works exactly as in the two-dimensional case. If we are in the situation of Proposition (2.4) (2) we get two inequalities and, from them, we also get $c = \tilde{c}$.

## §3. Moduli due to non-real eigenvalues.

### §3.a. The eigenvalue a or b is non-real.

Suppose that $W^u(p)$ and $W^s(q)$ have a quasi-transversal intersection as above and that the weakest contracting eigenvalue a at p and the weakest expanding eigenvalue b at q exist. Furthermore suppose that either a or b (or both) are non-real. One may assume that $\theta = \arg(a)$, $\psi = \arg(b)$ are in the interval $[0, \pi]$.

**Theorem (3.1).** For conjugate diffeomorphisms $\Phi$ and $\tilde{\Phi}$ as above

(1) $$\frac{\log |a|}{\log |b|} = \frac{\log |\tilde{a}|}{\log |\tilde{b}|}$$

(2) $$\theta = \tilde{\theta} \text{ and } \psi = \tilde{\psi} .$$

**Proof.** Assume that $\psi \in (0, \pi)$ and take a point r where $W^u(p)$ and $W^s(q)$ intersect quasi-transversally. As before this implies that $n-s+1 \leq u+2$ and that there are local coordinates $(x_1, \ldots, x_n)$ near r such that :

$$W^s(q) = \{x_1 = \ldots = x_{n-s} = 0\} ,$$

and

$$W^u(p) = \{x_{u+2} = \ldots = x_n = 0, \; x_1 = Q(x_{n-s+1}, \ldots, x_{u+1})\}$$

where one should read $x_1 = 0$ if $n-s+1 = u+2$, and Q is quadratic. Furthermore we have assumed that $W^{cu}(p) \pitchfork W^u(q)$ and $W^{cs}(q) \pitchfork W^u(p)$. We distinguish two cases :

Case 1: $n-s+1 = u+2$ .

In this case $T_r W^u(p) \cap T_r W^s(q) = 0$, and therefore for a small neighbourhood U of q,

$W^u(p) \cap U$ has countably many components which pile up at $W^s(q)$. The component of $W^u(p) \cap U$ which contains $\Phi^k(r)$ is called $W^u_k(p)$.

Assertion : If $(k_j.\psi)$ (mod $2\pi$) has $\ell$ limit points in $[0,2\pi)$ then $W^u_{k_j}(p) \cap U$ accumulates to $\ell$ smooth codimension-one submanifolds of $W^u(q)$ : $P_1,\ldots,P_\ell$. All these manifolds contain $W^{uu}(q)$.

Proof of Assertion : Take linearising coordinates $(y_1,\ldots,y_n)$ near q and let $\pi:U \to W^u(q)$ be the linear projection on $W^u(q)$ along $W^s(q)$. Since $T_rW^u(p) \cap T_rW^s(q) = 0$ and $\dim(T_rW^u(p) + T_rW^s(q)) = n-1$ one has that $\pi(W^u_k(p))$ is a smooth codimension-one submanifold of $W^u(q)$. Now in fact $\pi(W^u_k(p))$ is transversal to the two-dimensional eigenspace $W^c$ corresponding to the weakest expanding eigenvalue b (b is non-real, so $\dim(W^c) = 2$). This last fact follows from the assumption $T_rW^u(p) \pitchfork T_rW^{cs}(q)$ and therefore $\pi(T_rW^u_k(p))$ is transversal to $W^c$ as a subset of $W^u(q)$. But this certainly implies that $\pi(W^u_k(p))$ accumulates to the eigenspace of the strong-expanding eigenvalues $W^{uu}(q)$. Furthermore $\dim(T_rW^c \cap \pi(T_rW^u_k(p))) = 1$, and therefore the manifold $\pi(W^u_k(p))$ is rotated. From all this it follows that if $(k_i.\psi)$ (mod $2\pi$) converges then $W^u_{k_i}(p)$ converges to a plane P in $W^u(q)$, which contains $W^{uu}(q)$. This finishes the proof of the assertion.

Now assume that $k_j$ = integer part $(\frac{2\pi j}{\psi})$. Clearly then $W^u_{k_j}(p)$ converges to some manifold P in $W^u(q)$. If $\Phi$ and $\tilde{\Phi}$ are conjugate then $W^u_{k_j}(\tilde{p})$ must converge to the topological manifold h(P). But this is only possible if $k_j.\psi$ (mod $2\pi$) converges to some constant. By the choice of the sequence $k_j$, we have that $\psi = \tilde{\psi}$ .

Case 2: $n-s+1 < u+2$.
In this case $\dim(T_rW^u(p) \cap T_rW^s(q)) \geq 1$. Let N be a small cell-neighbourhood of r in U and let $W^u_k(p) = \Phi^k(W^u(p) \cap N) \cap U$. As before, assume that in U we have coordinates that linearise $\Phi$ and let $\pi:U \to W^u(q) \cap U$ be the corresponding projection along $W^s(q)$. We call a critical point of $\pi|W^u_k(p)$ a "fold" point and we say that $W^u(p)$ "folds" at $v \in W^u(q)$ if v is accumulated by "fold" (critical) points of $\pi|W^u_k(p)$, $k \in \mathbb{N}$. That is, v is a limit point of critical points of $\pi|W^u_{k_j}(p)$ for some sequence of integers $k_j \to \infty$.

Assertion: If $(k_j.\psi)$ (mod $2\pi$) has $\ell$ limit points in $[0,2\pi)$ then $W^u_{k_j}(p)$ is "folded" at $\ell$ smooth codimension-one manifolds of $W^u(q)$:$P_1,\ldots,P_\ell$.

These "folding manifolds" are topologically characterised in the following way : For $v \in W^u(q)$ then $v \notin P_1 \cup \ldots \cup P_\ell$ if and only if there is a cell-neighbourhood V of v such that for each $v' \in W^u(q) \cap V$ there are arbitrarily small cell-neighbourhoods V' of v' in V such that for j sufficiently big $W^u_{k_j}(p) \cap V'$ and $W^u_{k_j}(p) \cap V$ are homeomorphic.

Proof of Assertion : Since $\dim(T_r W^u(p) + T_r W^s(q)) = n-1$ and $W^u(p)$ intersects $W^s(q)$ quadratically at r, one has that the set of critical values of $\pi|W^u_k(p)$ is a codimension-one submanifold of $W^u(q)$. As in Case 1 one can show that if $(\psi . k_j) \pmod{2\pi}$ converges then the set of critical values of $\pi|W^u_{k_j}(p)$ converges to a plane P as $j \to \infty$. This plane P contains $W^{uu}(q)$ and has codimension one in $W^u(q)$. Using the normal form for the quasi-transversal intersection of $W^u(p)$ and $W^s(q)$ at r mentioned before Proposition (2.4), we get the desired topological characterisation of these "folding" manifolds.

As in Case 1, one can use the topological characterisation of the limit sets P to show that $\psi = \tilde{\psi}$.

## §3.b. Criticallity in the complex case.

Now we deal with the criticallity condition. Suppose that $W^u(p)$ and $W^s(q)$ have a quasi-transversal intersection as above and that $\log|a|/|b|$ is irrational and a and b are real. Furthermore suppose that there is a hyperbolic periodic point z such that $W^u(z)$ intersects some leaf of $F^{ss}(p)$ non-transversally. We also suppose that conditions (2), (3) of §2.c are satisfied but now we assume that c is not real.
As before let $\theta = \arg(c)$ and assume that $\theta \in [0,\pi]$.

Theorem (3.2). For conjugate diffeomorphisms $\Phi$, $\tilde{\Phi}$ as above we have :

(1) $$\frac{\log|a|}{\log|b|} = \frac{\log|\tilde{a}|}{\log|\tilde{b}|}$$

(2) $$\theta = \tilde{\theta}\ .$$

Proof. Let F be a leaf of $F^{ss}(p)$ such that F and $W^u(z)$ are non-transversal. By Theorem (2.1) the leaf F is mapped onto a similar leaf $\tilde{F}$ of $\tilde{F}^{ss}(p)$.

Since by assumption F is quasi-transversal to $W^u(z)$ and $W^{cu}(z)$ is transversal to F we can apply the same method as in Theorem (3.1) to the spaces $\Phi^{-k}(F)$. This will show that $\theta = \tilde{\theta}$.

## §4. Conclusion of the proof of the main theorem.

Let $\Omega(\Phi)$ be finite, hyperbolic and without cycles. Suppose that for each pair of periodic orbits p and q such that $W^u(p)$ and $W^s(q)$ have a non-transversal intersection, conditions (1) to (4) of Definition (1.1) are met (i.e., let $\Phi \in \mathcal{B}$). If all stable and unstable manifolds meet transversally, then $\Phi$ is stable (it is Morse-Smale), see [P], [PS], and so $\Phi$ has modulus of stability zero. Thus if $\Phi$ has modulus of stability one, there must be a pair of periodic orbits p, q whose unstable and stable manifolds have an orbit of quasi-transversal intersection. According to §2, such a non-transversal orbit of intersection must be unique and according to §3 the relevant eigenvalues (strictly weakest contracting and expanding, resp.) must be real. Finally, again in §2, the non-criticallity condition is proved when these eigenvalues are real. This concludes the proof of our main result.

## §5. The no-cycles condition.

Let $\Omega(\Phi)$ be finite and hyperbolic. Suppose that $\Phi$ had modulus of stability one.

Theorem. $\Omega(\Phi)$ has no cycles.

Proof. Suppose $\Omega(\Phi)$ has a cycle. Then we must have some orbit of non-transversal intersection of stable and unstable manifolds along the cycle for otherwise $\Omega(\Phi)$ would not be finite. In fact, we would have transversal homoclinic orbits. Now according to §2, there must be only one orbit $\gamma$ of non-transversal intersection for otherwise $\Phi$ would have modulus of stability bigger than one. But this is again a contradiction because in this case $\gamma \in \Omega(\Phi)$ and $\gamma$ is not periodic and thus $\Omega(\Phi)$ would not be finite.

References

[H.P.S.] M. Hirsch, C. Pugh & M. Shub : Invariant manifolds, Lecture Notes in Math., 583, Springer-Verlag, (1977).

[M] W. de Melo, Moduli of stability of two-dimensional diffeomorphisms, Topology 19, (1980), 9-21.

[M.P.] W. de Melo & J. Palis, Moduli of stability for diffeomorphisms, Conference at Northwestern Univ. on Global Theory of Dynamical Systems, Lecture Notes in Math., 819, Springer-Verlag, (1980), 318-339.

[N.P.T.] S. Newhouse, J. Palis & F. Takens, Stable families of diffeomorphisms, IMPA, preprint.

[P1] J. Palis, On Morse-Smale dynamical systems, Topology 8, (1969) 385-405.

[P2] J. Palis, A differentiable invariant of topological conjugacies and moduli of stability, Asterisque 51, (1978), 335-346.

[P.S.] J. Palis & S. Smale, Structural stability theorems, Proc. Symp. Pure Math. A.M.S. 14 (1970) 223-232.

[Ste] S. Sternberg, On the structure of local homeomorphisms of Euclidean n-spaces II, Amer. Journ. of Math. 80 (1958), 623-631.

[Str] S.J. van Strien, Saddle connections of arcs of diffeomorphisms, moduli of stability, this volume.

W. de Melo & J. Palis, Instituto de Matematica Pura e Aplicada, Rua Luiz de Camoes 68, Rio de Janeiro, R.J. Brazil.

S.J. van Strien, Mathematics Institute, University of Utrecht, Holland.

# Algebraic Kupka-Smale Theory.

J.W. Robbin.

The Kupka-Smale theorem ([3] and [5]) asserts that generically the critical points and periodic orbits of a vector field are hyperbolic and that (again generically) the stable and unstable manifolds of such critical elements intersect transversally. In this context the phrase "generically" means "for a residual (i.e. large) subset of the space of all vector fields". The proof (see e.g. [1]) involves making a sequence of locally supported perturbations and rests heavily on the fact that the space of all vector fields is closed under multiplication by "bump" functions.

We pose the question as to what extend the Kupka-Smale theorem remains true on much smaller spaces of vector fields than the space of all vector fields; e.g. the space of polynomial vector fields of a given degree. In such a context the proof becomes much harder for the effect of a perturbation cannot be localised. We give here some partial results : we restrict attention to critical points; periodic orbits are harder. (The reason is that the equation for a critical point of an algebraic vector field is algebraic while the equation for a periodic point is the projection of an analytic equation.)

Theorems 1 through 4 assert that generically vector fields of a certain kind have only hyperbolic critical points. The four cases considered are : polynomial vector fields of given degree in affine space; polynomial gradients of given degree in affine space; polynomial spherical vector fields of given degree; polynomial spherical gradients of given degree. Theorem 5 gives a sufficient condition for genericity of the property of transversal intersection of stable and unstable manifolds. The condition is hard to verify in particular cases and we are only able to apply it in a few (low dimensional or low degree) situations.

Most of the material on algebraic varieties which we need is proved in [7] and all the transversality theory we use can be found in [1].

Thanks to S. Smale, C. Glenton, and D. Passman for conversations/ encouragement.

§1. Let M be a smooth manifold, $\mathfrak{X}(M)$ be the vector space of smooth vector fields on M, and $\mathcal{Q}$ be an open subset of a finite dimensional vector subspace of $\mathfrak{X}(M)$. We define subsets $\mathcal{Q}_I \supset \mathcal{Q}_{II} \supset \mathcal{Q}_{III}$ of $\mathcal{Q}$ as follows :

$\mathcal{Q}_I$ is the set of all $\xi \in \mathcal{Q}$ such that

(I1) all critical points of $\xi$ are nondegenerate :

$\xi(p) = 0 \Rightarrow \det(D\xi(p)) \neq 0$ ;

(I2) $\xi$ has only finitely many critical points ; and

(I3) there is a neighbourhood $\mathcal{B}$ of $\xi$ in $\mathcal{Q}$ such that any $\xi' \in \mathcal{B}$ has the same number of critical points as $\xi$.

$\mathcal{Q}_{II}$ is the set of all $\xi \in \mathcal{Q}_I$ such that all critical points of $\xi$ are hyperbolic : $\xi(p) = 0 \Rightarrow \operatorname{spec}\ (D\xi(p)) \cap i\mathbb{R} = \emptyset$.

$\mathcal{Q}_{III}$ is the set of all $\xi \in \mathcal{Q}_{II}$ such that stable and unstable manifolds of critical points of $\xi$ intersect transversally : $\xi(p) = 0,\ \xi(q) = 0 \Rightarrow W^u(p) \pitchfork W^s(q)$.

REMARK 1.1 : If M is compact, (I2) and (I3) follow from (I1). The example $M = \mathbb{R}$ and $\mathcal{Q} = \{\xi(x) = ax + 1 : a \in \mathbb{R}\}$ shows that (I1) and (I2) do not imply (I3) in general; critical points can go to infinity.

REMARK 1.2 : By the implicit function theorem given $\xi_0 \in \mathcal{Q}_I$ we may find a neighbourhood $\mathcal{B}$ of $\xi$ in $\mathcal{Q}$ and maps $p_i : \mathcal{B} \to M$ $(i = 1, \ldots, r)$ such that for $\xi \in \mathcal{B}$ the critical points of $\xi$ are precisely $p_1(\xi), \ldots, p_r(\xi)$.

PROPOSITION 1.3 : $\mathcal{Q}_I$ and $\mathcal{Q}_{II}$ are open in $\mathcal{Q}$.
(See [1]).

THEOREM 1. Let $M = \mathbb{R}^m$ and $\mathcal{Q} \subset \mathfrak{X}(M)$ be the space of all polynomial vector fields $\xi = (\xi_1, \ldots, \xi_m)$ where each component $\xi_j$ is a polynomial of degree $\leq k$. Then $\mathcal{Q}_I$ and $\mathcal{Q}_{II}$ are dense in $\mathcal{Q}$.

THEOREM 2. Let $M = \mathbb{R}^m$ and $\mathcal{Q} \subset \mathfrak{X}(M)$ be the space of all vector fields $\xi = \operatorname{grad} h$ where h is a polynomial of degree $\leq k + 1$. Then $\mathcal{Q}_I$ and $\mathcal{Q}_{II}$ are dense in $\mathcal{Q}$.

THEOREM 3. Let $M = S^{m-1}$ and $\mathcal{Q} \subset \mathfrak{X}(M)$ be the space of all vector fields $\xi_{tan}$ where $\xi$

is a polynomial vector field on $\mathbb{R}^m$ with each component homogeneous of degree k and the tangential component $\xi_{tan}$ is given by :

$$\xi_{tan}(x) = \xi(x) - \langle \xi(x), x \rangle x$$

for $x \in S^{m-1}$. Then $\mathcal{A}_I$ and $\mathcal{A}_{II}$ are dense in $\mathcal{A}$.

THEOREM 4. Let $M = S^{m-1}$ and $\mathcal{A} \subset \mathfrak{X}(M)$ be the space of all vector fields $(\text{grad } h)_{tan}$ where h is a homogeneous polynomial of degree k+1. Then $\mathcal{A}_I$ and $\mathcal{A}_{II}$ are dense in $\mathcal{A}$.

Proof of Theorem 1 : Let $\mathcal{A}_{\mathbb{C}}$ denote the complexification of the real vector space $\mathcal{A}$. Consider elements $\zeta \in \mathcal{A}_{\mathbb{C}}$ as vector fields on $\mathbb{C}^m$.

For $\zeta \in \mathcal{A}_{\mathbb{C}}$ let $\tilde{\zeta}$ be its "homogeneisation of degree k". Thus $\tilde{\zeta} : \mathbb{C}^{m+1} \to \mathbb{C}^m$ is characterised by the conditions :

$$\tilde{\zeta}(\lambda \tilde{z}) = \lambda^k \tilde{\zeta}(\tilde{z})$$

and

$$\tilde{\zeta}(1, z_1, \ldots, z_m) = \zeta(z_1, \ldots, z_m)$$

for $\lambda \in \mathbb{C}$, $\tilde{z} = (z_0, z_1, \ldots, z_m) \in \mathbb{C}^{m+1}$. Consider $\tilde{\zeta}$ as a section of the appropriate vector bundle over $\mathbb{C}P^m$; viz. the bundle

$$(\mathbb{C}^{m+1} \setminus \{0\}) \times_{C^*} C^m \to \mathbb{C}P^m$$

where the action of $\mathbb{C}^*$ on $\mathbb{C}^m$ is given by $\lambda(z) = \lambda^{-k} z$ for $\lambda \in \mathbb{C}^*$, $z \in \mathbb{C}^m$.

Let $\mathcal{A}_{\mathbb{C}I}$ be the set of all $\zeta \in \mathcal{A}_{\mathbb{C}}$ such that $\tilde{\zeta}$ has no zero at infinity (i.e. $z_0 = 0$ and $\tilde{\zeta}$ is transverse to the zero section. Thus $\zeta \in \mathcal{A}_{\mathbb{C}}$ fails to be in $\mathcal{A}_{\mathbb{C}I}$ if and only if one of the two following over-determined systems of homogeneous equations has a nontrivi solution :

$$\tilde{\zeta}(0, z_1, \ldots, z_m) = 0 ;$$

$$\tilde{\zeta}(z_0, z_1, \ldots, z_m) = 0, \ \det((\partial \tilde{\zeta}_i / \partial z_j)_{1 \leq i, j \leq m}) = 0 .$$

By the homogeneous result theorem (§121 of [6] p.104) $\mathcal{a}_{\mathbb{C}}\backslash\mathcal{a}_{\mathbb{C}I}$ is an algebraic variety and hence $\mathcal{a}_{\mathbb{C}I}$ is, if non-empty, open, dense, and connected in $\mathcal{a}_{\mathbb{C}}$. But $\mathcal{a}_{\mathbb{C}I}$ is non-empty as it contains the element $\zeta$ given by :

$$\zeta_i(z_1,\ldots,z_m) = f(z_i)$$

where f is a polynomial of degree k with simple roots.

Now by transversality each vector field $\zeta \in \mathcal{a}_{\mathbb{C}I}$ has a finite number of zeros in $\mathbb{C}P^m$ the number of which is a locally constant function of $\zeta$. But $\mathcal{a}_{\mathbb{C}I}$ is connected so the number of zeros is the same for all $\zeta \in \mathcal{a}_{\mathbb{C}I}$. Also none of these is at infinity and the example has exactly $k^n$ zeros. Hence each $\zeta \in \mathcal{a}_{\mathbb{C}I}$ has exactly $k^n$ critical points in $\mathbb{C}^m$.

Now $\mathcal{a}_{\mathbb{C}I} \cap \mathcal{a}$ is dense in $\mathcal{a}$ for otherwise the proper algebraic variety $\mathcal{a}_{\mathbb{C}}\backslash\mathcal{a}_{\mathbb{C}I}$ would intersect the real space $\mathcal{a}$ in an open set which is absurd. Also $\mathcal{a}_{\mathbb{C}I} \cap \mathcal{a} \subset \mathcal{a}_I$ as the transversality condition on $\zeta$ implies that the critical points of $\zeta$ are nondegenerate. This proves $\mathcal{a}_I$ is dense in $\mathcal{a}$

Now chose $\xi_0 \in \mathcal{a}_I$ and by the remark 1.2 let $p_1(\xi),\ldots,p_r(\xi)$ be the critical points of $\xi$ near $\xi_0$. Define inductively :

$$\xi_j = \xi_{j-1} + \varepsilon_j\eta_j$$

where $\varepsilon_j \in \mathbb{R}$ and $\eta_j$ is given by

$$\eta_j(x) = x - p_j(\xi_{j-1}).$$

Thus at $p = p_j(\xi_{j-1})$ we have :

$$D\eta_j(p) = I$$

so

$$\text{spec } (D\xi_j(p)) = \text{spec } (D\xi_{j-1}(p)) + \varepsilon_j.$$

Choose $\varepsilon_j$ so as to make $p_j$ a hyperbolic critical point of $\xi_j$( all but at most m real values will do this) and so small as not to destroy the hyperbolicity of $p_1, \dots, p_{j-1}$ .

This completes the proof of Theorem 1; exactly the same argument proves Theorem 2 (note that the example was a gradient vector field).

Proof of Theorem 3 : Complexify and compactify as in Theorem 1 :

$$S_{\mathbb{C}}^{m-1} = \{z \in \mathbb{C}^m : \langle z,z\rangle = 1\}$$

$$\zeta_{tan}(z) = \zeta(z) - \langle z,z\rangle z$$

$$\bar{S}_{\mathbb{C}}^{m-1} = \{[z_0,z] \in \mathbb{C}P^m : \langle z,z\rangle = z_0^2\}$$

$$\hat{\zeta}(z_0,z) = \zeta(z) - z_0^{k-1} z$$

for $(z_0,z) \in \mathbb{C} \times \mathbb{C}^m = \mathbb{C}^{m+1}$ . (The idea is that the line joining any zero $[z_0,z]$ of $\hat{\zeta}$ to the origin $[1,0]$ intersects $S_{\mathbb{C}}^{m-1}$ in (zero or two) critical points of $\zeta_{tan}$.) Again interpret $\hat{\zeta}$ as a section of the vector bundle used in the proof of Theorem 1.

Let $\mathcal{A}_{\mathbb{C}I}$ be the set of all $\zeta \in \mathcal{A}_{\mathbb{C}}$ such that $\hat{\zeta}$ has no zeros at infinity, no zeros on the cone $\langle z,z\rangle = 0$, and is transverse to the zero section. Thus $\zeta \in \mathcal{A}_{\mathbb{C}}$ fails to be in $\mathcal{A}_{\mathbb{C}I}$ if and only if one of the three following over-determined systems of homogeneous equations has a nontrivial solution :

$$\zeta(z) = 0 \ ;$$

$$\zeta(z) - z_0^{k-1} = 0 \ , \ \langle z,z\rangle = 0 \ ;$$

$$\zeta(z) - z_0^{k-1} z = 0 \ , \ \det(D\zeta(z) - z_0^{k-1} I) = 0 \ .$$

As above $\mathcal{A}_{\mathbb{C}I}$ is, if non-empty, open, dense, and connected in $\mathcal{A}_{\mathbb{C}}$ so by transversality the number of zeros of $\hat{\zeta}$ is independent of the choice of $\zeta \in \mathcal{A}_{\mathbb{C}I}$ and by construction, no zero lies on $\langle z,z\rangle = 0$. It remains to show $\mathcal{A}_{\mathbb{C}I} \cap \mathcal{A} \subset \mathcal{A}_I$ (then $\mathcal{A}_I$ is dense in $\mathcal{A}$ as before) and to show $\mathcal{A}_{\mathbb{C}I}$ is non-empty by exhibiting an example. The former follows immediately from the following claim : Let $\zeta \in \mathcal{A}_{\mathbb{C}I}$ and $[z_0,z]$ be a zero of $\tilde{\zeta}$ normalised so that $\langle z,z\rangle = 1$ (i.e. $z \in S_{\mathbb{C}}^{m-1}$). Then z is a nondegenerate critical point of $\zeta_{tan}$.

To prove the claim introduce a covariant derivative by the formula

$$\nabla_v \zeta_{tan}(z) = D\zeta_{tan}(z)v = \langle D\zeta_{tan}(z)v, z\rangle z$$
$$= D\zeta(z)v - \langle D\zeta(z)v, z\rangle z - \langle \zeta(z), z\rangle v$$

for $(z,v) \in TS^{m-1}$ (i.e. $\langle z,z\rangle = 1$, $\langle z,v\rangle = 0$). Now if $\zeta_{tan}(z) = 0$ then

$$D\zeta_{tan}(z)v = \nabla_v \zeta_{tan}(z)$$

for $v \in T_z S_{\mathbb{C}}^{m-1}$. (Indeed the left side is intrinsically defined for any manifold, vector field, and critical point, and the equality holds for any covariant derivative.) Thus we must prove $v = 0$ from the assumptions that $(z,v) \in TS_{\mathbb{C}}^{m-1}$ and $\zeta_{tan}(z) = \nabla_v \zeta_{tan}(z) = 0$; i.e. that

$$\langle z,z\rangle = 1 ,$$
$$\langle z,v\rangle = 0 ,$$
$$\zeta(z) = \lambda z ,$$
$$(D\zeta(z) - \lambda)v = az$$

(where $\lambda = z_0^{k-1} = \langle \zeta(z), z\rangle$ and $a = \langle D\zeta(z)v, z\rangle$). By homogeneity

$$(D\zeta(z) - \lambda)z = bz$$

(where $b = k - \lambda$). Now $D\zeta(z) - \lambda$ is invertible so $b = k - \lambda \neq 0$ whence

$$(D\zeta(z) - \lambda)(bv - az) = 0$$

implies $v = 0$ as required.

We now show $\mathcal{Q}_{\mathbb{C}1}$ is non-empty. For $k = 1$ we have that $\zeta \in \mathcal{Q}_{\mathbb{C}1}$ where

$$\zeta_j(z) = a_j z_j \qquad (j = 1, \ldots, m)$$

and $a_1, \ldots, a_m$ are distinct and non-zero. For $k > 1$ we take

$$\zeta_j(z) = a_j^{1-k} z_j^k \qquad (j = 1, \ldots, m)$$

where the real numbers $a_1, \ldots, a_m$ are chosen so that

$$\Sigma(a_j\omega_j)^2 \neq 0$$

for every m-tuple $(\omega_1,\ldots,\omega_m) \neq (0,\ldots,0)$ of solutions of $\omega^k = \omega$. Then the zeros of $\hat{\zeta}$ are given by

$$z_0 = 1$$
$$z_j = a_j\omega_j \qquad (j = 1,\ldots,m)$$

so $\langle z, a\rangle \neq 0$ for such a zero by construction. This completes the proof that $\mathfrak{a}_I$ is dense in $\mathfrak{a}$.

To prove that $\mathfrak{a}_{II}$ is dense note that for $p \in S^{m-1}$ the element $\eta \in \mathfrak{a}$ given by

$$\eta(x) = -\langle p, x\rangle^k p$$

satisfies

$$\eta_{\tan}(p) = 0$$
$$\nabla_v \eta_{\tan}(p) = v\ .$$

The proof used in Theorem 1 goes through unchanged.

This completes the proof of Theorem 3; the same argument proves Theorem 4.

REMARK. Given $\zeta \in \mathfrak{a}_{\mathbb{C}}$ such that $\zeta(z) = 0$ has no non-trivial solution (e.g. $\zeta \in \mathfrak{a}_{\mathbb{C}I}$) we define two maps

$$\zeta_{\text{aff}} : \mathbb{C}^m \to \mathbb{C}^m : \qquad \zeta_{\text{aff}}(z) = \zeta(z)\ ;$$
$$\zeta_{\text{proj}} : \mathbb{C}P^{m-1} \to \mathbb{C}P^{m-1} : \qquad \zeta_{\text{proj}}([z]) = [\zeta(z)]\ .$$

Define

$$\text{fix}\ (\zeta_{\text{aff}}) = \{z \in \mathbb{C}^m : \zeta_{\text{aff}}(z) = z\}$$
$$\text{fix}\ (\zeta_{\text{proj}}) = \{[z] \in \mathbb{C}P^{m-1} : \zeta_{\text{proj}}([z]) = [z]\}$$
$$\text{crit}\ (\zeta_{\tan}) = \{z \in S_z^{m-1} : \zeta_{\tan}(z) = 0\}\ .$$

Now if $z \in \text{fix}\ (\zeta_{\text{aff}})$, $z \neq 0$ then $[z] \in \text{fix}\ (\zeta_{\text{proj}})$ and conversely every $[z] \in \text{fix}\ (\zeta_{\text{proj}})$

contains exactly k - 1 non-zero elements of $\mathrm{fix}(\zeta_{aff})$. Hence

$$\#\{\mathrm{fix}(\zeta_{aff})\} - 1 = (k-1).\ \#\{\mathrm{fix}(\zeta_{proj})\} .$$

(The -1 appears as $0 \in \mathrm{fix}(\zeta_{aff})$.) Now every $[z] \in \mathbb{C}P^{m-1}$ intersects $S_{\mathbb{C}}^{m-1}$ in precisely two points (if $\langle z,z\rangle \neq 0$) so for $\zeta \in G_{\mathbb{C}I}$ we have

$$\#\{\mathrm{crit}(\zeta_{tan})\} = 2.\ \#\{\mathrm{fix}(\zeta_{proj})\}$$

But the example shows that

$$\#\{\mathrm{fix}(\zeta_{aff})\} = k^n$$

so

$$\#\{\mathrm{fix}(\zeta_{proj})\} = \frac{k^n-1}{k-1} .$$

(This formula even holds for k = 1 if the right hand side is interpreted as a limit.)

I do not know if one can always find $\zeta$ real so that all the fixed points of $\zeta_{proj}$ are real. This is the case for k = 1, 2, 3 as the example given above shows. It is also the case when n = 2. Take

$$\zeta_1(x_1,x_2) = x_2^k$$
$$\zeta_2(x_1,x_2) = x_2^k f(\frac{x_1}{x_2})$$

where f is a polynomial so chosen that the equation

$$xf(x) = 1$$

has exactly $k + 1 = \frac{k^2-1}{k-1}$ simple roots.

§2. We now return to the general situation introduced at the beginning of §1; we will give a sufficient condition for $\mathfrak{a}_{III}$ to be residual.

To simplify notation we assume :

$$\mathfrak{a} = \mathfrak{a}_{II} .$$

Choose $\xi \in \mathfrak{a}$ and let

$$\varphi^t : M \to M \qquad (t \in \mathbb{R})$$

denote the flow of $\xi$. (If M is not compact $\varphi^t$ need not be everywhere defined but this doesn't affect the arguments.) Choose critical points p and q of $\xi$ and define sub-bundles $E^u \subset TM|W^u(p)$ and $E^s \subset TM|W^s(p)$ by :

$$E^u_x = T_x W^u(p) \qquad (x \in W^u(p)) ;$$
$$E^s_x = T_x W^s(q) \qquad (x \in W^s(q)) .$$

For $x \in W^u(p) \cap W^s(q)$ we denote by $F_x$ the quotient space given by :

$$F_x = T_x M/(E^u_x + E^s_x)$$

and we denote the natural projection onto the quotient by $\pi = \pi_x$ :

$$\pi_x : T_x M \to F_x .$$

Note that $F_x$ is zero-dimensional precisely when $W^u(p)$ and $W^s(q)$ intersect transversally at x.

For $x \in W^u(p) \cap W^s(q)$ we define a linear map

$$S = S_x : \mathfrak{X}(M) \to F_x$$

by :

$$S_\eta = \int_{-\infty}^{\infty} \pi_x(T\varphi^t)\eta(\varphi^{-t}(x))\, dt$$

for $\eta \in \mathfrak{X}(M)$. Thus

$$S\eta = S^+\eta + S^-\eta$$

where

$$S^+\eta = \int_0^\infty \pi_x (T\varphi^t)\eta(\varphi^{-t}(x))\, dt$$

and

$$S^-\eta = \int_{-\infty}^0 \pi_x (T\varphi^t)\eta(\varphi^{-t}(x))\, dt.$$

LEMMA. The integral $S^+\eta$ (and similarly $S^-\eta$) converges absolutely, so that $S\eta$ is well-defined.

Proof. First we construct a continuous vector bundle $K \subset TM|W^u(p)$ which is $\varphi^t$-invariant and complementary to $E^u$. Thus :

$$(T\varphi^t)K_x = K_y \ ;$$
$$(T\varphi^t)E^u_x = E^u_y \ ;$$
$$T_x M = K_x \oplus E^u_x$$

for $x \in W^u(p)$, $t \in \mathbb{R}$, $y = \varphi^t(x)$. This can be done as follows. First choose a cross section $\Sigma \subset W^u(p)\setminus\{p\}$ for the restriction of the flow $\{\varphi^t\}$ to the unstable manifold. Thus every orbit $\{\varphi^t(x) : t \in \mathbb{R}\}$ in $W^u(p)\setminus\{p\}$ intersects $\Sigma$ in exactly one point. (The cross section can be proved to exist by an appeal to Hartman's theorem that the flow $\{\varphi^t\}$ can be linearised near p or by construction of a Lyapunov function for $\varphi^t|W^u(p)$). Next choose an arbitrary vector bundle complement K to $E^u|\Sigma$ in $TM|\Sigma$ and extend K to $W^u(p)\setminus\{p\}$ by invariance. Finally, define $K_p$ to be the stable subspace of $T_p M$ :

$$K_p = E^s_p = T_p W^s(p)$$

and prove that $K_x$ is continuous at $x = p$ by using the fact that for large t the linear maps $(T_p\varphi^t)|K_p$ and $(T_p\varphi^t)^{-1}|E^u_p$ are contractions together with the fact $\lim_{t\to-\infty} \varphi^t(x) = p$ uniformly for $x \in \Sigma$ and the invariance of K.

Now for $x \in W^u(p)$ denote by :

$$\tilde{\pi}_x : T_xM \to K_x$$

the projection along $E^u_x$ and note that for $x \in W^u(p) \cap W^s(q)$ the projection $\pi_x : T_xM \to F_x$ factors through $\tilde{\pi}_x$ :

$$\pi_x = \pi_x \circ \tilde{\pi}_x .$$

The usual hyperbolic estimates show that the integral :

$$\tilde{S}\eta = \int_0^\infty \tilde{\pi}_x(T\varphi^t)\eta(\varphi^{-t}(x))\, dt$$

converges (for $\eta \in \mathfrak{X}(M)$, $x \in W^u(p)$); for this note the formula :

$$\tilde{\pi}_x(T\varphi^t) = (T\varphi^t)\tilde{\pi}_y$$

where $y = \varphi^t(x)$. As $\pi_x\tilde{S}\eta = S^+\eta$ this completes the proof of the lemma.

THEOREM 5. Assume that for all $\xi \in \mathcal{A}$, all pairs of critical points p,q of $\xi$, and all $x \in W^u(p) \cap W^s(q)$ the map S defined above maps the vector subspace $T_\xi\mathcal{A}$ of $\mathfrak{X}(M)$ surjectively onto $F_x$. Then $\mathcal{A}_{III}$ is dense in $\mathcal{A}$.

Proof : By standard arguments (see [1]) it is enough to prove a weaker assertion; viz. we assume that we are given particular hyperbolic critical points $p = p(\xi)$ and $q = q(\xi)$ for $\xi \in \mathcal{A}$ (which critical points depend smoothly on $\xi$) and we show that the set of $\xi$ for which $W^u(p)$ intersects $W^s(q)$ transversally is residual. For this we use Abraham's transversality density theorem of [1].

By [5] there are injective immersions :

$$\alpha : \underline{E}^u \to M$$
$$\beta : \underline{E}^s \to M$$

(with $\underline{E}^u$ and $\underline{E}^s$ vector spaces) which parameterise $W^u(p)$ and $W^s(q)$ :

$$\alpha(\underline{E}^u) = W^u(p) ;$$
$$\beta(\underline{E}^s) = W^s(q) .$$

Moreover, one can define flows :

$$\Phi^t : \underline{E}^u \to \underline{E}^u$$
$$\Phi^t : \underline{E}^s \to \underline{E}^s$$

such that :

$$\varphi^t \circ \alpha = \alpha \circ \Phi^t$$
$$\varphi^t \circ \beta = \beta \circ \Phi^t$$

for $t \in \mathbb{R}$ where $\{\varphi^t\}$ is the flow of $\xi$. Finally (see [1] and [4]) everything in sight depends smoothly on $\xi$; i.e. all the maps :

$$\mathcal{a} \times M \times \mathbb{R} \to M : (\xi, x, t) \mapsto \varphi^t(x)$$
$$\mathcal{a} \times \underline{E}^u \to M : (\xi, v) \mapsto \alpha(v)$$
$$\mathcal{a} \times \underline{E}^s \to M : (\xi, w) \mapsto \beta(w)$$
$$\mathcal{a} \times \underline{E}^u \times \mathbb{R} \to \underline{E}^u : (\xi, v, t) \mapsto \Phi^t(v)$$
$$\mathcal{a} \times \underline{E}^s \times \mathbb{R} \to \underline{E}^s : (\xi, w, t) \mapsto \Phi^t(w)$$

are smooth.

Now the condition that $W^u(p)$ and $W^s(q)$ intersect transversally is the same as the condition that the maps $\alpha : \underline{E}^u \to M$ and $\beta : \underline{E}^s \to M$ are transverse and this in turn is equivalent to the condition that the map :

$$\gamma : \underline{E}^u \times \underline{E}^s \to M \times M$$

given by :

$$\gamma(v, w) = (\alpha(v), \beta(w))$$

is transverse to the diagonal $\Delta$ in $M \times M$. In the lingo of [1] the map :

$$\mathcal{a} \to C^\infty(\underline{E}^u \times \underline{E}^s, M \times M) : \xi \mapsto \gamma$$

is a "representation of maps" and so by [1], page 48, it is enough to show that the evaluation map :

$$\mathrm{ev} : \mathcal{A} \times \underline{E}^u \times \underline{E}^s \to M \times M$$
$$\mathrm{ev}\,(\xi, v, w) = \gamma(v, w)$$

is transverse to $\Delta$.

For this we fix :

$$x = \alpha(v) = \beta(w) \in W^u(p) \cap W^s(q)$$

and compute :

$$\mathrm{Dev}\,(\xi, v, w)(\dot{\xi}, \dot{v}, \dot{w}) =$$
$$= (\dot{\alpha}(v) + D\alpha(v)\dot{v},\ \dot{\beta}(w) + D\beta(w)\dot{w})$$

for $\dot{\xi} \in T_\xi \mathcal{A} \subset \mathfrak{X}(M)$; $\dot{v} \in T_v \underline{E}^u = \underline{E}^u$; $\dot{w} \in T_w \underline{E}^s = \underline{E}^s$. (The term $\dot{\alpha}(v)$ arises from differentiating $\alpha(v)$ with v fixed in the direction $\dot{\xi} \in T_\xi \mathcal{A}$ ; similarly for $\dot{\beta}(w)$.)

We must show that the image of the linear map $\mathrm{Dev}(\xi, v, w)$ contains a vector space complement to the diagonal in $(T_x M) \times (T_x M)$. Now the image of the term $D\alpha(v)$ (resp. $D\beta(w)$) is precisely $T_x W^u(p) = E^u_x$ (resp. $T_x W^s(q) = E^s_x$) so the condition on $\mathrm{Dev}(\xi, v, w)$ reduces to the condition that the vector space :

$$\{\dot{\alpha}(v) - \dot{\beta}(w) \mid \dot{\xi} \in T_\xi \mathcal{A}\}$$

contains a complement to $E^u_x + E^s_x$. This can be written :

$$\{\pi_x(\dot{\alpha}(v) - \dot{\beta}(w)) \mid \dot{\xi} \in T_\xi \mathcal{A}\} = F_x \,.$$

Thus our theorem is proved once we have established the following.

LEMMA : For $\dot{\xi} \in T_\xi \mathcal{A}$ :

$$\pi_x(\dot{\alpha}(v) - \dot{\beta}(w)) = S_x \dot{\xi}$$

in fact :

$$\pi_x(\dot{\alpha}(v)) = S^+\dot{\xi}$$

and :

$$\pi_x(\dot{\beta}(w)) = -S^-\dot{\xi}.$$

Proof of Lemma : We differentiate the equation :

$$\alpha(v) = \varphi^t(\alpha(\Phi^{-t}(v))$$

in the direction $\dot{\xi}$ (with v fixed) to obtain :

$$\dot{\alpha}(v) = Z_1(t) + Z_2(t) + Z_3(t)$$

where

$$Z_1(t) = \dot{\varphi}^t(\alpha(\Phi^{-t}(v))) = \dot{\varphi}^t(\varphi^{-t}(x))$$
$$Z_2(t) = (T\varphi^t)\dot{\alpha}(\Phi^{-t}(v)) \ ;$$
$$Z_3(t) = D(\varphi^t \circ \alpha)(\Phi^{-t}(v))\dot{\Phi}^{-t}(v) \ .$$

Now $Z_3(t) \in E^u_x = T_xW^u(p)$ so :

$$\pi_x(Z_3(t)) = 0 \ .$$

Also :

$$\lim_{t\to\infty} \pi_x(Z_2(t)) = 0$$

as :

$$\pi_x(Z_2(t)) = \pi_x \circ \tilde{\pi}_x(Z_2(t))$$
$$= \pi_x(T\varphi^t)\tilde{\pi}_{y(t)}\dot{\alpha}(\Phi^{-t}(v)) \ ;$$
$$\lim_{t\to\infty} \Phi^{-t}(v) = \alpha^{-1}(p) \ ;$$

and :

$$\lim_{t\to\infty} (T\varphi^t) \circ \tilde{\pi}_p = 0 \ .$$

Thus :

$$\pi_x(\dot{\alpha}(v)) = \lim_{t\to\infty} \tilde{\pi}_x Z_1(t) .$$

But by [1] page 107:

$$\begin{aligned}\dot{\varphi}^t(y(t)) &= \int_0^t (T\varphi^{t-s})\dot{\xi}(\varphi^s(y(t)))\,ds \\ &= \int_0^t (T\varphi^{t-s})\dot{\xi}(\varphi^{s-t}(x))\,ds \\ &= \int_0^t (T\varphi^r)\dot{\xi}(\varphi^{-r}(x))\,dr .\end{aligned}$$

(Here $y(t) = \varphi^{-t}(x)$.) Thus :

$$\pi_x(\dot{\alpha}(v)) = \lim_{t\to\infty} \pi_x(\dot{\varphi}(y(t))) = S^+\dot{\xi}$$

as required. (The proof that $\pi_x(\dot{\beta}(w)) = S^-\dot{\xi}$ is the same.)

COROLLARY. In the context of Theorem 1 with $m = 2$, $\mathcal{a}_{III}$ is residual.

Proof : Let $\eta$ be the vector field obtained by rotating $\xi$ through $90^o$ :

$$\eta = (\eta_1, \eta_2) = (-\xi_2, \xi_1)$$

always points to the same side of the orbit; hence $S\eta \neq 0$ .

REMARK. This proof fails for Theorem 2 as $\eta$ is not a gradient. The analogous argument for Theorem 3 (with $m = 3$) also fails; one rotates through $90^o$ by taking the cross product with the normal, but this increases the degree k.

COROLLARY. In the context of Theorem 2 with $m = 2$ and $k = 2$, $\mathcal{a}_{III}$ is residual.

PROOF : By a beautiful theorem of Chicone [2] a saddle connection of a planar gradient of a cubic always lies on a straight line. Hence a suitable constant vector field $\eta$ always points to the same side of the saddle connection so that $S\eta \neq 0$. (One can prove this corollary without using Theorem 5; e.g. as in [2]).

## References.

1. R. Abraham & J. Robbin : Transversal Mappings and Flows; W.A. Benjamin, 1967.

2. C. Chicone : Quadratic gradients in the plane are generically Morse-Smale; J. Diff. Eq., 33 (1979), 159-166.

3. J. Kupka : Contribution à la théorie des champs génériques; Contributions to differential equations, 2 (1963), 457-484; 3 (1964), 411-420.

4. J. Robbin : Stable manifolds of semi-hyperbolic fixed points; Ill. J. Math. 15 (1979), 595-609.

5. S. Smale : Stable manifolds for differential equations and diffeomorphisms; Ann. Scuola Norm. Sup. Pisa (3), Vol. 17 (1963), pp.97-116.

6. B.L. Van der Waerden : Algebra Vol. II; Springer, Berlin/New York, 1960.

7. H. Whitney : Complex Analytic Varieties; Addison Wesley, 1972.

J. Robbin, Mathematics Department, University of Wisconsin, Madison, Wisconsin 53706, U.S.A.

# Differentiability of the Stable Foliation for the Model Lorenz Equations.

Clark Robinson*

Guckenheimer and Williams proved that their geometric model of the Lorenz equations have modulus two, i.e. there is a 2-parameter family of flows such that any $C^{14}$ perturbation of the geometric model Lorenz flow is topologically equivalent on a neighbourhood of the attractor to a member of this family, [1]. One of the steps of the proof is to show that any perturbation has a continuous foliation of stable manifolds. In this note we prove that any such $C^2$ perturbation has a differentiable foliation of stable manifolds. This result implies that the induced map of the interval to itself, which is the quotient of the Poincaré map, is also differentiable and not just continuous as proved by Guckenheimer and Williams. Next we prove this Poincaré map of the interval has the properties (condition 5 below) necessary to apply the topological equivalence result. In [1] it is merely claimed they are clear. Knowing that the Poincaré map of the interval is differentiable makes it easier to interpret and prove one of these conditions, mainly that the derivative goes to infinity as the point approaches the discontinuity (condition 5d below). Also the result that the map of the interval is "locally eventually onto" (condition 5e below and 9a in [1]) then follows from [6, p.78], while the stretching result for an arbitrarily small interval for a continuous map would require justification and may not be true. Lastly, the results of this paper clarify the differentiability assumptions needed on the perturbations : any $C^2$ (and not just $C^{14}$) perturbation of the geometric model Lorenz flow is topologically equivalent on a neighbourhood of the attractor to a member of the 2-parameter family of flows.

We prove this result by two approaches. The first approach is to show that any $C^2$ perturbation of the flow in $\mathbb{R}^3$ has a differentiable foliation of strong stable manifolds of points (rather than of orbits). It follows that for every perturbed flow there is a $C^1$ semi-flow on the branched manifold formed by taking the quotient by this foliation. (This parallels the work for the unperturbed flow in [6].) The Poincaré map of this semi-flow is a map of the interval to itself with a discontinuity at the point on the stable manifold of the critical point. It also has the other properties used in [1] to show all these perturbations are conjugate to a map taken from a two

* This research was partially supported by the NSF, Grant MCS-80-02177.

parameter family, i.e. that the geometric Lorenz equations have modulus two.

The second approach follows that in [1] and factors the Poincaré map of the flow in $\mathbb{R}^3$ into the composition of (1) the Poincaré map past the fixed point and (2) the Poincaré map from near the unstable manifold back to near the stable manifold. The Poincaré map past the fixed point is calculated using differentiable linearisation. It is necessary to assume the perturbation is $C^{14}$ to get a $C^1$ linearisation which is used to show the map of the interval is continuous and conjugate to one of the standard maps. To show the map of the interval is differentiable a $C^2$ linearisation is necessary for which the perturbed flow must be $C^{20}$.

The reason this analysis is included here even though it has stronger assumptions than the first approach is that it illustrates the type of assumptions and analysis needed to study foliations of maps with discontinuities. The differentiability of the foliation is delicate using the second approach because the Poincaré map is not uniformly $C^2$. Previous results in [4] and [5] has shown that the assumption that the map is $C^2$ is important to prove the differentiability of the foliation. Even though the second derivative of the Poincaré map becomes unbounded as the point approaches the discontinuity, the differentiability is preserved for these equations by the fact that a product of the second derivative and first derivative terms remains bounded. Compare with [7].

§1. Statement of results.

The actual Lorenz equations are

$$(1)\qquad \begin{aligned} \dot{x} &= -10x + 10y \\ \dot{y} &= 28x - y - xz \\ \dot{z} &= -\tfrac{8}{3}z + xy \end{aligned}$$

The eigenvalues of the fixed point at the origin are $a = -\frac{11}{2} + \frac{1}{2}(1201)^{1/2} \approx 11.83$, $-b = -\frac{11}{2} - \frac{1}{2}(1201)^{1/2} \approx -22.83$, and $-c = -\frac{8}{3} \approx -2.67$. By differentiable linearisation there is a neighbourhood of the origin where the equations are differentiably conjugate to

$$\dot{x} = ax$$
(2) $$\dot{y} = -by$$
$$\dot{z} = -cz$$

with $0 < c < a < b$. These equations take $S = \{(x, y, z) : z = 1, -1 \le x \le 1, -1 \le y \le 1\}$ onto two "cusped triangles" at $x = \pm 1$. The geometric model equations are taken (i) equal to (2) in this neighbourhood of the origin and (ii) defined outside this neighbourhood

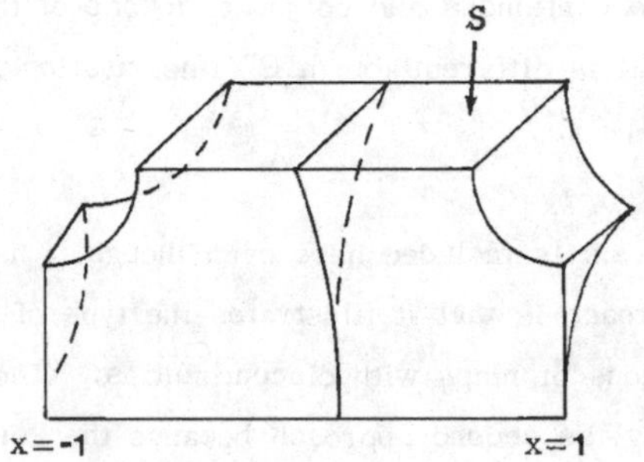

Figure 1.

such that these "cusped triangles" flow nonlinearly back to $z = 1$ so the image at time T is as indicated in figure 2 and the derivative of the flow on $x = \pm 1$ satisfies

(3) $$D\varphi^{T}(x, y, x) = \begin{pmatrix} 0 & 0 & \pm\eta \\ 0 & \zeta & 0 \\ \mp 1 & 0 & 0 \end{pmatrix}$$

with $\zeta$ and $\eta$ depending on the point and $0 < \zeta < 1$, $\eta > 1$, and $\zeta\eta \le 1$. Let $S_L = \{(x, y, 1) \in S : x < 0\}$ and $S_R = \{(x, y, 1) \in S : x > 0\}$.

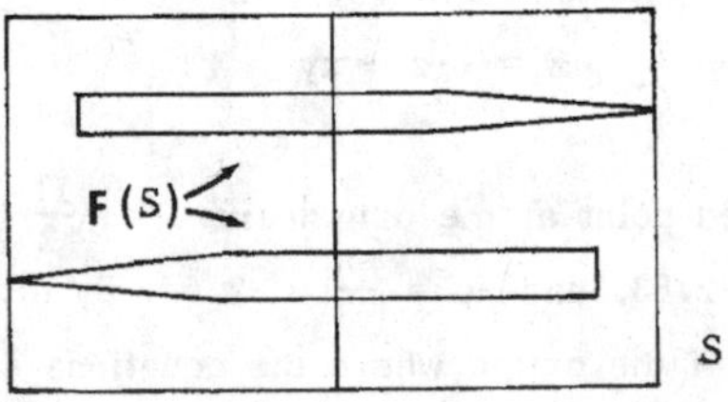

Figure 2.

The Poincaré map for equations (2) from $z = 1$ to $x = \pm 1$ is

$$y = y_0 |x_0|^u \tag{4}$$
$$z = |x_0|^s$$

where $u = \frac{b}{a} \approx 1.93$ and $S = \frac{c}{a} \approx 0.23$. Therefore the Poincaré map F from S at $z = 1$ back to itself is of the form $F(x, y) = (f(x), g(x, y))$. Here $S = I^2$ with $I = [-1, 1]$ is a square in $\mathbb{R}^2$ and also thought of as a transversal at $z = 1$. The important properties of $f: I \to I$ are as follows (see figure 3 and compare with condition 9 in [1]):

(5) (a) f has a single discontinuity at $x = d$

(b) f is nonuniformly $C^1$ on $I - \{d\}$ and $f'(x) \geq \lambda > 1$ for all $x \neq d$

(c) $f(d-) = 1$ (left limit), $f(d+) = -1$ (right limit), and $f(-1) < d < f(1)$

(d) $f'(x) \to \infty$ as $x \to d\pm$.

(e) f is locally eventually onto, i.e. for any subinterval $J \subset [-1, 1]$ there is an $n > 0$ such that $f^n(J) = [-1, 1]$.

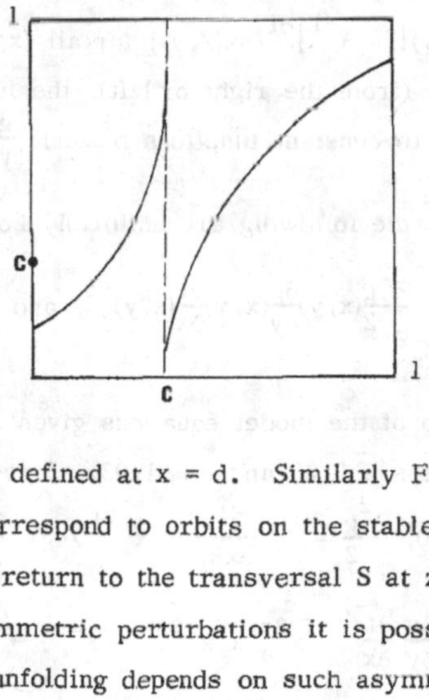


Figure 3.

Note that f is not actually defined at $x = d$. Similarly F is not actually defined on $\{d\} \times I$. These points correspond to orbits on the stable manifold of the fixed point of equations (2) which never return to the transversal S at $z = 1$. For the unperturbed flow $d = 0$, but for nonsymmetric perturbations it is possible for $d \neq 0$. In fact one of the parameters of the unfolding depends on such asymmetry of the graph, [1, p. 60]. The last property of the list, (5e), follows from the other properties as proved in [6, p. 78].

<u>Theorem 1</u>: Assume $b > a + c$ and $a > c$ in the definition of the geometric Lorenz equations. Then there is an open set $\mathfrak{h}$ of all $C^2$ flows which are $C^2$ near the

geometric model flow, such that every flow $\psi^t$ in $\mathfrak{n}$ has a $C^1$ foliation of strong stable manifolds of points in a neighbourhood of the attractor. Thus there is induced a branched manifold obtained by taking the quotient by these manifolds, see [6], and a $C^1$ semi-flow on this branched manifold. The induced Poincaré map $f:I \to I$ at $z = 1$ then satisfies properties (5). Therefore by [1] every $\psi^t$ in $\mathfrak{n}$ is topologically equivalent on a neighbourhood of the attractor to a flow from a two parameter family.

The proof is contained in Section 2.

The second approach, to prove the differentiability of the foliation and reduce to a map of an interval, first looks at the Poincaré map $F:S \to S$. To prove differentiability of the foliation, $F$ must satisfy the following conditions :

(6) (a) $F(x,y) = (f(x), g(x,y))$ is nonuniformly $C^2$ and $g$ is uniformly $C^1$ on $S - \{d\} \times I$

(b) $F$ is one to one on $S - \{d\} \times I$

(c) $|{}^{\partial g}/\partial y| < \frac{1}{2}$ and $|{}^{\partial g}/\partial x| \leq 1$ (actually what is needed is $|{}^{\partial g}/\partial x(x,y)| < \lambda^{-1}|{}^{\partial f}/\partial x(x,y)|$ for all $(x,y)$ where $\lambda > 1$).

(d) As $x \to d\pm$ (from the right or left), the functions $g|\{x\} \times I$ tend uniformly to constant functions $b^{\pm}$ and $\frac{\partial g}{\partial y}(x,y) \to 0$. Also as $x \to d\pm$ all the following are uniformly bounded : $\frac{\partial^2 g}{\partial y^2}(x,y)$,
$\frac{\partial^2 g}{\partial x \partial y}(x,y)$, $\frac{\partial^2 g}{\partial x^2}(x,y)\frac{\partial g}{\partial y}(x,y)\frac{\partial f}{\partial x}(x,y)^{-1}$, and $\frac{\partial^2 f}{\partial x^2}(x,y)\frac{\partial g}{\partial y}(x,y)\frac{\partial f}{\partial x}(x,y)^{-1}$.

<u>Remark</u> : In the Poincaré map of the model equations given above (with $d = 0$), $f(x) = x^s$ and $g(x,y) = yx^u$ where $s \approx 0.23$ and $u \approx 1.93$. Since $u < 2$, $g$ is not uniformly $C^2$, but as $x \to 0\pm$, $\frac{\partial g}{\partial y} = x^u \to 0$, $\frac{\partial^2 g}{\partial y^2} = 0$, $\frac{\partial^2 g}{\partial x \partial y} = x^{u-1} \to 0$, $\frac{\partial^2 g}{\partial x^2}\frac{\partial g}{\partial y}\frac{\partial f}{\partial x}^{-1} \sim yx^{u-2}x^u x^{-s+1} =$ $= yx^{2(u-1)+(1-s)} \to 0$, and $\frac{\partial^2 f}{\partial x^2}\frac{\partial g}{\partial y}\frac{\partial f}{\partial x}^{-1} \sim x^{s-2}x^u x^{-s+1} = x^{u-1} \to 0$. Thus assumption (6) is satisfied.

<u>Theorem 2</u> : Let $F:S \to S$ be a map satisfying properties (5) and (6). Let $G = J \circ F^*$ where $F^*$ satisfies (5) and (6) and is near $F$ and $J$ is $C^2$ and a $C^1$ perturbation of the identity. Then $G$ has a $C^1$ stable foliation near the stable foliation of $F$.

The proof of Theorem 2 is contained in Section 3. This theorem is applied to a perturbation of the geometric model Lorenz equations by taking a differentiable linearisation at the fixed point of the perturbed differential equations. To get a $C^k$ linearisation with the values $a \approx 11.83$, $b \approx 22.83$, and $c \approx 2.67$, the perturbation would need to be $C^r$ where $e^b e^{ak} e^{c(k-r)} < 1$ and $e^a e^{bk} e^{a(k-r)} < 1$ so need $r > 8.55 + k(5.43)$ and $r > 1 + k(2.93)$. Thus to apply $C^1$ linearisation used in [1], the perturbation would need to be $C^{14}$ and to apply the $C^2$ linearisation to get the assumption of Theorem 2 above, it would need to be $C^{20}$.

## §2. Proof of Theorem 1.

The existence and differentiability of the strong stable foliation is proved using estimates of the expansion and contraction in different directions. The existence uses the fact that the contraction in the y direction is stronger than the contraction in either other direction. The differentiability uses the fact this ratio of contractions times the maximal expansion is less than one. These estimates are verified for the unperturbed flow $\varphi^t$ and then the openness of the condition shows it holds for any perturbation $\psi^t$ which is $C^1$ close to $\varphi^t$.

Let S be the cross section at $z = 1$ as before and $\Sigma$ the cross sections at $x = \pm 1$ containing the cusped triangles of figure 1. The derivative $D\varphi^t(q)$ is not specified exactly as a trajectory passes from $\Sigma$ back to S (outside the neighbourhood of the origin), but if $p \in \Sigma$ and $\varphi^\tau(p) \in S$ then by (3) $D\varphi^\tau(p)$ does not stretch or contract in the wrong directions. For any point q in a neighbourhood of the attractor and $n > 0$, let $T(n,q)$ be the amount of time $\varphi^t(q)$ stays in the region with linear estimates (from S to $\Sigma$) for $0 \le t \le n$. There is a uniform $1 > \alpha > 0$ and N such that for $n \ge N$, $T(n,q) \ge \alpha n$. Also let $r = r(N,q)$ be the number of crossings from $\Sigma$ to S. Thus if q is between S and $\Sigma$ for $T = T(N,q)$, (i) if $e_2$ is the unit vector in the y direction $|D\varphi^N(q)e_2| \le \zeta^r e^{-Tb} \le e^{-Tb}$, and (ii) if v is a unit vector in the x-z plane then $e^{-Tc} \le |D\varphi^N(q)v| \le \eta^r e^{Ta}$. Thus for u and v unit vectors in the x-z plane
(iii) $|D\varphi^N(q)e_2|/|D\varphi^N(q)v| \le e^{-bT} e^{cT} \le e^{(-b+c)N\alpha} < \lambda < 1$ and
(iv) $|D\varphi^N(q)e_2| \cdot |D\varphi^N(q)u|/|D\varphi^N(q)v| \le (\zeta\eta)^r e^{-bT} e^{aT} e^{cT} \le e^{(-b+a+c)N\alpha} < \beta < 1$ since $\zeta\eta < 1$. For $\psi$ $C^1$ near enough to $\varphi$, its derivative $D\psi^N(q)$ will also satisfy (iii) and (iv) for all q in a neighbourhood. The bundles $\langle e_2 \rangle$ and $\langle e_1, e_3 \rangle$ are also nearly enough invariant (for q between S and $\Sigma$) to verify the cone condition for hyperbolicity

in the strong stable direction using (iii), [2, Theorem 4.8]. Then there exists strong stable foliation for $\psi^t$ tangent to these bundles. Note as usual the bundles and foliations are proved invariant for integer values of t and the uniqueness is used to prove invariance for all time. If $\psi$ is $C^2$ then condition (iv) allows the application of [2, Theorem 6.5] to prove the foliation is $C^1$.

By taking the quotient by the $C^1$ foliation there is a branched manifold with a $C^1$ semiflow $\rho^t(q)$ on it. The semi-flow induces a Poincaré map $f: S' \to S'$ where $f(q) = (\rho_1^\tau(q), \rho_2^\tau(q))$. The conditions on the derivative are the only nontrivial ones to check in property (5). (To simplify calculations) since the stable and unstable manifolds vary continuously under perturbation , a change of variables takes these manifolds back to $x = 0$ and $z = 0$ between S' and $\Sigma'$. The map f factors as $f = g \circ h$

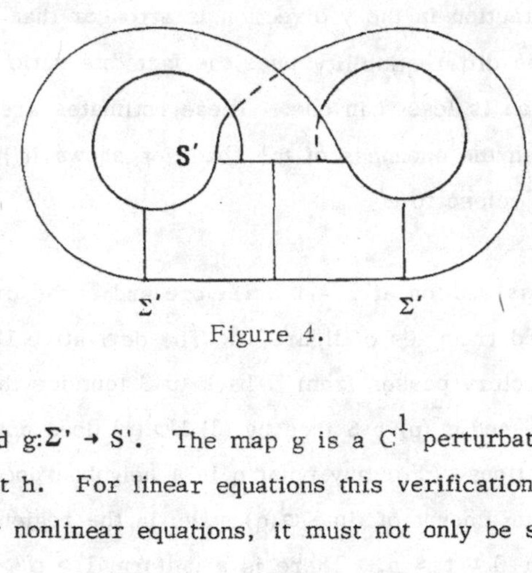

Figure 4.

where $h: S' \to \Sigma'$ and $g: \Sigma' \to S'$. The map g is a $C^1$ perturbation of the identity so it is enough to look at h. For linear equations this verification is straight forward using equations (4). For nonlinear equations, it must not only be shown that the vector $\partial\rho^\tau/\partial x$ is large but also that $V(h(x))$ has more negative slope than $\partial\rho^\tau/\partial x$ could possibly have. See figure 5. This fact may not be true for $C^1$ perturbations.

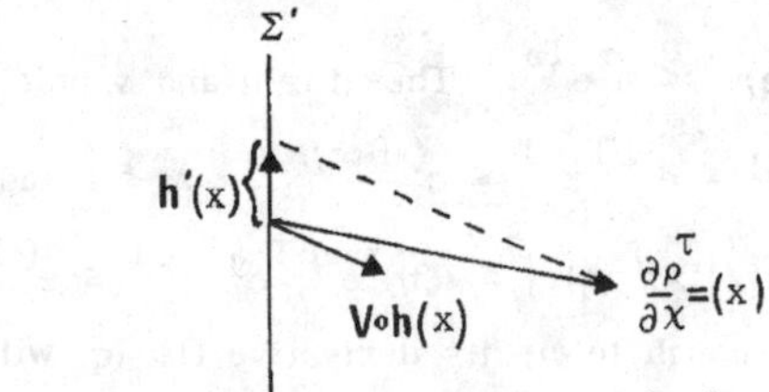

Figure 5.

Taking $\tau = \tau(x)$ such that $\rho^\tau(x,1) \in \Sigma'$, $h(x) = \rho_3^\tau(x)$. (Here x and z are used as coordinates on the branched manifold between S' and $\Sigma'$.) Assuming $x > 0$ to estimate the derivative ($x < 0$ is treated similarly with some sign changes), $h'(x) = V_3(h(x))\tau'(x) + \partial\rho_3^\tau/\partial x\ (x)$ where $V = (V_1, V_3)$ is the vector field for $\rho^t$. Since $\rho_1^\tau(x,1) = 1$, $0 = V_1(h(x))\tau'(x) + \partial\rho_1^\tau/\partial x\ (x,1)$ and $\tau'(x) = -\frac{\partial\rho_1^\tau}{\partial x} / V_1$. Thus

$$h'(x) = \frac{\partial\rho_1^\tau}{\partial x}(x,1) \left(\frac{-V_3 \circ h(x)}{V_1 \circ h(x)}\right) + \frac{\partial\rho_3^\tau}{\partial x}(x,1)$$

which is the projection of $\partial\rho^\tau/\partial x$ onto $e_3 = (0,1)$ along the vector $(V_1, V_3)$. See Figure 5. It is shown below that

$$V_1 \circ h(x) < a + \varepsilon,\ -V_3 \circ h(x) > (c - \varepsilon)h(x),\ h(x) > e^{(-c-\varepsilon)\tau}$$

$$\frac{\partial\rho_1^\tau}{\partial x}(x,1) > e^{(a-\varepsilon)\tau},\ \left|\frac{\partial\rho_3^\tau}{\partial x}(x,1)\right| < \varepsilon h(x) \frac{\partial\rho_1^\tau}{\partial x}(x,1) \quad .$$

Thus

$$\begin{aligned} h'(x) &\geq \left(\frac{c-\varepsilon}{a+\varepsilon} - \varepsilon\right) h(x) \frac{\partial\rho_1^\tau}{\partial x}(x,1) \\ &\geq \left(\frac{c-\varepsilon}{a+\varepsilon}\right) e^{(a-c-2\varepsilon)\tau} \\ &\geq \lambda > 1 \ . \end{aligned}$$

In the choice of $\lambda$, there is a hidden assumption on the geometric model equations that $\tau$ is large enough for all x on S such that $({}^c/a)e^{(a-c)\tau} > 1$. Also as $x \to 0$, $\tau(x) \to \infty$ so $h'(x) \to \infty$.

To check the estimates, $V_1 \circ h(x) = V_1(1,z) = a + \text{error} < a + \varepsilon$ for small error term. Next $V_3(x,0) \equiv 0$ and $-c + \varepsilon > \partial V_3/\partial z > -c - \varepsilon$, so $(-c+\varepsilon)z > V_3(x,z) > (-c-\varepsilon)z$ and $-V_3 \circ h(x) > (c-\varepsilon)h(x)$. From the differential inequality $\dot z > (-c-\varepsilon)z$, it follows that $h(x) = \rho_3^\tau(x,1) > e^{(-c-\varepsilon)\tau}$.

To extimate $\partial\rho_1^\tau/\partial x$, note that since $\psi$ is a $C^1$ perturbation of $\varphi$, $D\rho^\tau(x,z)$ preserves a cone about $e_1 = (1,0)$ and all these vectors have their first coordinate stretched by at least $e^{a-\varepsilon}$. Thus $\partial\rho_1^\tau/\partial x(x,1) \geq e^{(a-\varepsilon)\tau}$.

The estimate for $\partial\rho_3^\tau/\partial x$ is the most delicate and depends on the fact that $\rho^t$ can be taken to be $C^2$ between S' and $\Sigma'$. The flow $\varphi^t$ in $\mathbb{R}^3$ preserves the (x, z)-plane between S and $\Sigma$. This invariant manifold is 2-normally hyperbolic because $b > a + c > 2c$. (It is not globally invariant but this piece of the plane can be pasted into a globally invariant manifold by the usual trick.) By [7, Theorem 4.1f], for a $C^2$ perturbation $\psi^t$ there is a nearby $C^2$ invariant manifold B' on which restricted flow is $C^2$. The pair $(B', \psi^t|B')$ can be used for the branched manifold B and $\rho^t$ between S' and $\Sigma'$. By [2, Theorem 6.5], $\rho^t$ has a $C^1$ line field in the unstable direction on B', span $W(x,z) = \text{span}\,(1, W_3(x,z))$. (See remark 1 below for discussion of why this bundle is $C^1$ on B' and not all of $\mathbb{R}^3$.) In fact for $\psi^t$ $C^2$ near $\varphi^t$, $D\psi^t|TB'$ is $C^1$ near $D\varphi^t|\text{span}\,(e_1; e_3)$ so the bundle span W is $C^1$ near the bundle span $e_1$ for $\varphi^t$ by applying [7, Theorem 3.5] to the proof of [2, Theorem 6.5]; thus $|\partial W_3/\partial z| \le \varepsilon$. Also $W_3(x,0) \equiv 0$ so $|W_3(x,z)| \le \varepsilon z$. Then since $W_1 \equiv 1$,

$$\left|\frac{\partial\rho_3^\tau}{\partial x}(x,1)\right| = \left|W_3 \circ h(x)\, \frac{\partial\rho_1^\tau}{\partial x}(x,1)\right|$$

$$\le \varepsilon h(z)\, \frac{\partial\rho_1^\tau}{\partial x}(x,1)\ .$$

<u>Remarks.</u>

1. The unstable bundle is not necessarily $C^1$ on all of $\mathbb{R}^3$. The difference is that the Lipschitz size of $\psi^{-1}$ changes : $\text{Lip}(\psi^{-1}) \approx e^b$ on $\mathbb{R}^3$ and $\text{Lip}(\psi^{-1}|B') \approx e^c$. The contraction on lines in a cone about $e_1$ is about $e^{-a-c}$ in either space, so the relevant product $\text{Lip}(\psi^{-1}|B')e^{-a-c} \approx e^{-a-c+c} = e^{-a} < 1$ on B' while $\text{Lip}(\psi^{-1})e^{-a-c} \approx e^{b-a-c} > 1$ on all of $\mathbb{R}^3$.

2. The nondifferentiability result of [4] seems to imply that if the perturbation $\psi$ is only $C^1$ then h'(x) might not go to infinity as x goes to zero.

§3. <u>Proof of Theorem 2.</u>

The method of proof is the graph transform. Given a nearly vertical trial foliation $\mathcal{F}$ of S, $G^{-1}(\mathcal{F})$ defines a foliation everywhere except $\{c\} \times I$. Adding this one leaf yields a new foliation $\mathcal{F}^1$ of S. As checked in [1], $\mathcal{F}^1$ is a continuous foliation even at $\{c\} \times I$. Below it is shown that it is $C^1$ on all of S. With repeated transformations adding $\{c\} \times I$ to $G^{-1}(\mathcal{F}^{n-1})$ to yield a new foliation $\mathcal{F}^n$, the contraction mapping principle can be applied to show $\mathcal{F}^n$ converges $C^0$ to a foliation $\mathcal{F}^\infty$

which is the invariant stable manifold foliation. The fibre contraction theorem, [2, Theorem 1.2], applied to the graph transform of the tangent lines to the foliation shows the convergence in $C^1$. Away from the leaf $\{c\} \times I$ the proof is the same as [2, Theorem 6.3]. The new feature in this situation is checking the continuity and differentiability (zero derivative) at this special leaf caused by the discontinuity of the map F.

We start to give the details of the proof by writing out the derivative of G. Let $p = (x, y)$ be a point of S and

$$\begin{aligned} DG_p &= DJ_{F^*(p)} DF^*(p) \\ &= \begin{pmatrix} a_p & c_p \\ b_p & d_p \end{pmatrix} \\ &= \begin{pmatrix} 1+\varepsilon & \varepsilon \\ \varepsilon & 1+\varepsilon \end{pmatrix} \begin{pmatrix} a^*_p & 0 \\ b^*_p & d^*_p \end{pmatrix} \\ &= \begin{pmatrix} (1+\varepsilon)a^*_p + \varepsilon b^*_p & \varepsilon\, d^*_p \\ \varepsilon a^*_p + (1+\varepsilon)b^*_p & (1+\varepsilon)d^*_p \end{pmatrix} \end{aligned}$$

where $\varepsilon$ stand for a small term which is either positive or negative. By assumptions (5) and (6), as $p \to \{c\} \times I$ it follows that $a^*_p \to \infty$ and $d^*_p \to 0$. Also $|d^*_p| \le \frac{1}{2}$, $|a^*_p| \ge \lambda > 1$, and $|b^*_p| \le 1$ everywhere. Let $\{\mu_p\}$ be a set of slopes of trial tangent lines (as functions from y to x) with $\mu_p \le 1$ and let $\{M_p\} = \Gamma\{\mu_p\}$ be the set of slopes of the graph transformed foliation. Letting u and v be vectors in the y direction, the formula for $M_p$ in terms of $\mu_{G_p}$ is as follows :

$$\begin{aligned} \begin{pmatrix} \mu & v \\ & v \end{pmatrix} &= \begin{pmatrix} a & c \\ b & d \end{pmatrix} \begin{pmatrix} Mu \\ u \end{pmatrix} \\ &= \begin{pmatrix} (aM+c)u \\ (bM+d)u \end{pmatrix} \end{aligned}$$

$$v = (bM+d)u$$
$$\mu v = \mu(bM+d)u = (aM+c)u$$
$$(a-\mu b)M = (\mu d-c)$$

and finally

$$M = \Gamma(\mu) = (a-\mu b)^{-1}(\mu d-c) .$$

In this equation everything is evaluated at p except $\mu$ which is evaluated at G(p).

If $|\mu_p| \leq 1$ everywhere then

$$\begin{aligned}|M| &= |(\mu d-c)(a-\mu b)^{-1}| \\ &= |[\mu(1+\varepsilon)d^*-\varepsilon d^*][1+\varepsilon a^*+\varepsilon b^*-\mu\varepsilon a^*+\mu(1+\varepsilon)b^*]^{-1}| \\ &\leq |d^*|(1-2|\varepsilon|)[|a^*|(1-2|\varepsilon|)-(1+2|\varepsilon|)|b^*|]^{-1} \\ &\leq \frac{1}{2}(1-2|\varepsilon|)[\lambda(1-2|\varepsilon|)-(1+2|\varepsilon|)]^{-1} \\ &< 1\end{aligned}$$

is $\varepsilon$ is small enough. Therefore $\{M_p\}$ is a new nearly vertical set of tangent lines. Also as $p \to \{c\} \times I$, $a_p \to \infty$ and $d_p \to 0$ so $|M_p| \to 0$. Therefore $\{M_p\}$ extends continuously so $M_p = 0$ for $p = (c, y)$, i.e. to the tangent lines of $\{c\} \times I$. This shows $\Gamma$ preserves the set $\mathfrak{J}$ of continuous trial tangent lines $\{\mu_p\}$ with $|\mu_p| \leq 1$ everywhere and $\mu_p = 0$ for $p = (c, y)$. To show $\Gamma$ is a contraction on $\mathfrak{J}$, assume $M = \Gamma(\mu)$ and $M' = \Gamma(\mu')$. Then

$$\begin{aligned}|M-M'| &= \left|\frac{\mu d-c}{a-\mu b} - \frac{\mu' d-c}{a-\mu' b}\right| \\ &\leq \left|\frac{d(\mu-\mu')}{a-\mu b}\right| + |\mu' d-c| \circ \left|\frac{b(\mu-\mu')}{(a-\mu b)(a-\mu' b)}\right| \\ &\leq |\mu-\mu'| \left\{\frac{|d|}{|a-\mu b|} + \frac{|\mu' d-c|}{|a-\mu b|} \circ \frac{|b|}{|a-\mu' b|}\right\} .\end{aligned}$$

For $\varepsilon$ small and $|\mu|, |\mu'| \leq 1$ it follows that the term $\{\ \} < 1$ so $\Gamma$ is a contraction. (Note $|b|/|a-\mu' b| \leq 1$.) The fixed point of $\Gamma$ is the tangent line field of the invariant foliation.

The fibre contraction method used to prove differentiability in [2] is to take the derivative of $\Gamma$ and show for each fixed $\mu$ it is a contraction on trial derivatives. Because all the $\mu$'s vanish on $\{c\} \times I$, their partials with respect to y is zero at those points. Imposing the additional condition that the partial derivative with respect to x is also zero at points of $\{c\} \times I$, the set of trial derivatives $\mathcal{D}$ are continuous functions $H: TS \to \mathbb{R}$ which vanish on $TS|\{c\} \times I$. If $\Gamma(\mu) = M$, then a trial derivative H of $\mu$ gets transformed by

$$\psi_\mu(H) = -(a-\mu b)^{-2}(Da-bHDG-\mu Db)(\mu d-c) \\ +(a-\mu b)^{-1}(HDGd+\mu Dd-Dc) .$$

This equation is obtained by taking the derivative of $\Gamma$ at $\mu$. This equation defines $\psi_\mu(H)$ off $\{c\} \times I$, and it is extended to be zero on $\{c\} \times I$. For each fixed $\mu$, a calculation shows that $\psi_\mu$ is a contraction, $|\psi_\mu(H)-\psi_\mu(H^*)| \le \beta|H-H^*|$ with $\beta < 1$. To insure $\psi_\mu(H)$ is in $\mathcal{D}$, it is necessary to check that $\psi_\mu(H)$ is continuous at $\{c\} \times I$, i.e. that $\psi_\mu(H)(p) \to 0$ as $p \to \{c\} \times I$. First note that

$$\begin{aligned} |a-\mu b| &\ge (1-\varepsilon)\frac{\partial f}{\partial x}(p) - \varepsilon\frac{\partial g}{\partial x}(p) - \varepsilon\frac{\partial f}{\partial x}(p) - (1+\varepsilon)\frac{\partial g}{\partial x}(p) \\ &\ge (1-2\varepsilon)\frac{\partial f}{\partial x}(p) - (1+\varepsilon) \\ &\ge (1-\lambda^{-1}-3\varepsilon)\frac{\partial f}{\partial x}(p) \end{aligned}$$

since $|\partial g/\partial x\ (p)| \le 1$ and $|\partial f/\partial x\ (p)| \ge \lambda > 1$. Next

$$\begin{aligned} |\mu d-c| &= |\mu(1+\varepsilon)\frac{\partial g}{\partial y} - \varepsilon\frac{\partial g}{\partial y}| \\ &\le (1+2\varepsilon)\frac{\partial g}{\partial y} \end{aligned}$$

and

$$\begin{aligned} |b(a-\mu b)^{-1}| &\le |\varepsilon\frac{\partial f}{\partial x} + (1+\varepsilon)\frac{\partial g}{\partial x}|(1-\lambda^{-1}-3\varepsilon)\frac{\partial f}{\partial x}^{-1} \\ &\le (1+2\varepsilon)(1-\lambda^{-1}-3\varepsilon)^{-1} \end{aligned}$$

is bounded. Also $|DG| \le 2\frac{\partial f}{\partial x}(p)$ since this term dominates the derivative. Therefore in the first term of $\psi_\mu(H)$,

$$|(a-\mu b)^{-2} bH|DG|.|\mu d-c||$$
$$\leq |DG|(a-\mu b)^{-1}[b(a-\mu b)^{-1}]|\mu d-c| \; |H|$$
$$\leq 2(1-\lambda^{-1}-3\varepsilon)^{-1}\lambda^{-1}(1+2\varepsilon)(1-\lambda^{-1}-3\varepsilon)^{-1}(1+2\varepsilon)\frac{\partial g}{\partial y} \; |H|$$
$$\to 0$$

as $p \to \{c\} \times I$ because $\{H\} \in \mathfrak{D}$ is uniformly bounded on S. The second term can be shown to go to zero in a similar way. The only remaining terms are in fact the most difficult : $-(a-\mu b)^{-2}(Da-\mu Db)(\mu d-c)$. The difficulty is that $Da = (1+\varepsilon)D(\partial f/\partial x) + \partial f/\partial x\ D(1+\varepsilon) + \varepsilon D(\partial g/\partial x) + \partial g/\partial x\ D(\varepsilon)$ and $Db = \varepsilon D(\partial f/\partial x) + \partial f/\partial x\ D(\varepsilon) + (1+\varepsilon)D(\partial g/\partial x) + \partial g/\partial x\ D(1+\varepsilon)$ contain the terms $\partial^2 f/\partial x^2$ and $\partial^2 g/\partial x^2$ which are unbounded. (Note here $\varepsilon$ stands for several different $C^1$ functions whose derivatives are bounded.) Concentrating on the term containing $\partial^2 f/\partial x^2$ (the term with $\partial^2 g/\partial x^2$ is treated similarly),

$$|(a-\mu b)^{-2}(1+\varepsilon)\frac{\partial^2 f}{\partial x^2}(\mu d-c)|$$
$$\leq (1-\lambda^{-1}-3\varepsilon)^{-2}\frac{\partial f}{\partial x}^{-2}(1+\varepsilon)\frac{\partial^2 f}{\partial x^2}(1+2\varepsilon)\frac{\partial g}{\partial y}$$
$$\leq (1-\lambda^{-1}-3\varepsilon)^{-2}(1+\varepsilon)(1+2\varepsilon)\frac{\partial f}{\partial x}^{-1}\left[\frac{\partial f}{\partial x}\frac{\partial^2 f}{\partial x^2}\frac{\partial g}{\partial y}\right]$$
$$\to 0$$

as $p \to \{c\} \times I$ because $\partial f/\partial x^{-1} \to 0$ and the term inside [ ] is bounded by assumption (6). This completes the check that $\psi_\mu(H)$ is continuous at $\{c\} \times I$.

Therefore $(\Gamma,\psi):\mathfrak{J} \times \mathfrak{D} \to \mathfrak{J} \times \mathfrak{D}$ is a fibre contraction and by [2, Theorem 1.2] given any $C^1$ $\mu$ and $H = D\mu$ the sequence $(\mu^n, H^n) = (\Gamma(\mu^{n-1}), \psi_{\mu^{n-1}}(H^{n-1}))$ converges to some invariant $(\mu^\infty, H^\infty)$. Below it is proved by induction on n that $\mu^n$ is differentiable and $D\mu^n = H^n$ even on $\{c\} \times I$. By uniform convergence it follows that $\mu^\infty$ is differentiable and $D\mu^\infty = H^\infty$.

Now assume $\mu^{n-1}$ is differentiable and $D\mu^{n-1} = H^{n-1}$. From the form of the graph transform it follows that $D\mu^n = D\Gamma(\mu^{n-1}) = \psi_{\mu^{n-1}}(D\mu^{n-1}) = \psi_{\mu^{n-1}}(H^{n-1}) = H^n$ off $\{c\} \times I$. Now $\mu^n$ is a continuous function which vanishes on $\{c\} \times I$, which is $C^1$ off

$\{c\} \times I$, and whose derivative goes to zero as $p \to \{c\} \times I$. Let $h(x,y)$ be the continuous function which equals $\partial\mu^n/\partial x$ off $x = c$ and equals zero for $x = c$. Then

$$\int_c^x h(t,y)dt = \lim_{r\to c} \int_r^x h(t,y)dt$$

$$= \lim_{r\to c} \mu^n(x,y) - \mu^n(r,y)$$

$$= \mu^n(x,y) \quad .$$

Thus $\mu^n$ has a partial derivative with respect to x everywhere which is continuous. Clearly $\partial\mu^n/\partial y(c,y)$ exists and equals zero. Because $\mu^n$ has continuous partial derivatives, it is $C^1$.

## References.


1. J. Guckenheimer & R.F. Williams, Structural stability of Lorenz attractors, Publ. IHES, no. 50 (1979), pp.59-72.

2. M. Hirsch & C. Pugh, Stable manifolds and hyperbolic sets, Proceedings of Symposia in Pure Math 14, Amer. Math. Soc., 1970, pp.133-163.

3. E. Lorenz, Deterministic nonperiodic flow, J. Atmospheric Sciences 20 (1963) pp.130-141.

4. J. Palis, C. Pugh & R.C. Robinson, Nondifferentiability of invariant foliations, Dynamical Systems - Warwick 1974, ed. A. Manning, Lecture Notes in Math., 468, Springer-Verlag, Berlin-Heidelberg-New York, 1975.

5. C. Robinson & L.S. Young, Nonabsolutely continuous foliations for an Anosov diffeomorphism, to appear in Inventiones Math.

6. R.F. Williams, The structure of Lorenz attractors, Publ. IHES, no. 50 (1979), pp.73-99.

7. A. Katok & J-M. Strelcyn, Invariant manifolds for smooth maps with singularities, preprint.


R.C. Robinson : Department of Mathematics, Northwestern University, Evanston, Illinois 60201, USA.

## On the bifurcations creating horseshoes.

Sebastian J. van Strien*.

### §1. Introduction.

One of the basic problems in bifurcation theory is to understand the way in which horseshoes are created. The simplest example of a horseshoe is provided by the Smale horseshoe, see [Sm], realised as a basic set of a diffeomorphism of the 2-sphere $S^2$. The construction of this diffeomorphism proceeds as follows. First define a map $\Phi_1$, from a rectangle Q = ABCD in $\mathbb{R}^2$ into itself, as in figure (1.1). Then extend this

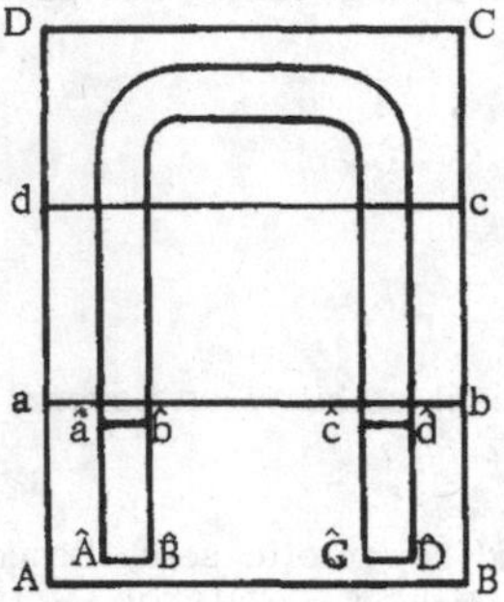


Figure (1.1)

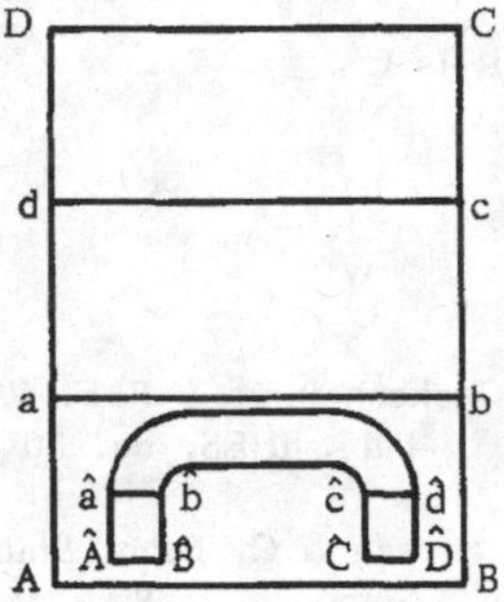


Figure (1.2)

map to a diffeomorphism of $S^2$. If $\Phi_1$ is chosen properly the set $\Lambda = \bigcap_{n\in\mathbb{Z}} \Phi_1^n$ (abcd) is a hyperbolic basic set of $\Phi_1$ and $\Phi_1|\Lambda$ is conjugate to the shift map on all biinfinite sequences of 0's and 1's. A rather more trivial diffeomorphism of $S^2$ is provided by the same construction except that the image of Q lies in ABba, as in figure (1.2).

Extend this to $S^2$ to get a diffeomorphism $\Phi_0$. $\Phi_0$ can be chosen such that the non-wandering set of $\Phi_0$ consists of a sink in ABba and a source at $\infty$.

Now consider a path $\Phi_\mu, \mu \in [0,1]$ of diffeomorphisms of $S^2$ such that $\Phi_0$ and $\Phi_1$ are the diffeomorphisms defined as above. Then it is a natural and important problem of bifurcation theory to understand the structure of the bifurcations which occur in the family as $\mu$ changes from 0 to 1.

---

* Research supported by Netherlands Organisation for the Advancement of Pure Research (ZWO). This is a rewritten version of a Utrecht preprint (November 1979).

For such a family $\Phi_\mu$, Hénon [Hé] did computer experiments. For a certain value, $\mu_0$, the picture of the orbits of $\Phi_{\mu_0}$ suggests that $\Phi_{\mu_0}$ has a strange attractor. Furthermore, a result of Newhouse [Ne] indicates that for certain parameter values $\mu_i$ the maps $\Phi_{\mu_i}$ have infinitely many sinks. This phenomena might be an explanation of the pictures the computer experiments gave.

In any case this complicated behaviour suggests that the task of understanding the complete bifurcation structure is probably hopeless. Therefore I shall look at the very special case when the diffeomorphisms $\Phi_\mu : Q \to Q$ are very near to singular maps, i.e. when the image of Q is very thin for all $\mu \in [0,1]$.

Here it will be shown that in this case one can learn at least something from the structure theory of one-dimensional endomorphisms (e.g. [JR, Mil, M.T.]), using some analytical estimates. In fact consider one-dimensional endomorphisms $f: \mathbb{R} \to \mathbb{R}$ with a graph as in figure (1.3). In Theorem A it will be shown that an open class of such one-dimensional endomorphisms is nearly Axiom A.

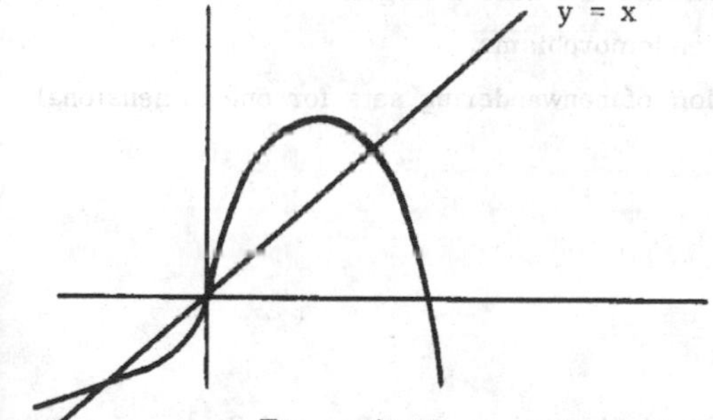

Figure (1.3)

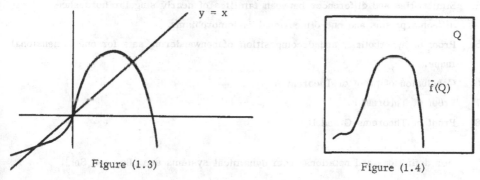

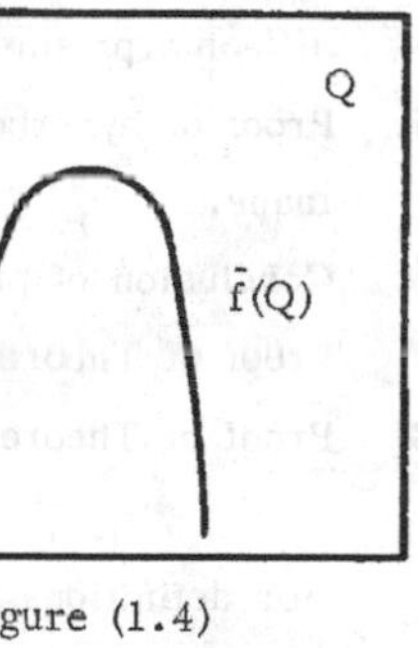

Figure (1.4)

In fact, the nonwandering set of such endomorphisms decompose into sets $\Omega_i$ (possibly countably many), all of which, with the possible exception of one, are hyperbolic. There is precisely one attracting set, and it is this set that might not be hyperbolic. In Theorem A a quite detailed description of the dynamics of these maps is given. As the statement of this Theorem is very long I have found it necessary to subdivide it.

In Theorem B it will follow that families of one-dimensional endomorphisms have nice bifurcations.

Now one can consider a one-dimensional endomorphism $f: \mathbb{R} \to \mathbb{R}$ also as a singular map $\bar{f}:Q \to Q$ :

$$\bar{f}(x,y) = (y, f(y))$$

as in figure (1.4). Diffeomorphisms $\Phi:Q \to Q$ which are $C^1$ near singular maps are called nearly-singular horseshoe diffeomorphisms. In Theorems C and D a comparison is made between the bifurcations of families of nearly-singular horseshoe diffeomorphisms and one-dimensional endomorphisms.

The key result of this paper is the hyperbolicity and the decomposition of the nonwandering set for one-dimensional endomorphisms obtained in Theorem A.

The organisation of this paper is as follows :

§2. Statement of results for one-dimensional maps. Theorem A.
§3. Statement of result for families of one-dimensional maps. Theorem B.
§4. Similarities and differences between families of nearly singular horseshoe diffeomorphisms and one-dimensional endomorphisms.
§5. Proof of hyperbolicity and decomposition of nonwandering sets for one-dimensional maps.
§6. Conclusion of proof of Theorem A.
§7. Proof of Theorem B.
§8. Proof of Theorems C and D.

For definitions and notations from dynamical systems see [Bo, and Sm].

Finally I wish to thank my supervisors David Rand and Floris Takens, and also J. Guckenheimer, M. Misiurewicz, S. Newhouse, H. Nijmeijer, and the referee, for their stimulating remarks.

## §2. Statement of results for one-dimensional maps.

A very important concept in the theory of dynamical systems is hyperbolicity. In general, little is understood about the structure of non-hyperbolic dynamical systems and hyperbolicity seems to be necessary for the persistance of the nonwandering set under perturbations of the dynamical system. But for most explicitly given dynamical systems it is extremely difficult to check whether the nonwandering set $\Omega(\Phi)$ is hyperbolic. Even for quadratic maps of the plane little is known, (see [Hé]).

For a rather large open class of one-dimensional endomorphisms, it turns out that the nonwandering set of such endomorphisms decomposes into sets $\Omega_i$ (possibly countably many), all of which, with the possible exception of one, are hyperbolic. To check whether a one-dimensional endomorphism is in this class involves only a trivial computation. I will show that the dynamics of these maps are also known in great detail.

§2.a Decomposition and Hyperbolicity.

The class of one-dimensional endomorphisms we will consider is the class $\mathfrak{a}$ of $C^3$ functions $f:\mathbb{R} \to \mathbb{R}$ with

(1) $f(+1) = f(-1) = -1, \quad f'(-1) > 1$

(2) $f(x)$ has one unique maximum at a point $c$, $c \in (-1,1)$ and $f'(x) \neq 0$ for $x \neq c$. Also $f''(c) \neq 0$.

(3) $f$ has one hyperbolic attracting fixed point $x_0 \in (-\infty, -1)$.

Additionally we need a rigidity property on the map f and it is this property that will give us hyperbolicity.

(4) The Schwarzian derivative

$$S(f)(x) = \frac{f'''(x)}{f'(x)} - \frac{3}{2} \cdot \left(\frac{f''(x)}{f'(x)}\right)^2$$

is negative for $x \in [-1,1]\setminus\{c\}$.

Unless otherwise stated we assume that maps f are contained in this open class $\mathfrak{a}$. The condition (4) was introduced by D. Singer [Si]. Two crucial properties of the Schwarzian derivative are that

i) $S(f) < 0$ implies $S(f^n) < 0$ for all $n \in \mathbb{N}$.

ii) $|f'|$ has no positive local minima, if $S(f) < 0$ .

Therefore the graph of $f \in \mathfrak{a}$ is as in figure (2.1).

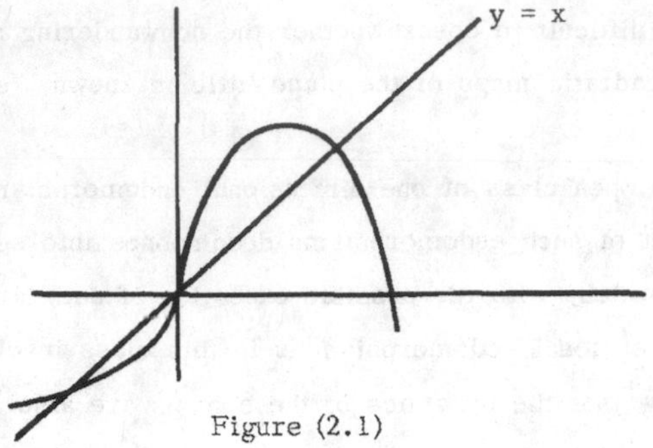


Figure (2.1)

Proposition (2.1), see [Si]. If $f \in \mathcal{G}$ then there exists at the most one $z \in (-1,1)$ such that $f^n(z) = z$ and $|(f^n)'(z)| \leq 1$, for some $n > 0$.

We shall see that such an orbit O(z) "attracts". But let us first give the following definition. A closed set $\Lambda$ with $f(\Lambda) = \Lambda$ is hyperbolically repelling if there are numbers $C \in (0,\infty)$ and $\tau \in (1,\infty)$ such that for $x \in \Lambda$

$$|(f^k)'(x)| \geq C.\tau^k$$

for all $k \in \mathbb{N}$. If this is the case, we shall sometimes express this by saying that $f:\Lambda \to \Lambda$ is called hyperbolically repelling.

Theorem A(a). (Decomposition and hyperbolicity.) Let $\Omega(f)$ be the nonwandering set of f. $\Omega(f)$ has a nearly hyperbolic $\Omega$-decomposition : for some $p \in \mathbb{N} \cup \{\infty\}$

$$\Omega(f) = \bigcup_{j=0}^{p} \Omega_j$$

where

1) $\Omega_j$ is closed, $f(\Omega_j) = \Omega_j$.
2) $\Omega_j$ is hyperbolically repelling for $0 \leq j < p$.
3) $\Omega_p$ might not be hyperbolic.

So we have a decomposition of the nonwandering set, which is similar to those of Axiom A systems. The difference is that here we can have an infinite number of basic sets and that one basic set need not be hyperbolic. The next Theorem gives conditions for the finiteness of the decomposition and states that the non-hyperbolic set $\Omega_p$ is an attractor.

A compact invariant set Z is called an <u>attractor</u> if the set $B = \{x : f^n(x) \to Z \text{ as } n \to \infty\}$ is a neighbourhood of Z or more generally if the Lebesque measure m(B) of B is strictly bigger than m(Z).

<u>Theorem A(b).</u> In the above decomposition

1) $\Omega_p$ is the unique attractor. Moreover, the set $B := \{x : f^n(x) \to \Omega_p \text{ as } n \to \infty\}$ is an open, dense set in (-1,1) if $p < \infty$ and is residual in (-1,1) if $p = \infty$. Furthermore B has full Lebesgue measure in (-1,1).

2) If there is a periodic orbit O(z) with eigenvalue $\lambda$, with $|\lambda| \le 1$, then

   a) $p < \infty$

   b) The set $\tilde{B} := \{x | f^n(x) \to O(z)\}$ is open and dense in (-1,1). Also it has full Lebesgue measure in (-1,1).

3) Call the period of such an attracting orbit n. This orbit is either hyperbolic, i.e. $|\lambda| < 1$, or quasi-hyperbolic in this sense :

   a) if $\lambda = 1$, then $(f^n)''(z) \neq 0$, i.e. O(z) attracts from one side,

   b) if $\lambda = -1$, then $(f^{2n})''(z) = 0$ and $(f^{2n})'''(z) < 0$, i.e. O(z) attracts locally from both sides.

## §2.b <u>Dynamics of $f : \Omega_j \to \Omega_j$.</u>

The dynamics of $f : \Omega_j \to \Omega_j$ is very well understood.

<u>Theorem A(c).</u> <u>(Dynamics of $f : \Omega_j \to \Omega_j$, $0 \le j < p$.)</u> In the above decomposition

1) $\Omega_0 \subset Per(f)$ and $\Omega_0$ is finite, (in fact $\Omega_0 = \{x_0, -1\}$).

2) $\Omega_j = Per_j \cup C_j$, $0 < j < p$ where $Per_j$ is a finite subset of Per(f) and $C_j$ is a Cantor set.

3) $f : C_j \to C_j$ is a transitive subshift of finite type.

So these sets $\Omega_j$ are similar to the basic sets of Axiom A diffeomorphisms. If $p < \infty$, then there are four possibilities for $f : \Omega_p \to \Omega_p$. The attracting set contained in $\Omega_p$ can be :

1(a) a periodic orbit, isolated in $\Omega(f)$, with eigenvalue $\lambda$, $-1 \leq \lambda < 1$,

1(b) a Cantor set,

2(a) a periodic orbit, non-isolated in $\Omega(f)$, with eigenvalue $\lambda = 1$.

2(b) a finite union of intervals.

<u>Theorem A(d).</u> <u>(The four possibilities for the dynamics of $f: \Omega_p \to \Omega_p$, if $p < \infty$.)</u>

<u>Case 1:</u> If the topological entropy $h(f|\Omega_p)$ of $f:\Omega_p \to \Omega_p$ is zero, then one of the following two cases holds.

(a) The attracting set is a periodic orbit, which is isolated in $\Omega(f)$. Then $\Omega_p$ is a finite subset of Per(f) and there is at the most one non-hyperbolic periodic in $\Omega_p$. For some $n \geq 0$ and some $N_p > 0$, $\Omega_p$ consists of repelling orbits, precisely one for each period $N_p \cdot 2^i$, for each $0 \leq i \leq n$, and precisely one attracting periodic orbit O(z) of period $N_p \cdot 2^n$ or $N_p \cdot 2^{n+1}$. (For the eigenvalue $\lambda$ of this orbit O(z) one has $-1 \leq \lambda < 1$.)

(b) The attracting set is a Cantor set $C_p$. Then $\Omega_p = \mathrm{Per}_p \cup C_p$. $C_p$ is an invariant attracting Cantor set, namely the orbit of the critical point c. $\partial \mathrm{Per}_p = C_p$. All periodic orbits are hyperbolically repelling, but clearly $C_p$ is non-hyperbolic. $\mathrm{Per}_p$ consists of periodic orbits, precisely one of period $M_p \cdot 2^i$ for each $i > 1$.

<u>Case 2 :</u> If $h(f|\Omega_p) > 0$ then one of the following two cases holds.

(a) The attracting set is a periodic orbit, which is not isolated in $\Omega(f)$. Then $\Omega(f) = \mathrm{Per}_p \cup C_p$. $\mathrm{Per}_p$ and $C_p$ are exactly as in Theorem A(c). The Cantor set $C_p$ contains exactly one periodic orbit which is quasi-hyperbolic as in Theorem A(b), case 3(a).

(b) The attracting set is a finite union of intervals. Then $\Omega_p$ contains infinitely many periodic orbits, all of which are hyperbolically repelling. $\Omega_p = \mathrm{Per}_p \cup I_p$ where $\mathrm{Per}_p$ is a finite set of periodic points isolated in $\Omega(f)$ and the set $I_p$ is the finite union of intervals. $I_p$ is the attracting set. $f:\Omega_p \to \Omega_p$ is conjugate to a piece-wise linear map, (see §6).

If $p = \infty$, then there is just one case.

<u>Theorem A(e).</u> <u>(Dynamics of $f:\Omega_p \to \Omega_p$, if $p = \infty$).</u> If $p = \infty$, then $\Omega_\infty = \overline{O(c)}$, the

closure of the orbit of the critical point. This set is a Cantor set and $h(f|\Omega_\infty) = 0$.

§2.c No-cycle property.

The previous properties of f, such as the decomposition and the hyperbolicity, did not take into consideration how all these sets $\Omega_j$ are related. We will show that we have a filtration of the manifold, as for Axiom A systems.

**Theorem A(f).** (No cycles and filtrations.) We can choose the decomposition $\Omega(f) = \bigcup_{j=0}^{p} \Omega_j$ such that there exists a sequence of closed intervals $M_j$, $M_j \subset M_{j-1}$, for $0 \le j \le p$ and j finite, with

1) $M_j$ is a finite union of closed intervals.

2.a) $f(M_j) \subset \mathrm{int}(M_j)$, $0 \le j < p$.

b) If p is finite, then $f(M_p) \subset M_p$. If $f: \Omega_p \to \Omega_p$ is as in case 1 or case 2(a) of Theorem A(d), then $f(M_p) \subset \mathrm{int}(M_p)$.

3) $\Omega_j = \bigcap_{n\ge 0} f^n(M_j \setminus M_{j+1})$ for $0 \le j < p$.

4) $\Omega_p = \Omega(f) \cap (\bigcap_{0\le i<p+1} M_i)$ .

Furthermore one has that :

5) the topological entropy $h(f|\Omega_j)$ decreases : $h(f|\Omega_{j+1}) < h(f|\Omega_j)$ for $1 \le j < p$. (Since $\Omega_0$ is finite, $h(f|\Omega_o) = 0$.)

6) $\Omega_i \cap \Omega_j = \emptyset$ for $0 \le i, j \le p$ and $i \ne j$. Possibly $\Omega_{p-1} \cap \Omega_p \ne \emptyset$, if p is finite.

§2.d Relation with other papers.

A decomposition of $\Omega(f)$ of this form, for maps not necessarily satisfying the Schwarzian derivative condition, was already given by L. Jonker and D. Rand in [J.R. part I].

The new content of Theorem A if that when $S(f) < 0$, one also obtains (i) hyperbolicity of the sets $\Omega_j$, $0 \le j < p$ and (ii) the conjugacies (rather than semi-conjugacies) of $f|\Omega_j$ with subshifts of finite type, for $1 \le j < p$.

The analytical estimates needed to prove the hyperbolicity of $\Omega_j$ are related to work of J. Guckenheimer ([Gu]), M. Misiurewicz ([Mis.2]) and L. Nusse ([Nu]) see also [C.E.].

## §3. Statement of Results for familes of one-dimensional maps.

In this paper we are particularly interested in bifurcations of maps, so let us consider the class $P\mathcal{G}$ of $C^1$ families of maps in $\mathcal{G}$, i.e. maps $f:[0,1] \times \mathbb{R} \to \mathbb{R}$ such that for $f_t(x) = f(t,x)$, $f_t \in \mathcal{G}$ and $\frac{d^{i+j}}{dt^i dx^j} f(t,x)$ exists and is continuous for $t = 0,1$ and $j = 0,1,2,3$.

In the previous Theorem A we saw that for $f_t$ only one basic set can be non-hyperbolic. This non-hyperbolic set is in cases 1(a), 2(a) a periodic orbit. We will restrict our attention to a set of parameter values P defined as follows. P is the interior of the set of parameter values $t \in [0,1]$ such that $f_t$ has an attracting periodic orbit, i.e. a point $z \in (-1,1)$ with $f^n(z) = z$ and $-1 \leq (f^n)'(z) < 1$.

Let $\mathfrak{U}$ be a connected component of P. Theorem B says that the bifurcations of $\Omega(f_t)$, for $t \in \mathfrak{U}$, are quite simple.

Theorem B(a). (Elementary bifurcations for $t \in \mathfrak{U}$.) For $t \in \mathfrak{U}$, let $\Omega(f_t) = \Omega_0(t) \cup \Omega_1(t) \cup \ldots \cup \Omega_{p(t)}(t)$ be the generalised $\Omega$-decomposition of $f_t$ as above. Then $p(t) = p$ is a finite constant and the topological entropy $h(f_t)$ is constant for $t \in \mathfrak{U}$. For $t,t' \in \mathfrak{U}$ and $j < p$, $f_t|\Omega_j(t)$ is topologically conjugate to $f_{t'}|\Omega_j(t')$.

Let $O(z(t)) \subset \Omega_p(t)$ be the attracting periodic orbit. Then z(t) may be chosen to vary continuously with t, and z(t) has period $N_p \cdot 2^{n(t)}$, where $N_p$ is a constant and $|n(t) - n(s)| \leq 1$ for $|t-s|$ small and $t,s \in \mathfrak{U}$. Only bifurcations of O(z (t)) occur.

Let us draw the set $\Omega_p(t)$ for the case that $N_p = 3$ in a bifurcation diagram. The solid curves represent the attracting periodic orbit O(z(t)) and the dotted lines represent the repelling orbits, from which the orbit O(z(t)) splits off when the orbit O(z(t)) doubles in period. Actually it is not known whether $t \mapsto n(t)$ is monotone for some class of one-dimensional maps $\{f_t\}$.

So we see that the bifurcations of $\Omega(f_t)$ for $t \in \mathfrak{U}$ are well understood. Let $\mathfrak{U} = (h_1, h_2)$. What happens if $t \downarrow h_1$ or $t \uparrow h_2$?

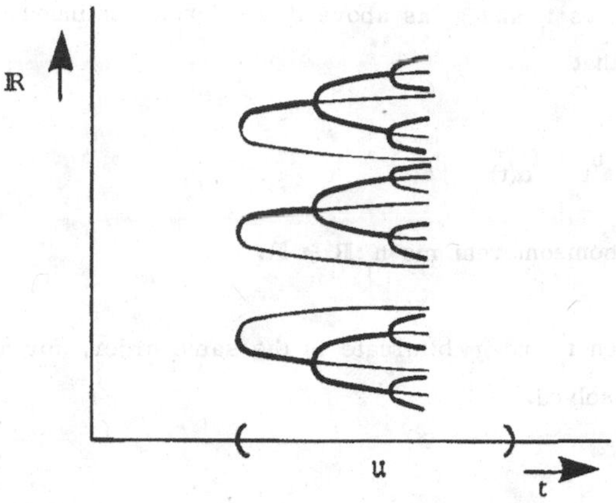


Figure (3.1)

Theorem B(b). (Non-elementary bifurcations for $t = h_1$ or $t = h_2$.) For $t \downarrow h_1$ or $t \uparrow h_2$ one of the following two cases holds :

Case 1 : $n(t) \to \infty$. In this case the attracting set of $f_{h_i}$ is a Cantor set, and $f_{h_i}: \Omega_p(h_i) \to \Omega_p(h_i)$ is as in case 1(b) of Theorem A(d).

Case 2 : $n(t) = 0$ for $t \approx h_i$, $t \in \mathfrak{U}$. In this case the attracting set is a quasi-hyperbolic periodic orbit, but this orbit is non-isolated in $\Omega_{p-1}(h_i)$. $f_{h_i}: \Omega_{p-1}(h_i) \to \Omega_{p-1}(h_i)$ is as in case 2(a) of Theorem A(d). $\Omega_p(h_i)$ is empty.

The other situations i) $p(t) = \infty$ and ii) $p(t) < \infty$ and $f_t: \Omega_p(t) \to \Omega_p(t)$ as in case 2(b) in Theorem A(d) occur often:

Theorem B(c). If $h(f_0) \neq h(f_1)$ then there exist an infinite number of parameter values $\mu$ such that i) $p(\mu) = \infty$ or ii) $p(\mu) < \infty$ and $f_\mu: \Omega_p(\mu) \to \Omega_p(\mu)$ as in case 2(b).

Finally, the bifurcation pattern of all one-dimensional endomorphisms is essentially the same. Consider a family $f_t$ as above such that $Per(f_0)$ consists only of fixed points and such that $f_1[-1,1] \supset [-1,1]$. Then the following Theorem follows from [Gu] and [J.R., part II].

Theorem B(d). For families $f_t$ and $g_t$ as above there exist continuous surjections $\alpha,\beta:[0,1] \to [0,1]$ such that

$$h_t \circ f_{\alpha(t)} = g_{\beta(t)} \circ h_t$$

where $h_t$ is a family of homeomorphisms $h_t:\mathbb{R} \to \mathbb{R}$.

The question of when $f_t$ and $g_t$ bifurcate in the same order, for an open class of families in PG is still unsolved.

## §4. Families of nearly singular horseshoe diffeomorphisms.

### §4.a Similarities between nearly singular horseshoe diffeomorphisms and one-dimensional endomorphisms.

In the previous section it was shown that families of one-dimensional maps bifurcate in a well understood way. Here we show that families of diffeomorphisms near singular families bifurcate approximately in the same way, although we will also show that there are big differences.

Let PG be as in §3. Define a singular family $\bar{f}_t:\mathbb{R}^2 \to \mathbb{R}^2$ corresponding to a one-dimensional endomorphism family $\{f_t\}$ as follows :

$$\bar{f}_t(x,y) = (y, f_t(y)) .$$

Clearly the endomorphism $f_t:\mathbb{R} \to \mathbb{R}$ has the same dynamics as the singular map $\bar{f}_t:\mathbb{R}^2 \to \mathbb{R}^2$.

Let E (resp. Emb) be the class of $C^3$-maps (resp. embeddings) for $\mathbb{R}^2$ to $\mathbb{R}^2$, and give A, E, and Emb the uniform $C^3$-topology. Let PE (resp. PEmb) be the class of $C^1$-families of maps in E (resp. in Emb), similar to PG. Put a metric on PE:

$$d(\Phi,\Psi) = \sup_{t\in[0,1]} \|\Phi_t - \Psi_t\|_{C^3} .$$

Remark that $\{\bar{f}_t\} \in$ PE and that P was defined as the interior of the set of parameters for which $f_t$ has an attracting periodic orbit with eigenvalue $\lambda$, $-1 \le \lambda < 1$. For $\delta > 0$ let :

$$P(\delta) = \{t \in [0,1] \mid (t-\delta, t+\delta) \subset P\},$$

i.e. $P(\delta)$ is slightly smaller than P.

Families $\{\Phi_t\}$ and $\{\Psi_t\} \in PE$ are called <u>$\Omega$-conjugate on $Q \subset [0,1]$</u>, if there exists a homeomorphism $\rho: Q \to Q$ and a family of conjugating homeomorphisms $h_t: \Omega(\Phi_t) \to \Omega(\Psi_{\rho(t)})$ such that $\Psi_{\rho(t)} \circ h_t = h_t \circ \Phi_t$, for $t \in Q$. $\{\Phi_t\}$ is <u>$\Omega$-semi-conjugate on $\Phi$</u> to $\{\Psi_t\}$ if $\rho$ and $h_t$ are as above, except that $h_t$ need not be a homeomorphism, but only a bijection from $Per(\Phi_t)$ to $Per(\Psi_{\rho(t)})$ and a continuous surjection from $\Omega(\Phi_t)$ to $\Omega(\Psi_{\rho(t)})$.

<u>Theorem C.</u> There is a residual subset of PG such that for each $\{\bar{f}_t\}$ in this set there exists a continuous function $\delta(\varepsilon)$ for $\varepsilon \geq 0$, with $\delta(0) = 0$, such that for each $\varepsilon > 0$ the following holds :

1) For each $\{\Phi_t\} \in$ PEmb with $d(\Phi, \bar{f}) < \varepsilon$, $\{\Phi_t\}$ and $\{\bar{f}_t\}$ are <u>$\Omega$-semi conjugate on $P(\delta(\varepsilon))$.</u>

2) For $\{\Phi_t^{(i)}\} \in$ PEmb with $d(\Phi^{(i)}, f) < \varepsilon$, $\{\Phi_t^{(1)}\}$ and $\{\Phi_t^{(2)}\}$ are $\Omega$-conjugate on $P(\delta(\varepsilon))$.

This theorem says that if a family of diffeomorphisms is near to a singular family, then the set of parameter values $P(\delta)$ on which the $\Omega$-bifurcations are the same is nearly as big as the set P. Schematically this Theorem can be explained as follows. In figure (5.1) the family $\{f_t\}$ is represented by a solid horizontal line and the nearby diffeomorphism family $\{\Phi_t\}$ by a dotted line. For convenience take a component $\mathfrak{U}$ of P. For $t \in \mathfrak{U}(\delta)$ these families have the same bifurcations.

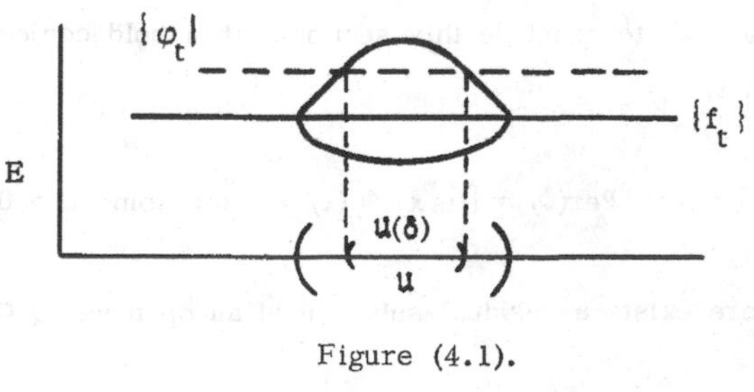


Figure (4.1).

§4.b Differences between nearly-singular diffeomorphisms and endomorphisms.

Now there are certainly no families $\{\Phi_\mu\} \in$ PEmb near $\{\bar{f}_\mu\}$ such that these families are $\Omega$-conjugate for all $\mu \in [0,1]$. In fact this follows from a result of Newhouse [Ne]:

Let $\Lambda$ be a hyperbolic basic set for a $C^r$ diffeomorphism $\Phi: M \to M$ with $r \geq 2$ and M a compact two dimensional manifold. A non-degenerate tangency z of $W^u(x,\Phi)$ and $W^s(y,\Phi)$ for $x, y \in \Lambda$ will be called a non-degenerate homoclinic tangency for $\Lambda$ of $\Phi$. $\Lambda$ is a wild hyperbolic set if each $\Psi \in C^r$ near $\Phi$ has the property that $\Lambda(\Psi)$ has a non-degenerate homoclinic tangency.

Using the results of Newhouse we will prove :

Theorem D. Let $\{f_\mu\} \in$ PG, with $h(f_0) \neq h(f_1)$. For each family of diffeomorphisms $\{\Phi_\mu\}$, sufficiently near $\{\bar{f}_\mu\}$ in the PE topology, there exist countably many parameter values $\mu_i$, neighbourhoods $W(\mu_i)$ of the diffeomorphism $\Phi_{\mu_i}$, and residual subsets $\tilde{W}(\mu_i)$ of $W(\mu_i)$, such that :

1) Each $\Psi \in \tilde{W}(\mu_i)$ has a wild hyperbolic set.
2) Each $\Psi \in \tilde{W}(\mu_i)$ has infinitely many sinks.

Since singular maps $\bar{f}_\mu$ can have at most one sink, diffeomorphisms $\Psi \in W(\mu_i)$ cannot be conjugate to maps $\bar{f}_\mu$. Therefore there exists no neighbourhood V of $\{\bar{f}_\mu\}$ such that maps $\{\Phi_\mu\} \in$ V and $\{\bar{f}_\mu\}$ are semi-conjugate.

§4.c A conjecture.

A real understanding of the difference between diffeomorphism families and families of singular maps seems an extremely difficult, but important task.

Yet, I would like to conclude this section with a bold conjecture about this relation. Let for $\{\Phi_\mu\} \in$ PE,

$$\mathrm{Per}(\Phi) = \{(\mu, x) \mid \Phi_\mu^n(x) = x \text{ for some } n > 0\} .$$

Conjecture. There exists a residual subset B of an open set $Q \subset$ PE such that

1) The singular family $\{\bar{f}_\mu\}$, corresponding to the quadratic family

$f_\mu = -\mu . x^2 + \mu - 1$, $\mu \in (0,2]$, is contained in B.

2) For each $\Phi$, $\Phi \in B$, there exists a bijection $h:Per(\Phi) \to Per(\Psi)$.

Theorem D does not contradict this conjecture. It merely indicates that $h:Per(\Phi) \to Per(\Psi)$ cannot be of the form $h(\mu,x) = (\rho(\mu), h_\mu(x))$. The conjecture only states that the families $\{\Phi_\mu\}$ and $\{\Psi_\mu\}$ have the same bifurcations, but these bifurcations can occur in a different order.

## §5. Proof of Theorem A(a) : Decomposition and Hyperbolicity for f ε G.

The construction of the decomposition of the nonwandering set is due to L. Jonker and D. Rand, see [J.R., part I]. In order to be thorough, and because I want to prove the stronger properties for maps satisfying the Schwarzian derivative condition $S(f) < 0$, I shall give the construction here again. This decomposition uses the kneading theory of Milnor and Thurston, [M.T.].

### §5.a Some kneading theory, [M.T., J., J.R.].

Milnor and Thurston consider the class C of $C^0$ functions $f:[a,b] \to [a,b]$ such that $f(a) = f(b) = \{a,b\}$ and such that f has only one turning point c. The aim of the kneading theory of Milnor and Thurston is to relate the dynamics of maps $f \in C$ to the dynamics of piecewise linear maps $F_s:[-1,1] \to [-1,1]$, $F_s(x) = s - 1 - s|x|$.

graph of f ε C　　　　graph of $F_s$

Figure (5.1)

The key result of Milnor and Thurston is the following Theorem. Let h(f) be the topological entropy of f and $s(f) = \exp h(f)$.

Theorem (5.1), see [Mil, M.T., and J.R., part I].

i) Let $f \in C$ then $1 \le s \le 2$ .

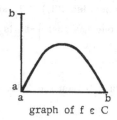

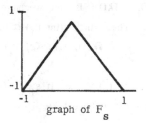

If $f([a,b]) = [a,b]$ then $s(f) = 2$.

ii) Let $s = s(f) > 1$, then $f$ is semi-conjugate to the piecewise linear map $F_s : [-1,1] \to [-1,1]$, i.e. there is a continuous monotone surjection $K_f$ (not necessarily strictly monotone) such that

$$\begin{array}{ccc} [a,b] & \xrightarrow{\ f\ } & [a,b] \\ K_f \downarrow & & \downarrow K_f \\ [-1,1] & \xrightarrow{\ F_s\ } & [-1,1] \end{array}$$

commutes.

iii) If moreover $F_s^n(0) = 0$, i.e. if $f^n(K^{-1}(0)) \subset K^{-1}(0)$, for some $n \in \mathbb{N}$, then $K_f^{-1}(0)$ is an interval. For $n$ minimal with $F_s^n(0) = 0$,

$$f^n | K^{-1}(0) \in \mathcal{C}\ .$$

In this case the entropy decreases :

$$h(f| \bigcup_{i\geq 0} f^i(K^{-1}(0))) < h(f|[a,b]).$$

## §5.b The decomposition of $\Omega(f)$.

The decomposition of the nonwandering set is done inductively using Theorem (5.1).

Let $\Omega_0 = \{x_0, -1\}$, $B_1 = [-1,1]$, $\tilde{M}_1 = \text{int}(B_1)$ and $n_1 = N_1 = 1$. If $f(B_1) \supset B_1$, then let $\Omega_1 = \Omega(f) \cap B_1$ and we have then $\Omega(f) = \Omega_0 \cup \Omega_1$. Otherwise $f(B_1) \subsetneqq B_1$ so put $f_1 = f|B_1$. Then since $f(\text{int } B_1) \subset \text{int } B_1$ and since $f'(-1) > 1$ one has that $\{-1\}$ is an isolated point in $\Omega(f_1)$, i.e. $\Omega(f_1) \cap \tilde{M}_1$ is a closed, invariant set.

If $s_1 = s(f_1) = 1$ or if $s_1 = s(f_1) > 1$ and $F^n_{s_1}(0) \neq 0$, $\forall n \geq 0$ then the induction step breaks off and we take

$$\Omega_1 = \Omega(f) \cap \tilde{M}_1\ .$$

If however $s_1 = s(f_1) > 1$ and $F_{s_1}^{n_2}(0) = 0$ for some $n_2 \in \mathbb{N}$, then we can repeat the construction. Let $B_2 = K_{f_1}^{-1}(0)$ and for $n_2$ minimal consider $f_2 = f_1^{n_2}|B_2$. Furthermore let $N_2 = n_1 \cdot n_2$ and $\tilde{M}_2 = \text{int} \bigcup_{0 \leq i < N_2} f^i(B_2)$. Since $f_2(B_2) \subset B_2$ the set $\tilde{M}_1 \setminus \tilde{M}_2$ is invariant, and since $f_1(B_1) \subsetneqq B_1$,

$$\Omega_1 = \Omega(f) \cap (\tilde{M}_1 \setminus \tilde{M}_2)$$

is a closed, invariant set, ($\{-1\}$ was an isolated point in $\Omega(f_1)$).

So suppose by induction that $f_j : B_j \to B_j$ is in class C and that the induction assumptions (5.2) hold :

Induction step (5.2).

(5.2(a)) $\quad f_j(B_j) \subsetneq B_j$

(5.2(b)) $\quad s_j = s(f_j) > 1$ and

(5.2(c)) $\quad F_{s_j}^{n_{j+1}}(0) = 0$ for some $n_{j+1} \in \mathbb{N}$, (take $n_{j+1}$ minimal).

Then according to Theorem (5.1), there exists a smaller interval $B_{j+1} \subset B_j$ ($B_{j+1} = K_{f_j}^{-1}(0)$) with $f_{j+1} = f_j^{n_{j+1}} | B_{j+1} \in C$. Let $N_{j+1} = n_1 \cdot \ldots \cdot n_{j+1}$ and $\tilde{M}_{j+1} = \mathrm{int}(\bigcup_{0 \le i < N_{j+1}} f^i(B_{j+1}))$. As before we have that $\Omega_j = \Omega(f) \cap (\tilde{M}_j \setminus \tilde{M}_{j+1})$ is a closed invariant set, provided $\partial B_j$ is isolated in $\Omega(f) \cap \tilde{M}_j$, which we will prove in Proposition (5.4)(a).

If the induction breaks off, $f_j(B_j) \subsetneq B_j$ but $s(f_j) = 1$ or $s_j = s(f_j) > 1$ and $F_{s_j}^n(0) \neq 0$, $\forall n > 0$, then let

$$\Omega_j = \Omega(f) \cap \tilde{M}_j \quad .$$

Finally, if the induction breaks off and $f_j(B_j) = B_j$, (i.e. $s(f_j) = 2$), then let

$$\Omega_j = \Omega(f) \cap Cl(\tilde{M}_j).$$

If the induction step (5.2) does not break off, then let $p = \infty$. If it does break off, then let $p \in \mathbb{N}$ be such that $f_j : B_j \to B_j$ satisfies (5.2) for $1 \le j < p$ and such that (5.3) holds:

(5.3) $\quad$ <u>The map $f_p : B_p \to B_p$ satisfies case 1 or 2 :</u>

Case 1 : $\quad s(f_p) = 1$ .

Case 2(a) : $s(f_p) > 1$ and $F_{s_p}^{n_{p+1}}(0) = 0$, i.e. $f_p : B_p \to B_p$ does satisfy (5.2) but for the fixed point $x \in \partial B_{p+1}$ of the map $f_{p+1} : B_{p+1} \to B_{p+1}$ one has $f'_{p+1}(x) = 1$. In the next Proposition we will prove that in this case $s(f_{p+1}) = 1$, $\Omega(f_{p+1}) = \{x, f(x), \ldots, f^{N_{p+1}}(x) = x\}$ i.e. $\Omega(f_{p+1}) \subset \Omega(f_p)$.

Case 2(b) : $s(f_p) > 1$ and $F_{s_p}^{n}(0) \neq 0$, for all $n > 0$.

These cases correspond to the several possibilities for $\Omega_p$ in Theorem A(d). In the next section we will prove that the sets $\Omega_j$, $0 \leq j < p$ are hyperbolic. In order to prove that we will need the following result :

Proposition (5.4).

(a) For $1 \leq j < p$ one has for the fixed point $x \in \partial B_{j+1}$ of $f_{j+1} : B_{j+1} \to B_{j+1}$, $f'_{j+1}(x) > 1$.

(b) If $f'_{j+1}(x) = 1$ for the fixed point $x \in \partial B_{j+1}$ of $f_{j+1}$ then one has $f''_{j+1}(x) \neq 0$, $\Omega(f_{j+1}) = \{x\}$ and $s(f_{j+1}) = 1$.

Proof :

i) Since $f_{j+1} \in \mathcal{C}$, $f'_{j+1}(x) > 0$.

ii) Suppose that $0 < f'_{j+1}(x) < 1$. Then x is an attracting periodic point of $f_j : B_j \to B_j$. But $F_{s_j}$ has no attracting periodic orbits, so the semi-conjugacy $K_{f_j}$ is locally constant near x. However this is a contradiction with $x \in \partial B_{j+1} = \partial K_{f_j}^{-1}(0)$.

iii) Now suppose that $f'_{j+1}(x) = 1$. Let $[a,b] = K_{f_j}^{-1}(0)$. Since $f_{j+1} \in \mathcal{C}$, either $f_{j+1}(a) = a$ or $f_{j+1}(b) = b$. We will treat the former case, i.e. the case that $x = a$, and show that $f''_{j+1}(a) < 0$ and that $f_{j+1}$ has just one fixed point.

As above in ii) $f''_{j+1}(a) > 0$ or $f''_{j+1}(a) = 0$ and $f'''_{j+1}(a) < 0$ are in contradiction with $a = x \in \partial B_{j+1}$. But for $f''_{j+1}(a) = 0$ we have $S(f_{j+1})(a) = f'''_{j+1}(a)$ and since $S(f) < 0$ implies $S(f^n) < 0$ this implies $f'''_{j+1}(a) < 0$. Linking these two observations together proves $f''_{j+1}(a) < 0$.

Moreover, $S(f) < 0$ also implies that $|f'|$ has no local positive minima. Hence $f_{j+1} : [a,b] \to [a,b]$ has only one fixed point $\{a\}$, from which $\Omega(f_{j+1}) = \{a\}$ and $s(f_{j+1}) = 1$ follows. Q.E.D.

Later we also will consider the sets $\Lambda_j$: Since $f^{N_{j+1}}(B_{j+1}) \subset B_{j+1}$ the set of points $\Lambda_j$ in the closure of $\tilde{M}_j$ that are never mapped into $\mathrm{int}(B_{j+1})$ i.e.

$$\Lambda_j = C\ell(\tilde{M}_j) \setminus \bigcup_{n\geq 0} f^{-n}(\mathrm{int}\, B_{j+1})$$

is closed and invariant. .

§5.c Hyperbolicity of the sets $\Omega_j$, $0 < j < p$.

We will now prove the hyperbolicity of the sets $\Omega_j$, $0 < j < p$. Let $q = j + 1$,

$$T_1 = \mathrm{int}(B_q) \quad \text{and}$$
$$\Lambda = [-1,1] \setminus \bigcup_{k\geq 0} f^{-k}(T_1)$$

Since $\Omega_j \subset \Lambda$ it suffices to prove that $\Lambda$ is a hyperbolic set :

Lemma A : $\Lambda$ is a hyperbolic Cantor set, i.e. there exists a $\tau > 1$ and $C > 0$ such that $|(f^k)'(x)| \geq C.\tau^k$ for all $x \in \Lambda$ and $k \in \mathbb{N}$.

The proof of this Lemma goes as follows. Let $I_n$ be a maximal interval such that $f^k(I_n) \cap T_1 = \emptyset$, for $k = 0,1,2,\ldots,n-1$, i.e. a component of $[-1,1] \setminus \bigcup_{k=0}^{n-1} f^{-k}(T_1)$, and let $F_n$ be the union of all such components $I_n$. Since the singular point c is contained in $T_1$, this implies that $f^k|I_n$ is a homeomorphism for $k = 1,1,\ldots,n$. I shall first prove a generalisation of a Theorem by Henry [He], using the condition on the Schwarzian derivative of f.

Lemma B : There exists a $\tau < 1$ such that Lebesgue measure $\ell(F_n)$ of $F_n$ satisfies $\ell(F_n) < \tau^n$ for $n \in \mathbb{N}$.

Suppose that $f^n(I_n) \cap T_1 \neq \emptyset$. We will show that there are two maximal intervals $I^1_{n+1}$, $I^2_{n+1} \subset I_n$ such that $f^n(I^i_{n+1}) \cap T_1 \neq \emptyset$, $i = 1,2$. Let $J_n \supset I_n$ be the maximal interval such that $f^n|J_n$ is a homeomorphism. $f^n(J_n \setminus I_n)$ consists of two disjoint intervals. Now by assumption $f^{N_q}(T_1) \subset T_1$ and therefore by the maximality of $I_n$, these two intervals $f^n(J_n \setminus I_n)$ are contained in disjoint components $f^k(T_1)$ and $f^\ell(T_1)$ of $\bigcup_{0\leq i<N_q} f^i(T_1)$, in such a way that the boundary points separating $f^n(J_n \setminus I_n)$ from $f^n(I_n)$ are boundary points of $f^k(T_1)$ and $f^\ell(T_1)$ respectively. For a schematic picture see

figure (5.2).

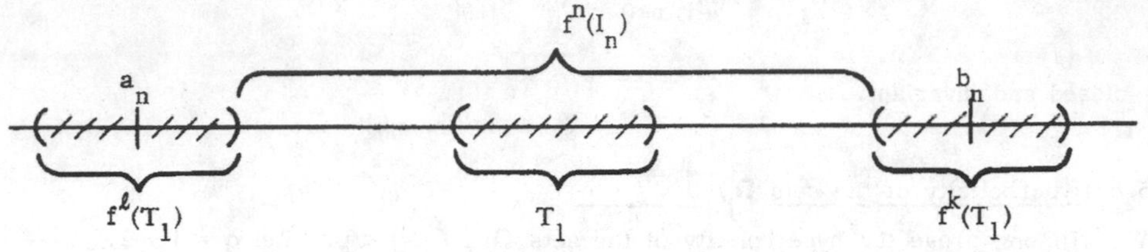


Figure (5.2)

From this it is clear that there are intervals $I^1_{n+1}, I^2_{n+1} \subset I_n$ with $f^n(I^i_{n+1}) \cap T_1 = \emptyset$. Furthermore let $T_2$ be the maximal component of $[-1,1] \setminus \bigcup_{0<i<N_q} f^i(T_1)$ containing $T_1$. Then we have $f^n(I_n) \supset T_2$. Now we are ready to prove the following Proposition.

<u>Proposition (5.4).</u> There exists a $\tilde{\tau} > 1$ such that if $f^n(I_n) \cap T_1 \neq \emptyset$, then

$$\frac{\ell(I_n)}{\ell(I^1_{n+1} \cup I^2_{n+1})} \geq \tilde{\tau}$$

for any $n \in \mathbb{N}$ and any interval $I_n$ as above.

<u>Proof</u> : Define $\tilde{f}^n = f^n|J_n$. Then

$$\frac{\ell(I_n)}{\ell(I^1_{n+1} \cup I^2_{n+1})} \geq \frac{\ell\, \tilde{f}^{-n}(T_2)}{\ell\, \tilde{f}^{-n}(T_2 \setminus T_1)} = 1 + \frac{\ell\, \tilde{f}^{-n}(T_1)}{\ell\, \tilde{f}^{-n}(T_2 \setminus T_1)} \geq 1 + \frac{\ell\, \tilde{f}^{-n}(T_1)}{\ell\, \tilde{f}^{-n}(T_2)} .$$

We need two Lemmas in order to prove that $\dfrac{\ell\, \tilde{f}^{-n}(T_1)}{\ell\, \tilde{f}^{-n}(T_2)} \geq \tilde{\tilde{C}}$, for some $\tilde{\tilde{C}}$ (independently of n).

<u>Lemma (5.5).</u> Suppose $f^n(I_n) \cap T_1 \neq \emptyset$ and let $[a_n, b_n] = f^n(J_n)$. There is a $\delta > 0$, which does not depend on $I_n$ and n such that

$$T_1 \subset T_2 \subset [a_n + \delta, b_n - \delta] .$$

<u>Proof</u> : The boundary points of $T_2$ are contained in $\bigcup_{0<i<N_q} f^i(T_1)$ and the critical values

$a_n, b_n$ of $f^n|J_n$ are also critical values of $f^n|_{\bigcup_{0\le i<N_q} f^i(T_1)}$. Now there are two cases to consider for the map $f_q = f^{N_q}|T_1 \in C$.

i) $f_q(T_1) \subsetneq T_1$ , ii) $f_q(T_1) = T_1$

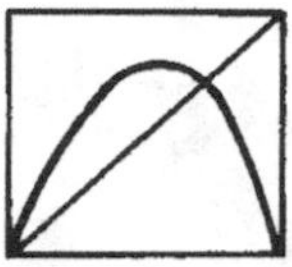

case i)

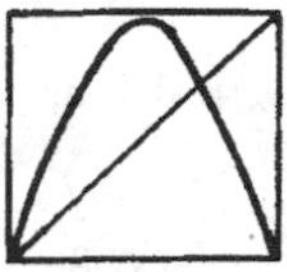

case ii)

Figure (5.3)

In the first case we have that the critical values of $f_q^k$, for all $k > 0$, stay bounded away from $\partial T_1$, since we have that the fixed point $x \in \partial T_1$ of $f_q$ satisfies $f_q'(x) > 1$. This proves the result in this case. In case ii) the result is obvious.

Lemma (5.6). Take $\tilde{f}^n = f^n|J_n$ as above and let $[a_n, b_n] = f^n(J_n)$.

For all $\delta > 0$ there exists a $\tilde{C} > 0$ such that

$$\frac{\ell\, \tilde{f}^{-n}(L_1)}{\ell\, \tilde{f}^{-n}(L_2)} \ge \left(\frac{\ell(L_1)}{\ell(L_2)}\right)^{\tilde{C}}$$

i) for all intervals $L_1, L_2$ with $L_1 \subset L_2 \subset [a_n + \delta, b_n - \delta]$.
ii) for all intervals $L_1 = [u, v]$ and $L_2 = [u, w]$ with $L_1 \subset L_2 \subset [a_n, b_n - \delta]$. This constant $\tilde{C}$ does not depend on $n > 0$.

Proof : Define $g:(a_n, b_n) \to \mathbb{R}$ by $g(t) = (\tilde{f}^{-n})'(t)$. Clearly either $g(t) > 0$, $\forall t \in (a_n, b_n)$ or $g(t) < 0$, $\forall t \in (a_n, b_n)$. Assume we are in the former case.

Claim 1. $g.g'' \ge \frac{3}{2}.(g')^2$ and $\frac{g(t)}{g'(t)} \to 0$ for $t \to a_n$ or $t \to b_n$.

Proof of Claim 1. A simple calculation shows that :

$$g''(t) = -(g(t))^3.(Sf^n)(f^{-n}(t)) + \frac{3}{2} \cdot \frac{(g'(t))^2}{g(t)} .$$

But $S(f) < 0$ and as one can easily check this implies $S(f^n) < 0$, $\forall n > 0$. Therefore $g(t).g''(t) \geq \frac{3}{2}.(g'(t))^2$.

$$\frac{g(t)}{g'(t)} = \frac{(f^n)'(f^{-n}(t))}{(f^n)''(f^{-n}(t))} \to 0$$

for $t \to a_n$ or $t \to b_n$, because then $(f^n)'(f^{-n}(t)) \to 0$ and because $f''(c) \neq 0$.

<u>Claim 2.</u> For $L_1, L_2$ and g as above

$$\frac{\int_{L_1} g}{\int_{L_2} g} \geq \left(\frac{\ell(L_1)}{\ell(L_2)}\right)^{\tilde{C}} \qquad \text{for some } \tilde{C} > 0.$$

<u>Proof of Claim 2 :</u>

$$\left(\frac{g}{g'}\right)' = 1 - \frac{g.g''}{(g')^2} \leq -\frac{1}{2} .$$

Using $\frac{g(t)}{g'(t)} \to 0$ as $t \to b$, integrate this inequality over $[t,b]$ and rewriting yields :

$$\frac{g'(t)}{g(t)} \leq \frac{2}{b-t} , \text{ i.e.}$$

$$\frac{g'(t)}{g(t)} \leq \frac{2}{\delta} \quad \text{for } t \leq b - \delta .$$

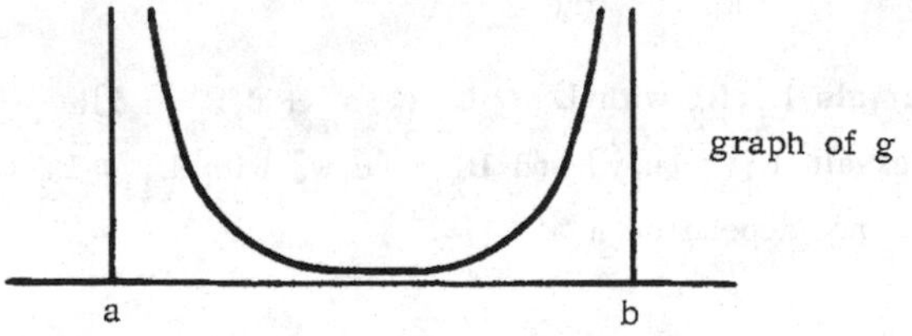


Figure (5.4)

Assume $L_1$ and $L_2$ are as in case (b) of Lemma (5.6). Case (a) is similar. For simplicity of notation assume 0 is the common boundary point u of $L_1$ and $L_2$. (Everything is invariant under translation.) Define

$$I(s) = \int_0^s g \quad \text{for } s \in L_2 .$$

Claim 3. For some $\tilde{C}$ (independently of g)

$$\left|\frac{s.I'(s)}{I(s)}\right| \leq \tilde{C}$$

for all $s \in L_2$, $s > 0$.

It suffices to prove this Claim since then

$$\ell n(I)\Big|_{s/_r}^{s} = \int_{s/_r}^{s} \left|\frac{I'}{I}\right| \leq \tilde{C} . \int_{s/_r}^{s} \frac{du}{|u|} = \tilde{C} . \ell n(r)$$

and therefore

$$\frac{I(s)}{I(\frac{s}{r})} \leq r^{\tilde{C}} \quad , \text{ i.e. taking } r = \frac{\ell(L_2)}{\ell(L_1)} \, ,$$

$$\frac{\ell\, \tilde{f}^{-n}(L_1)}{\ell\, \tilde{f}^{-n}(L_2)} = \frac{\int_{L_1} g}{\int_{L_2} g} \geq \left(\frac{\ell(L_1)}{\ell(L_2)}\right)^{\tilde{C}} .$$

So we only have to prove Claim 3.

Proof of Claim 3 : Define

$$J(t) = \tilde{C} . \int_0^t g - g(t).t \quad .$$

Then $J(0) = 0$ and

$$J'(t) = \tilde{C} . g(t) - g'(t).t - g(t) = (\tilde{C}-1).g(t).(1 - \frac{1}{\tilde{C}-1} . \frac{g'(t)}{g(t)} . t)$$

which is positive for $0 < t < b - \delta$ provided $\frac{1}{\tilde{C}-1} \leq \frac{\delta}{2(b-\delta)}$. Q.E.D.

Conclusion of the proof of Proposition (5.4). Taking $L_1 = T_1$ and $L_2 = T_2$, Lemmas (5.5) and (5.6) show that

$$\frac{\ell\, \tilde{f}^{-n}(T_1)}{\ell\, \tilde{f}^{-n}(T_2)} \geq \tilde{\tilde{C}} \, ,$$

where $\tilde{\tilde{C}} = \left( \dfrac{\ell(T_1)}{\ell(T_2)} \right)^{\tilde{C}}$ and $\tilde{C} = 1 + \frac{1}{\delta}$. This completes the proof of Proposition (5.4).

Proof of Lemma B. Take a sequence $I_n$ as above.

Claim 4. There are $k_1 < k_2 < k_3 \quad \dots$, with $k_{i+1} < k_i + N_q$ (where $q = j + 1$) and

$$f^{k_i}(I_{k_i}) \cap T_1 \neq \emptyset .$$

Proof : Let us first assume that $q = 2$. Then $T_1 = K_{f_1}^{-1}(0)$ and $N_2 = n_2$, where $n_2$ is the minimal number such that $F_{s_1}^{n_2}(0) = 0$. Let $[a_1, a_2] = K_{f_1}(I_n)$. $[a_1, a_2]$ is a maximal interval such that $F_{s_1}^n |[a_1, a_2]$ is a homeomorphism. Therefore $a_1 \neq a_2$ and the extreme values $F_{s_1}^n(a_1)$ and $F_{s_1}^n(a_2)$ of $F_{s_1}^n$ will have period $N_2$, since $F_{s_1}^{n_2}(0) = 0$. But $F_{s_1}$ has slope bigger than one and therefore $F_{s_1}^{n+N_2}$ cannot be a homeomorphism.

For $q = 3$ one uses the argument for $K_{f_2}$ a.s.o. by induction.

From the claim above and from proposition (5.4) it then follows that for some sequence $k_1 < k_2 < k_3 \dots$ with $k_{i+1} < k_i + N_q$,

$$\ell(F_{k_{i+1}}) < (\tilde{\tau})^{-1} \cdot \ell(F_{k_i}) .$$

From this Lemma B follows.

Proof of Lemma A. $F_n$ was defined to be the union of all components $I_n$ for $[-1,1] \setminus \bigcup_{k=0}^{n-1} f^{-k}(T_1)$, and $\Lambda = \bigcap_{n \geq 0} F_n$. Since $\ell(F_n) \to 0$, $\Lambda$ is a Cantor set.

Take $x \in \Lambda$ and a sequence of intervals $I_n, J_n$ as above with $x \in I_n \subset J_n$, $\forall n$. By Lemma B, $\ell(I_n) \leq \tau^n$. By Claim 4 from above there exist $k_i$, $k_1 < k_2 < k_3 \dots$ with $k_{i+1} < k_i + N_q$ such that $f^{k_i}(I_{k_i}) \cap T_1 \neq \emptyset$ from which it follows that

$$f^{k_i}(I_{k_i}) \supset T_2 .$$

Therefore there is a sequence of points $x_{k_i} \in I_{k_i}$ with

$$|(f^{k_i})'(x_{k_i})| \geq \frac{\ell(T_2)}{\ell(I_{k_i})} \geq \ell(T_2).(\frac{1}{\tau})^{k_i}. \qquad (*)$$

For $\alpha, \beta \in I_n$,

$$|1 - \frac{f'(\alpha)}{f'(\beta)}| = |\frac{1}{f'(\beta)}| . |f'(\alpha) - f'(\beta)| \leq C.\tau^n$$

for some constant $C < \infty$, because $\ell(I_n) < \tau^n$, f is $C^2$ and $I_n \cap T_1 = \emptyset$. ($T_1$ contains the unique critical point of f.) From this it follows that the product

$$|\frac{(f^n)'(x)}{(f^n)'(x_n)}| = \prod_{k=1}^{n} |(1-(1-\frac{f'(f^k(x))}{f'(f^k(x_n))}))|$$

stays bounded away from zero :

$$|\frac{(f^n)'(x)}{(f^n)'(x_n)}| \geq \bar{k} > 0 ,$$

for all $n > 0$. From this and (*) the Lemma follows.

§5.d <u>Proof of Theorem A(b).</u> <u>Proof of 1) : $\Omega_p$ is the unique attractor.</u>

First assume that $p < \infty$. According to §5.c the set of points $\Lambda$ never mapped into $B_p$ has measure zero, and is the complement of a cantor set. Therefore the set $B := \{x | f^n(x) \to \Omega_p\}$ is an open and dense set in $[-1,1]$ and has full measure.

If $p = \infty$, the result follows by considering a countable intersection of such sets.

<u>Proof of 2(a)</u> : The existence of an attracting periodic orbit O(z) implies $p < \infty$.

Because $n_j > 1$ for $1 \leq j < p$, it follows that the period of periodic points in $B_j$ are a multiple of $n_1 \cdot \ldots \cdot n_j > 2^{j-1}$. Since an attracting periodic orbit is contained in $\Omega_p$, ($\Omega_j$, $0 < j < p$ is hyperbolically repelling), p must be finite.

<u>Proof of 2(b)</u> : The set $\tilde{B} := \{x | f^n(x) \to O(z)\}$ is open and dense and has full Lebesgue measure in $(-1,1)$.

There are two cases for $f_p : B_p \to B_p$ :

i) $s(f_p) = 1$ or ii) $s(f_p) > 1$.

Later, in Theorem A(d), we will show that for case i) $\Omega(f_p)$ is finite. From the proof of (1) above, the result follows.

For case ii) we had that $f_{p+1}: B_{p+1} \to B_{p+1}$ did exist, but for the fixed point $x \in \partial B_{p+1}$ for $f_{p+1}$ one had $f'_{p+1}(x) = 1$. Therefore it is impossible that the set

$$\Lambda = [-1,1] \setminus \bigcup_{k \geq 0} \operatorname{int}(f^{-k}(B_{p+1}))$$

is hyperbolic. In fact Lemma (5.5) is not true in this case. The number $\delta$ depends on n, $\delta_n \downarrow 0$ and therefore $\tilde{C}(n) \to \infty$ and $\tilde{\tilde{C}}(n) \downarrow 0$. But since $f_{p+1}$ is $C^2$ one can prove that $\delta_n$ goes to zero quite slowly, such that in fact $\prod_{n>0}(1+\tilde{\tilde{C}}(n)) \to \infty$. From this it follows that $\ell(\Lambda) = 0$. For more details about this see [Gu, Mis. 2, C.E. and Nu]. From $\ell(\Lambda) = 0$, it follows that the set "$\tilde{B}$" has full measure in $(-1,1)$.

Proof of 3 : $O(z)$ is hyperbolic or quasi-hyperbolic.

In Theorem A(d) it will be proved that if $\lambda = 1$, then $f: B_p \to B_p$ is as in case 2(a), so the result follows from Proposition (5.4). If $O(z)$ has period n and $(f^n)'(z) = -1$, then $(f^{2n})'(z) = 1$ and $(f^{2n})''(z) = 0$. From $S(f) < 0$ it then follows that $(f^{2n})'''(z) < 0$.

## §6. Proof of Theorem A(c), (d).

### §6.a Proof of Theorem A(c), dynamics of $f: \Omega_j \to \Omega_j$.

According to Theorem (4.1) for $0 < j < p$ there is a semi-conjugacy $K_j$ such that

$$\begin{array}{ccc} B_j & \xrightarrow{f_j} & B_j \\ {\scriptstyle K_j}\downarrow & & \downarrow{\scriptstyle K_j} \\ [-1,1] & \xrightarrow{F_{s_j}} & [-1,1] \end{array}$$

figure (5.1)

commutes. Since $S(f) < 0$ this result can be strengthened considerably : Let

$$\tilde{\Lambda}_j = B_j \setminus \bigcup_{k \geq 0} f^{-k}(B_{j+1})$$

and remember that

$$\Lambda_j = B_j \setminus \bigcup_{k\geq 0} f^{-k}(\text{int } B_{j+1}) .$$

Clearly $\Lambda_j = C\ell(\tilde{\Lambda}_j)$ .

Theorem (6.1). For $1 \leq j < p$

i) $K_j|\tilde{\Lambda}_j$ is injective.

ii) $\Omega(f_j) = C_j \cup Per_j$, where $C_j$ is an invariant basic set and $Per_j$ is a finite set of periodic points.

iii) $f_j:C_j \to C_j$ is a subshift of finite type.

Proof of Theorem (6.1)(i). Suppose $K_j$ is constant on some interval U. If $F^n_{s_j}(K_j(U)) \neq 0$, $\forall n > 0$ then $U \subset \Lambda_j$. But since $\Lambda_j$ is a Cantor set, this is impossible. But if $F^n_{s_j}(K_j(U)) = \emptyset$, then $U \cap \tilde{\Lambda}_j = \emptyset$.

Proof of Theorem (6.1)(ii). The decomposition $\Omega_j = C_j \cup Per_j$ follows from i) and

Proposition (6.2) : see [M.T., J.R.].

For $\sqrt{2} < s^n \leq 2$, the nonwandering set $\Omega(F_s)$ of $F_s$ decomposes into (n+1)-basic sets, namely $\{-1\}$, periodic orbits of period $2^k$, one for each $k = 0,1,2,\ldots,n-2$ and a basic set X, consisting of the union of $2^{n-1}$ closed intervals $X^i$, $0 \leq i < 2^{n-1}$.

$$f(X^i) = X^{i+1(\text{mod } 2^{n-1})}, \text{ and}$$
$$0 \in X^0 .$$

Proof of Theorem (6.1)(iii). We need to introduce some symbolic dynamics (kneading theory). Let $f:[a,b] \to [a,b]$ be a map in C with an extremum in c as before. For $x \in [a,b]$ and $i = 0,1,2,\ldots$ let

$$\theta_i(x) = 0, \text{ if } f^j(x) = c \text{ for some } 0 \leq j \leq i$$
$$\theta_i(x) = 1, \text{ if } f^{i+1} \text{ is orientation preserving near x, and}$$
$$\theta_i(x) = -1, \text{ otherwise.}$$

Let $\theta_f(x)$ denote the formal power series $\sum_{i=0}^{\infty} \theta_i(x).t^i$. This power series is called the

<u>kneading coordinate</u> of x. The map $x \to \theta_f(x)$ is monotone (if $\mathbb{Z}[[t]]$ is ordered lexicographically), see [Mil., Jo.]. The limit $\lim_{y \downarrow x} \theta_f(y)$ in the (t)-adic topology is denoted by $\theta_f(x^+)$ and $\theta_f(x^-) = \lim_{y \uparrow x} \theta_f(y)$. $\theta_f(c^-)$ is called the <u>kneading invariant</u> of f. For $\theta = \sum_{i \geq 0} \theta_i . t^i$, let

$$|\theta| = \begin{cases} \theta & \text{if } \theta \geq 0 \\ -\theta & \text{if } \theta \leq 0 \,, \end{cases}$$

with respect to the lexicographic ordering and let

$$\sigma : \mathbb{Z}[[t]] \to \mathbb{Z}[[t]] \quad \text{be}$$

$$\sigma(\sum_{i \geq 0} \theta_i t^i) = (\sum_{i \geq 0} \theta_{i+1} t^i) . \theta_0 \,.$$

The kneading coordinate function $\theta_f(x)$ is a semi-conjugacy between f and $\sigma$ :

$$\begin{array}{ccc} [a,b] & \xrightarrow{f} & [a,b] \\ \downarrow \theta_f & & \downarrow \theta_f \\ \mathbb{Z}[[t]] & \xrightarrow{\sigma} & \mathbb{Z}[[t]] \end{array}$$

For $0 < j < p$, $f_j(x_1) = f_j(x_2)$ for the two points $x_1, x_2 \in B_{j+1}$ and therefore $\theta_{f_j}(x_1) = -\theta_{f_j}(x_2)$. Let

$$v_j = |\theta_{f_j}(x_1)| = |\theta_{f_j}(x_2)| \,.$$

Since $f_j^{n_{j+1}} | B_{j+1} \in \mathcal{C}$ this power series $v_j$ has period $n_{j+1}$, i.e. if $v_j = \sum \tilde{v}_i t^i$ then $\tilde{v}_i = \tilde{v}_{i+n_{j+1}}$, $\forall i \geq 0$. First we prove :

<u>Proposition (6.3).</u>

Let $S_j = \{\alpha = \sum_{i \geq 0} \alpha_i t^i \,|\, \alpha_i \in \{-1, 1\}$ and $|\sigma^k(\alpha)| \geq v_j . (\text{mod } t^{n_{j+1}}),\ \forall i,k \geq 0\}$

Then $\theta_{f_j}(\Lambda_j) = S_j$ and $\theta_{f_j} : \Lambda_j \to S_j$ is a homeomorphism.

$$\begin{array}{ccc} \Lambda_j & \xrightarrow{f_j} & \Lambda_j \\ \downarrow \theta_{f_j} & & \downarrow \theta_{f_j} \\ S_j & \xrightarrow{\sigma} & S_j \end{array}$$

Proof :

i) $\theta_{f_j}|\Lambda_j$ is continuous : that $x \mapsto \theta_{f_j}(x)$ is discontinuous at x, implies that for some $n > 0$, $f^n(x) = 0$ and therefore $x \notin \Lambda_j$.

ii) $\theta_{f_j}|\Lambda_j$ is strictly monotone : Since $x \mapsto \theta_{f_j}(x)$ is monotone, $x \notin \mathrm{int}(B_{j+1})$ implies that $|\theta_{f_j}(x)| \geq v_j$. Therefore $x \in \Lambda_j$ implies that $|\sigma^k(\theta_{f_j}(x))| \geq v_j$, $\forall k \geq 0$. Hence, if $\theta_{f_j}(x) = \theta_{f_j}(y)$ with $x, y \in \Lambda_j$ then the convex hull $\langle x,y \rangle$ is contained in $\Lambda_j$. But this contradicts the fact that $\Lambda_j$ is a Cantor set.

iii) Since $\theta_{f_j}|\Lambda_j$ is strictly monotone one has in ii) $x \in \Lambda_{j+1}$ if and only if $|\sigma^k(\theta_{f_j}(x))| \geq v_j$, $\forall k \geq 0$.

Conclusion of the proof of Theorem (6.1)(iii). From Theorem (5.1)(ii) we have that $\Omega_j = C_j \cup Per_j$, where $C_j$ is a basic set for $f_j$. Therefore $\theta_{f_j}(C_j) \subset S_j$ is a basic set for $\sigma$. To check whether $\alpha \in \mathbb{Z}[[t]]$ is in $S_j$, only conditions up to order $t^{n_{j+1}}$ have to be taken into consideration. Therefore

$$\sigma: \theta_{f_j}(C_j) \to \theta_{f_j}(C_j)$$

is an indecomposable subshift of finite type.

§6.b Proof of Theorem A(d), dynamics of $f: \Omega_p \to \Omega_p$, if $p < \infty$.

Case 1 : $h(f|\Omega_p) = 0$.

Case 1(a) : $f_p : B_p \to B_p$ has an attracting periodic orbit.

Periodic point x and y of a map $f \in \mathcal{C}$ are called equivalent if $\theta_f(x) = \theta_f(y)$, i.e. if $f^n$ restructed to the convex hull $\langle x,y \rangle$ of x and y is a homeomorphism for all $n \in \mathbb{N}$. For maps $f \in \mathcal{C}$, with $h(f) = 0$, Jonker and Rand, [J.R., part I], show that $\Omega(f)$ contains precisely the number of equivalence classes of periodic points as given in Theorem A(d). But if x and y are equivalent periodic points, then there is an attracting point $z \in \langle x,y \rangle$. According to Proposition (2.1) there is at the most one such an attracting periodic orbit O(z). From [J.R.] it follows that, since $h(f_p) = 0$, the periodic orbit cannot have eigenvalue 1.

Case 1(b) : $f_p : B_p \to B_p$ has no attracting periodic point. Again the result follows from

[J.R.] and Proposition (2.1). In fact Misiurewicz proved that the kneading coordinate $x \mapsto \theta_f(x)$ is strictly monotone for such maps, and that all such maps are topologically conjugate.

Case 2 : $h(f|\Omega_p) > 0$ .

Case 2(a) : In this case $F_{s_p}^{n_{p+1}}(0) = 0$ but for the fixed point $x \in \partial B_{p+1}$ of the map $f_{p+1}: B_{p+1} \to B_{p+1}$ one has $f'_{p+1}(x) = 1$. Therefore $\Omega_p$ is not a hyperbolic set, but according to §5.d $\Omega_p$ is a cantor set. Also as in §6.a $f_p: \Omega_p \to \Omega_p$ is a subshift of finite type.

Case 2(b) : In this case $F_{s_p}^{n}(0) \neq 0$, $\forall n > 0$. We will show that $f_p: B_p \to B_p$ has no periodic attracting orbit O(z). If such an orbit O(z) exists, then it is easy to see that the positive iterates of the critical point c converge to O(z), [J.R., Si.]. Therefore in that case $F_{s_p}^{n}(K(c)) = F_{s_p}^{n}(0) \to K(O(z))$, which is impossible if $F_{s_p}^{n}(0) \neq 0$.

In [Gu] it is proved that when $f \in \mathcal{G}$, and f has no attracting periodic orbit, that then $x \mapsto \theta_g(x)$ is strictly monotone. Since $F_{s_p}^{n}(0) \neq 0$, $\theta_f(x) = \theta_{F_s}(K_f(x))$, and therefore $K_{f_p}: B_p \to [-1,1]$ is strictly monotone. This proves case 2(b).

§6.c Proof of Theorem A(e), dynamics of $f: \Omega_p \to \Omega_p$, if $p = \infty$. If $p = \infty$ then f also has no attracting periodic point and again $x \mapsto \theta_f(x)$ is strictly monotone. But if $\langle x, y\rangle \subset \bigcap_{n=1}^{N} \tilde{M}_n$, then $f^n|\langle x,y\rangle$ is a homeomorphism for $n = 1,2,\dots,N$. Therefore $\bigcap_{n=1} \tilde{M}_n$ is a Cantor set. Together with [J.R.] this proves Theorem A(e).

§6.d Proof of Theorem A(f). The filtration $M_j$ follows from the inductive decomposition in §5.b of $\Omega(f)$ : If $f_j|B_j$ satisfies

i) $f_j(B_j) \subsetneq B_j$ and
ii) the fixed point $x \in \partial B_j$ satisfies $f'_j(x) > 1$,

then there is a slightly smaller set $\tilde{B}_j$ such that $f_j(\tilde{B}_j) \subsetneq \text{int}\, \tilde{B}_j$. Put

$$M_j = \bigcup_{0 \le i < N_j} f^i(\tilde{B}_j) \quad .$$

The other assertions follow from §5.

## §7. Proof of Theorem B.

Milnor and Thurston prove that for a continuous family of maps $f_t \in \mathcal{C}$, $t \mapsto s(f_t)$ is continuous. Since for all $t \in \mathfrak{U}$, 0 is a periodic point for $F_s(t)$, $s(t)$ must be constant for all $t \in \mathfrak{U}$.

Also the intervals $B_j(t)$ from §5.b depend continuously on t, since for $1 \leq j < p$, the fixed point in $\partial B_j(t)$ for $f_{j,t}$ is hyperbolically repelling for $t \in \mathfrak{U}$, From this it follows easily that the numbers p, $n_j$, $s_j$ and $N_j$ are constant for $t \in \mathfrak{U}$.

Furthermore it follows from Proposition (6.3) that for $1 \leq j < p$, $f_t : \Omega_j(t) \to \Omega_j(t)$ is conjugate to a fixed subshift $\sigma_j : S_j \to S_j$ for all $t \in \mathfrak{U}$. The existence of conjugating homeomorphisms $h_j : \Omega_j(t) \to \Omega_j(t')$ for $t, t' \in \mathfrak{U}$ and $0 < j < p$, follows.

$\Omega_p(t)$ is a finite subset of Per(f) for $t \in \mathfrak{U}$ and this set can only change if there is a periodic point with eigenvalue $\lambda$ and $\lambda = 1$ (fold bifurcation) or $\lambda = -1$ (quasi-harmonic, flip or pitchfork bifurcation, where the period of the attractor doubles). For these bifurcations see [Br, M.P.T. and So.]. But since there is precisely one attracting periodic orbit $O(z(t))$ of $f_t$, for $t \in \mathfrak{U}$, Theorem B(a) follows.

Theorem B(b) follows from Theorem A(d). The Theorems B(c) and B(d) follow from [Mil., M.T. and J.R., part II].

## §8. Proof of Theorem C, similarities between diffeomorphisms and one-dimensional endomorphisms.

### §8.a Hyperbolic sets of endomorphisms, a perturbation result.

Let $\Phi : M \to M$ be a $C^1$ map on a smooth compact manifold M, and $\Lambda$ a maximal invariant compact hyperbolic set. First we will prove a perturbation result. Remark that $\Phi$ may not be an embedding.

Theorem (8.1), see [H.P. and H.P.S.].

Given $\varepsilon > 0$, there exists a compact neighbourhood V of $\Lambda$ and a neighbourhood Q of $\Phi$ in $C^1(M,M)$, with the following properties :

(1) If $\psi \in Q \cap \mathrm{Emb}^1(M,M)$ (the class of $C^1$-embeddings), then $\psi$ has a unique

maximal invariant set $\Lambda(\psi) \subset V$ and there is a unique continuous surjection $h:\Lambda(\psi) \to \Lambda$ such that

$$\begin{array}{ccc} \Lambda(\psi) & \xrightarrow{\psi} & \Lambda(\psi) \\ h\downarrow & & \downarrow h \\ \Lambda & \xrightarrow{\Phi} & \Lambda \end{array}$$

commutes and the $C^0$ distance $d(h, id_\Lambda) < \varepsilon$.

(2) Furthermore h is a bijection from $Per(\psi)$ to $Per(\Phi)$ and depends continuously on $\psi \in Q$.

(3) If $\psi_i \in Q \cap Emb^1(M, M)$, $i = 1,2$, then there exists a unique homeomorphism $h:\Lambda(\psi_1) \to \Lambda(\psi_2)$ such that $\psi_2 \circ h = h \circ \psi_1$.

Proof. As in Theorem (7.1) of [H.P.] one gets a neighbourhood Q of $\Phi$ and a neighbourhood V of $\Lambda$ such that $\psi \in Q$ has a maximal invariant set $\Lambda(\psi)$ in V. (In Proposition (4.8) in [H.P.] the contracting part need not be invertible). The existence of a continuous map $h:\Lambda(\psi) \to \Lambda$, if $\psi$ is an embedding, is proved as in Theorem (7.3) in [H.P.].

But an inverse of h cannot be constructed in general. Therefore first construct the local stable manifold $W^s_\varepsilon(\Lambda(\Phi))$ of $\Lambda(\Phi)$. For this use Theorem (5.1) of [H.P.S.] which allows for singular behaviour. Then, in order to prove that h is a surjection and a bijection between the periodic points it suffices to make a homeomorphism

$$\tilde{h}:W^s_\varepsilon(\Lambda(\Phi)) \to W^s_\varepsilon(\Lambda(\psi)) ,$$

which preserves leaves, i.e. for all $x \in \Lambda$ :

$$\tilde{h}(W^s_\varepsilon(\Phi, h(x))) \subset W^s_\varepsilon(\psi, x) .$$

This is done as in Theorem (7.4) in [H.P.]. Q.E.D.

§8.b Proof of Theorem C. To avoid pathological bifurcations we assume that the famil

$\{f_t\}$ satisfies certain genericity conditions. For more details, see for example [Br., So., and N.P.T.].

Assumption (8.2). (Generic flip bifurcation).

If $f_{t_0}$ has a periodic point z with period n and with eigenvalue -1, then $\frac{d}{df}\frac{d}{dx} f^{2n}_{t_0}(z) \neq 0$.

From Theorem A(b) it follows that for such an orbit $(f^{2n})''(z) = 0$ and $(f^{2n})'''(z) < 0$, i.e. O(z) is an attracting orbit. From this it follows that, as t changes, the attracting orbit O(z(t)) of $f_t$ doubles its period and splits off a repelling orbit. This is called a flip or pitchfork bifurcation, see the figure in §3.

Now we are in a position to prove Theorem C. Take a singular family $\{\bar{f}_t\}$ such that $f_t$ satisfies Assumption (8.2), and take a component $\mathfrak{U}$ of P. For $t \in \mathfrak{U}$ there is a $p \in \mathbb{N}$ such that there is a decomposition as in Theorem A :

$$\Omega(\bar{f}_t) = \Omega_0(t) \cup \dots \cup \Omega_p(t) .$$

$\Omega_i(t)$, for $0 \le i < p$ is hyperbolic and $\Omega_p(t)$ contains a periodic orbit O(z(t)) with eigenvalue $\lambda(t)$, $-1 \le \lambda(t) < 1$.

a) According to Theorem A(f) there is a filtration of $\mathbb{R}^2$ : there are sets

$$M_0(t) \supset M_1(t) \supset \dots \supset M_p(t)$$

where $M_j(t)$ are compact neighbourhoods of segments of the curve $\bar{f}_t(\mathbb{R}^2)$ in $\mathbb{R}^2$. Furthermore

$$\bar{f}_t(M_j(t)) \subset \mathrm{int}(M_j(t)), \text{ for } 0 \le j \le p, \text{ and}$$

$$\Omega_j(t) = \bigcap_{n\in\mathbb{N}} \overline{\bar{f}^n_t(M_j(t)\setminus M_{j+1}(t))}$$

for $0 \le j < p$. $\Omega_p(t)$ is a finite set.

b) Take $t_0 \in \mathfrak{U}$. For $0 \le j < p$ take a small neighbourhood $V_j(t_0)$ of $\Omega_j(t_0)$ as in Theorem (8.1). For some $k_j$ :

$$\Omega_j(t_0) \subset \bigcap_{-k_j<n<k_j} \overline{\bar{f}^{\,n}_{t_0}(M_j \setminus M_{j+1})} \subset V_j(t_0) .$$

For diffeomorphisms $\Phi_{t_0} : \mathbb{R}^2 \to \mathbb{R}^2$ uniformly $C^0$ near $\bar{f}_{t_0}$

$$\bigcap_{-k_j<n<k_j} \Phi^n_{t_0} \overline{(M_j \setminus M_{j+1})} \subset V_j(t_0) ,$$

and $\Phi_{t_0}(M_j(t_0)) \subset \operatorname{int} M_j(t_0)$. Therefore

$$\Omega(\Phi_{t_0}) \subset \left(\bigcup_{j=0}^{p-1} V_j\right) \cup M_p(t_0) .$$

According to Theorem (8.1) there exists for $\Phi_{t_0}$ in a $C^1$ neighbourhood $Q(t_0)$ of $\bar{f}_{t_0}$, a maximal hyperbolic set $\tilde{\Omega}_j(t_0)$, for $0 \le j < p$. $\Phi_{t_0} : \tilde{\Omega}_j(t_0) \to \tilde{\Omega}_j(t_0)$ and $f_{t_0} : \Omega_j(t_0) \to \Omega_j(t_0)$ are semi-conjugate. This can be done for each $t_0 \in \mathfrak{U}$, but for $t_0$ near $\partial\mathfrak{U}$ the neighbourhood $Q(t_0)$ might "shrink".

c) Now we consider the finite set $\Omega_p(t_0)$. If the attracting periodic orbit $O(z(t_0))$ is hyperbolic, then there is a $C^1$ neighbourhood $Q(t_0)$ of $f_{t_0}$ and a neighbourhood $V$ of $\Omega_p(t_0)$ such that for $\Phi \in Q(t_0)$ :

$$\Omega(\Phi) \cap V \cong \Omega_p(f_{t_0}) .$$

Since $\Phi(M_p(t_0)) \subset M_p(t_0)$, and since $f^n_{t_0}(x) \to \Omega_p(f_{t_0})$ for $x \in M_p(t_0)$, for $\Phi$ sufficiently near $f_{t_0}$ :

$$\Omega(\Phi) \cap M_p \cong \Omega_p(f_{t_0}) .$$

If the periodic orbit $O(z(t_0))$has eigenvalue $\lambda = -1$ then it follows from assumption (8.2) that for a family $\{\Phi_t\}$ which is sufficiently $C^3$ near $\{f_t\}$, for $t$ near $t_0$, there exists a reparametrisation $t \mapsto \rho(t)$ such that

$$\Omega(\Phi_t) \cap V \cong \Omega(\bar{f}_{\rho(t)}) ,$$

for $t$ near $t_0$. Since $O(z(t_0))$ is weakly attracting $((f^{2n}_{t_0})'''(z(t_0)) < 0)$ we have also

$$\Omega(\Phi_t) \cap M_p \cong \Omega_p(\bar{f}_{\rho(t)}) .$$

d) From b) and c) it follows that for $d(\Phi, f) < \varepsilon$ there exists a number $\delta(\varepsilon)$ such that $\{\Phi_t\}$ and $\{\bar{f}_t\}$ are $\Omega$-semi-conjugate on $P(\delta(\varepsilon))$. The number $\delta(\varepsilon)$ can be chosen to vary continuously on $\varepsilon$, such that $\delta(0) = 0$.

§8.c Proof of Theorem D.

We have $\{f_\mu\} \in PC$ and $h(f_0) \neq h(f_1)$. According to Theorem B, there exists therefore a countable number of parameter values $\mu_i$ such that the following is true. For $g = f_{\mu_i}$ the decomposition of the nonwandering set of Theorem A is such that :

$$\Omega(g) = \Omega_0 \cup \Omega_1 \cup \ldots \cup \Omega_{p-1} \cup \Omega_p$$

with $p \geq 2$, and such that furthermore the following condition is satisfied. For the number $N_p$ from §5,

$$g^{N_p} : \Omega_p \to \Omega_p$$

is conjugate to the piecewise linear map $F:[-1,1] \to [-1,1]$ defined by $F(x) = 1 - 2 . |x|$, or equivalently conjugate to $Q(x) = 1 - 2 . x^2$ This implies that $\Omega_p \cap \Omega_{p-1} \neq \emptyset$ and that $\Omega_p$ consists of $N_p$ disjoint intervals.

Let $Q = [-2,2] \times [-2,2]$ and let us draw $g(Q)$, in figure (8.1). The segment <a, b> of $g(Q)$ is one of the components of $\Omega_p$, and the hyperbolic basic set $\Omega_{p-1}$ also contains a and c. We have $g^{N_p}(a) = a$, $g^{N_p}(b) = a$, $g^{N_p}(c) = b$.

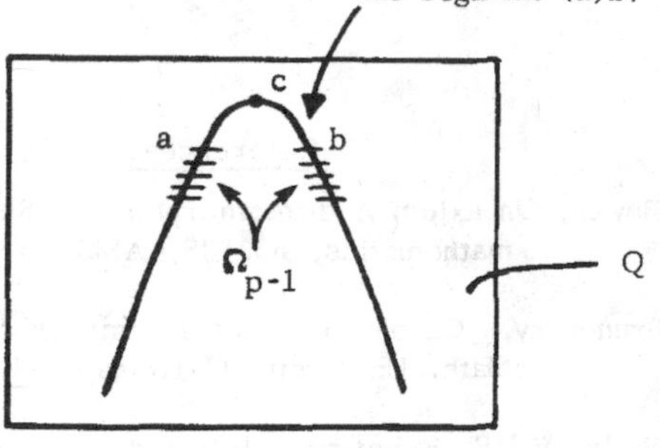


Figure (8.1).

The map g is singular, (i.e. non-invertible), but even so there exist invariant manifolds $W^s(a)$ and $W^u(a)$ for the periodic point a : see Theorem (5.1) of [H.P.S.]. $W^u(a)$ contains the segment <a, b> of g(Q) and $W^s(a)$ contains the horizontal line passing through c. So we have that $W^u(a)$ and $W^s(a)$ have a quadratic tangency in c, and $a \in \Omega_{p-1}$. $\Omega_{p-1}$ is a hyperbolic basic set.

Since $h(f_0) \neq h(f_1)$ there exist $t_i^1$ and $t_i^2$ such that

1) for $t \in \langle t_i^1, t_i^2 \rangle$, $\Omega_{p-1}(t)$ is a hyperbolic basic set, and a(t) is a periodic point in $\Omega_{p-1}(t)$ for $f_t$.

2) $W^u(a(t_i^1))$ and $W^s(a(t_i^1))$ have no intersections.

3) $W^u(a(t_i^2))$ and $W^s(a(t_i^2))$ have two intersections.

We are now going to apply this to families of diffeomorphisms. For a family of diffeomorphisms $\{\Phi_t\}$, sufficiently $C^1$-near $\{\bar{f}_t\}$, $\Phi_t$ has a hyperbolic basic set $\tilde{\Omega}_{p-1}(t)$ and a periodic point $\tilde{a}(t)$ in $\tilde{\Omega}_{p-1}(t)$, with period $N_p$, for $t \in \langle t_i^1, t_i^2 \rangle$. Also $\det|d\Phi_t^{N_p}(\tilde{a}(t))| < 1$.

According to Theorem (5.1) the manifolds $W^s(\tilde{a}(t))$ and $W^u(\tilde{a}(t))$ depend continuously on $\Phi_t$ in the $C^r$ sense. Therefore there is a $\tilde{t}_i \in \langle t_i^1, t_i^2 \rangle$ such that $W^s(\tilde{a}(\tilde{t}_i))$ and $W^u(\tilde{a}(\tilde{t}_i))$ have a quadratic tangency. By possibly perturbing this family $\{\Phi_t\}$ slightly, one can assume that there is a $\varepsilon_i > 0$ such that $W^s(\tilde{a}(t))$ and $W^u(\tilde{a}(t))$ have i) no intersection for $\tilde{t}_i - \varepsilon_i < t < \tilde{t}_i$, ii) two intersections for $\tilde{t}_i < t < \tilde{t}_i + \varepsilon$, and iii) a quadratic tangency for $t = \tilde{t}_i$.

Now apply Theorem 3 from [Ne].

## References.

[Bo] R. Bowen, On axiom A diffeomorphisms, Regional conference series in mathematics, no. 35, AMS.

[Br] P. Brunovsky, On one-parameter families of diffeomorphisms, Comment. Math. University Carolinae, 11, (1970), pp. 559-582.

[C.E.] P. Collet & J.P. Eckmann, Iterated maps on the interval as dynamical systems, Birkhäuser, Boston, 1980.

[Gu] J. Guckenheimer, Sensitive dependence to initial conditions for one dimensional maps, preprint IHES (1979).

[Hé] M. Hénon, A two-dimensional mapping with a strange attractor, Comm. Math. Phys. 50, (1976), pp. 69-78.

[He] B.R. Henry, Escape from the unit interval under the transformation $x \mapsto \lambda.x.(1-x)$, Proc. Am. Math. Soc., 41, (1973), pp. 146-150.

[H.P.] M. Hirsch & C. Pugh, Stable manifolds and hyperbolic sets, Proc. Symp. in Pure Math., Vol. 14, AMS, Providence, R.I. (1970), pp. 133-164.

[H.P.S.] M. Hirsch, C. Pugh & M. Shub, Invariant manifolds, Lecture Notes in Math, 583, Springer Verlag, (1977).

[Jo] L. Jonker, Periodic orbits and kneading invariants, Proc. London Math. Soc. 39 (1979), pp. 428-450.

[J.R.] L. Jonker & D.A. Rand, Bifurcations in one dimension, I : The nonwandering set; II : A versal model for bifurcations, Warwick Preprint, 1978. To appear in Inventiones Math.

[Mil] J. Milnor, The theory of kneading, (Handwritten Notes), (1976).

[M.T.] J. Milnor & W. Thurston, On iterated maps of the interval I and II, Preprint Princeton University & the Institute for Advanced Study, Princeton, (1977).

[Mis 1] M. Misiurewicz, Structure of mappings of an interval with zero entropy, Preprint IHES, (1978).

[Mis 2] M. Misiurewicz, Absolutely continuous measures for certain Markov maps of an interval, Preprint IHES, (1979).

[Ne] S. Newhouse, The abundance of wild hyperbolic sets and non-smooth stable sets for diffeomorphisms, Publ. Math. IHES, 50, (1979), pp. 101-152.

[N.P.T.] S. Newhouse, J. Palis & F. Takens, Stable arcs of diffeomorphisms, Bull. Am. Math. Soc., 82, (1976) and to appear.

[Nu] L. Nusse, No chance to get lost, Preprint Utrecht, The Netherlands, unpublished.

[Si] D. Singer, Stable orbits and bifurcations of maps of the interval, SIAM, Journal of Appl. Math., 35, (1978), pp. 260-267.

[Sm] S. Smale, Differentiable dynamical systems, Bull. Am. Math. Soc., 73, (1967), pp. 747-817.

[So] J. Sotomayor, Generic bifurcations of dynamical systems, Proc. of Symp. at Salvador, Academic Press, New York, (1973), pp. 561-582.

S.J. van Strien : Mathematics Institute, University of Utrecht, Utrecht, The Netherlands.

# Saddle Connections of Arcs of Diffeomorphisms : Moduli of Stability

Sebastian J. van Strien.

## I. Introduction.

Consider a generic arc of diffeomorphisms $\{\Phi_\mu\}_{\mu\varepsilon(-1,1)}$ on a compact manifold M. Suppose that there are two hyperbolic periodic saddle points $p_\mu$ and $q_\mu$ such that $W^u(p_\mu)$ and $W^s(q_\mu)$ intersect non-transversally for $\mu = \mu_0$; for the two-dimensional case see Figure (I.1).

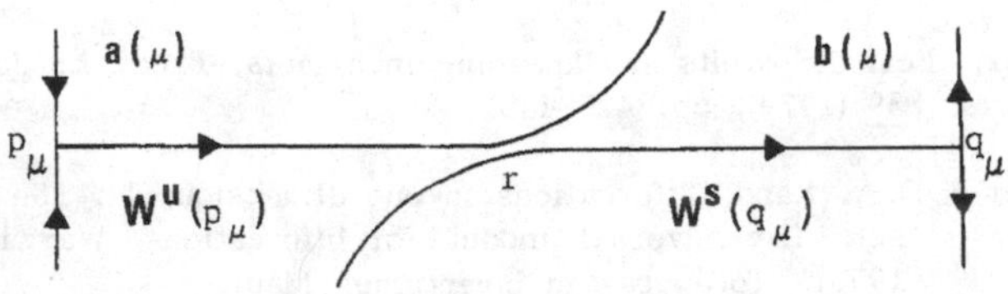


Figure (I.1).

Let $\varphi = \Phi_{\mu_0}$ and let $a(\mu)$ be the weakest contracting eigenvalue at $p_\mu$ and $b(\mu)$ be the weakest expanding eigenvalue at $q_\mu$. Palis et al. proved that for such diffeomorphisms $\varphi$ the number

$$P(\varphi) := \frac{\log|a(\mu_0)|}{\log|b(\mu_0)|}$$

is a topological invariant, (see [P, M, M.P. and N.P.T.]). This means that if $\varphi$ and $\tilde{\varphi}$ as above are conjugate then $P(\varphi) = P(\tilde{\varphi})$. Such a number is called a modulus of stability.

Here we look at arcs of diffeomorphisms and prove that if two arcs of diffeomorphisms are conjugate and if the conjugacy depends continuously on the parameter, then in addition to $P(\varphi)$ another modulus condition appears. In fact, if two arcs $\{\Phi_\mu\}$ and $\{\tilde{\Phi}_\mu\}$ are conjugate in this sense, then :

$$\begin{aligned} a(\mu_0) &= \tilde{a}(\tilde{\mu}_0) \ , \\ b(\mu_0) &= \tilde{b}(\tilde{\mu}_0) \ . \end{aligned} \tag{I.1}$$

In another paper we will construct conjugacies, depending continously on the parameter, between two arcs which satisfy these two moduli conditions. Actually de Melo proves that two arcs $\{\Phi_\mu\}$ and $\{\tilde{\Phi}_\mu\}$ are conjugate if $P(\varphi) = P(\tilde{\varphi})$, but certainly the conjugacies he constructs do not depend continuously on the parameter, unless the equations in (I.1) are satisfied. So we see that we get a new modulus, just because we want the conjugacies to depend continuously on the parameter. This situation also occurs in the saddle-node bifurcation, see [N.P.T.].

We wish to thank Floris Takens for suggesting the problem and for his encouragement.

## II. Statement of Results, for dim (M) = 2.

Let $\mathcal{C}$ be the class of smooth arcs of $C^\infty$ diffeomorphisms on a two dimensional $C^\infty$ manifold as in the introduction. For an arc $\{\Phi_\mu\} \in \mathcal{C}$ denote the contracting eigenvalue at $p_\mu$ by $a(\mu)$ and the expanding eigenvalue at $q_\mu$ by $b(\mu)$. For a similar arc $\{\tilde{\Phi}_\mu\}$ denote the analogue of $p_\mu, \mu_0, W^u(p_\mu)$ and $a(\mu)$ by $\tilde{p}_\mu, \tilde{\mu}_0, W^u(\tilde{p}_\mu)$ and $\tilde{a}(\mu)$. Furthermore assume :

(II.1) Quasi-transversal intersection).
For $\mu = \mu_0$, $W^u(p_\mu)$ and $W^s(q_\mu)$ have a quasi-transversal intersection (i.e. second order contact) on at least one orbit $O(r)$ and this tangency unfolds generically. (In fact it suffices that for some sequence $\mu_i \to \mu_0$ the separatrices $W^u(p_{\mu_i})$ and $W^s(q_{\mu_i})$ intersect near r transversally or not at all.)

Palis [P] proved that for the diffeomorphism $\varphi = \Phi_{\mu_0}$ in such arcs $\{\Phi_\mu\}_{\mu\in(-1,1)}$ the number $P(\varphi)$ defined by :

$$P(\varphi) := \frac{\log |a(\mu_0)|}{\log |b(\mu_0)|} \qquad \text{(II.1)}$$

is a topological invariant. We will generalise this to arcs of diffeomorphisms. Arcs $\{\Phi_\mu\}$ and $\{\tilde{\Phi}_\mu\}$ are called mildly conjugate at $\mu_0$ if for some orientation preserving homeomorphism $\xi : (-1,1) \to (-1,1)$ and some homeomorphism $h_\mu$, not necessarily continuous in $\mu$, one has $h_\mu \circ \Phi_\mu = \tilde{\Phi}_{\xi(\mu)} \circ h_\mu$ for $\mu$ near $\mu_0$. Often it is more natural to require that the homeomorphisms $h_\mu$ depend continuously on $\mu$. In that case the arcs are called conjugate at $\mu_0$. The arcs are called semi-locally mildly conjugate (resp. conjugate) at $\mu_0 \times r$ if $(\mu, x) \mapsto (\xi(\mu), h_\mu(x))$ is only defined on a neighbourhood U of the closure of $\mu_0 \times (\cup\, \Phi^i(r))$.

Theorem A.

For (semi-locally) conjugate arcs $\{\Phi_\mu\}$ and $\{\widetilde{\Phi}_\mu\}$ as above assume that the topological invariant $P(\varphi)(= P(\widetilde{\varphi}))$ is irrational. Then :

$$a(\mu_0) = \widetilde{a}(\widetilde{\mu}_0) \quad ,$$
$$b(\mu_0) = \widetilde{b}(\widetilde{\mu}_0) \quad . \tag{II.2}$$

Melo [M] proves that $P(\varphi) = P(\widetilde{\varphi})$ and certain global assumptions on $\varphi$ and $\widetilde{\varphi}$ imply that $\varphi$ and $\widetilde{\varphi}$ are conjugate. In another paper will also prove that analogously (II.2) implies that the arcs $\{\Phi_\mu\}$ and $\{\widetilde{\Phi}_\mu\}$ are conjugate.

Remark (II.2) In fact it is sufficient to assume in Theorem A that $(x,\mu) \to \Phi_\mu(x)$ is $C^2$.

Remark (II.3) Theorem A gives only a semi-local topological invariant. If we want that two arcs are globally conjugate we can get an infinite number of moduli : Take an arc of diffeomorphisms as in Figure (II.1)

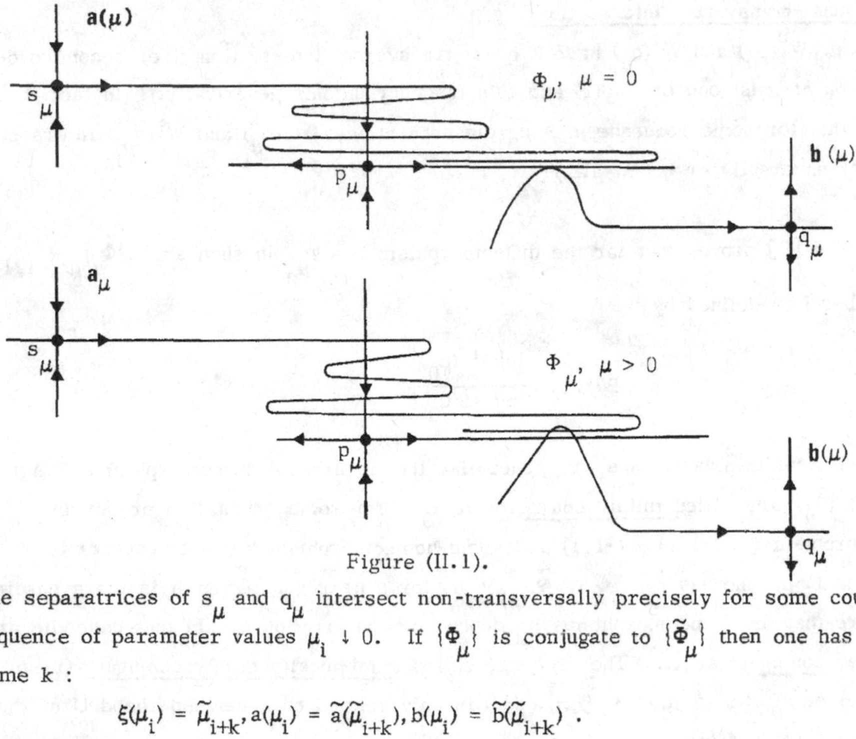


Figure (II.1).

The separatrices of $s_\mu$ and $q_\mu$ intersect non-transversally precisely for some countable sequence of parameter values $\mu_i \downarrow 0$. If $\{\Phi_\mu\}$ is conjugate to $\{\widetilde{\Phi}_\mu\}$ then one has for some k :

$$\xi(\mu_i) = \widetilde{\mu}_{i+k}, a(\mu_i) = \widetilde{a}(\widetilde{\mu}_{i+k}), b(\mu_i) = \widetilde{b}(\widetilde{\mu}_{i+k}) \; .$$

Since in general $\mu \to a(\mu)$ and $\mu \to b(\mu)$ are non-constant functions these equations give rise to a countable number of modulus conditions.

Notation (II.4) From now on we use the following notation for sequences of numbers $\alpha_i, \beta_i$ :

$\alpha_i \lesssim \beta_i$ means that $\alpha_i/\beta_i$ is bounded.
$\alpha_i \simeq \beta_i$ means that $\alpha_i/\beta_i$ converges to some positive finite number.
$\alpha_i \lesssim \beta_i$ means that $\alpha_i/\beta_i$ has a supremum not bigger than one and finally
$\alpha_i \cong \beta_i$ means that $\alpha_i/\beta_i$ converges to one.

## III. Statement of results for dim (M) > 2.

Suppose that one has

(III.1) (Linearisability) : The diffeomorphism $\Phi_\mu$ is $C^1$-linearisable near $p_\mu$ and $q_\mu$ for $\mu$ near $\mu_0$ .

(III.2) (Quasi-transversal tangency) : $W^u(p_\mu)$ and $W^s(q_\mu)$ has a quasi-transversal intersection for $\mu = \mu_0$ at a point r. An intrinsic definition of this generic property is given in [N.P.T.]. It means that for each value of $\mu$ near $\mu_0$ there is a $C^2$ local chart $\varphi_\mu$ near the intersection point r such that the mapping $(x,\mu) \to \varphi_\mu(x)$ is $C^2$. Furthermore

$$\varphi_\mu(W^s(q_\mu)) = \{x_1 = \ldots = x_{n-s} = 0\}, \text{ and}$$
$$\varphi_\mu(W^u(p_\mu)) = \{x_{u+2} = \ldots = x_n = 0,\ x_1 = f_\mu(x_{n-s+1}, \ldots, x_{u+1})\},$$

if $n - s + 1 \leq u + 1$, and otherwise

$$\varphi_\mu(W^u(p_\mu)) = \{x_{u+2} = \ldots = x_n = 0,\ x_1 = \varepsilon_\mu\} .$$

Here $s = \dim W^s(q_\mu)$, $u = \dim W^u(p_\mu)$ and $f_\mu = Q + \varepsilon_\mu$ where Q is a non-degenerate homogeneous quadratic function and $\varepsilon_\mu$ a constant depending continuously on $\mu$ with $\varepsilon_0 = 0$. Assume that $\varepsilon_{\mu_i} \neq 0$ for some sequence $\mu_i \to 0$.

(III.3) (Real weakest eigenvalues) : $\Phi_\mu$ has a real, strictly weakest contracting eigenvalue $a(\mu)$ at $p_\mu$, i.e. this eigenvalue has multiplicity one and for any other eigenvalue $\lambda(\mu) \neq a(\mu)$ at $p_\mu$ with $|\lambda(\mu)| < 1$ one has $|\lambda(\mu)| < a(\mu)$. Similarly assume the weakest expanding eigenvalue $b(\mu)$ at $q_\mu$ exists and is real.

(III.4) (Regular-tangency) : let $W^{cu}(p_\mu)$ be an invariant manifold tangent at $p_\mu$ to the direct sum of the expanding eigenspace and the weakest contracting eigenspace of $\Phi_\mu$ at $p_\mu$. These manifolds are not unique, but all these manifolds $W^{cu}(p_\mu)$ are tangent to each other at $W^u(p_\mu)$. Therefore it makes sense to demand that $W^{cu}(p_\mu)$ is transversal to $W^s(q_\mu)$ near $(r, \mu_0)$. Similarly assume that $W^u(p_\mu)$ intersects $W^{cs}(q_\mu)$ transversally. See [H.P.S.] and [N.P.T.].

In [N.P.T.] it is proved that for $\varphi = \Phi_{\mu_0}$ as above the number $P(\varphi)$ is a topological invariant. Here we prove :

Theorem B.

For conjugate families $\{\Phi_\mu\}$ and $\{\tilde{\Phi}_\mu\}$ as above with the topological invariant $P(\varphi)(= P(\tilde{\varphi}))$ irrational and $a(\mu_0)$ and $b(\mu_0)$ real, one has :

$$a(\mu_0) = \tilde{a}(\tilde{\mu}_0) \quad \text{and} \quad b(\mu_0) = \tilde{b}(\tilde{\mu}_0).$$

The proof of Theorem B depends very much on the fact that the distance of a point $x \in U$ to $W^u(p_\mu)$ and $W^s(q_\mu)$ is strongly related to the distance of $h_\mu(x)$ to $W^u(\tilde{p}_{\xi(\mu)})$ and $W^s(\tilde{q}_{\xi(\mu)})$. Using this we have nearly finished the proof of our Theorem B for $a(\mu)$ and $b(\mu)$ possibly complex. Also we expect to solve by related techniques the conjecture of Floris Takens from [T].

## IV. Proof of Theorem A (dimension 2).

Assume that the families $\{\Phi_\mu\}$ and $\{\tilde{\Phi}_\mu\}$ are conjugate and that the conjugacy depends continuously on the parameter. Topological invariance of $P(\varphi)$ is equivalent to

$$a = \tilde{a}^{\delta} \quad , \quad b = \tilde{b}^{\delta} \quad , \tag{IV.1}$$

for some $\delta$, where $a = |a(\mu_0)|$, $b = |b(\mu_0)|$, $\tilde{a} = |\tilde{a}(\tilde{\mu}_0)|$ and $\tilde{b} = |\tilde{b}(\tilde{\mu}_0)|$. For simplicity assume $\mu_0 = \tilde{\mu}_0 = 0$.

First assume that there is a sequence of parameter values $\mu_i \to 0$ such that $W^u(p_{\mu_i})$ and $W^s(q_{\mu_i})$ have two intersections near r, and therefore bound a compact set $D_{\mu_i}$ as in figure (IV.1). The case where $W^u(p_{\mu_i})$ and $W^s(q_{\mu_i})$ have no intersections near r will be considered later on.

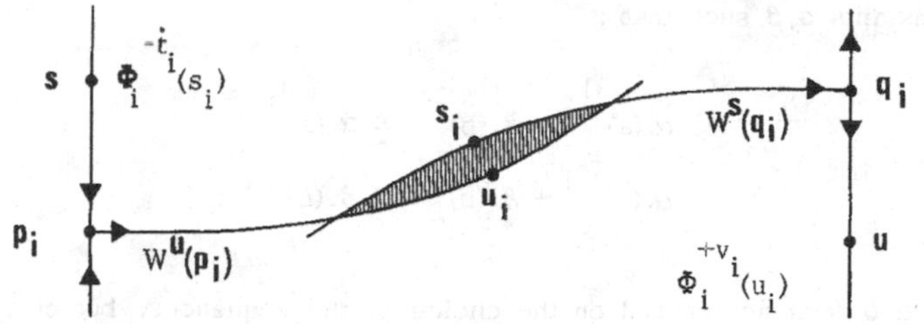


Figure (IV.1).

For simplicity we sometimes denote $\mu_i$ by i, $p_{\mu_i}$ by $p_i$ and $\Phi_{\mu_i}$ by $\Phi_i$, and let :

$$d_{\mu,u}(x) = d(x, W^u(p_\mu)) ,$$
$$d_{\mu,s}(x) = d(x, W^s(q_\mu)) .$$

where d is some distance function. The functions $d_{i,s}|D_i$ and $d_{i,u}|D_i$ have the same maximum value $d_i$, for each i, (remember "i" stands for "$\mu_i$"). For x on the curve $c_i = \{x \in D_i | d_s(x) + d_u(x) = d_i\}$ one has $\tilde{d}_s(h_i(x)) + \tilde{d}_u(h_i(x)) \leq \tilde{d}_i$, where $\tilde{d}_i = \tilde{d}_{\xi(\mu_i)}$ . We will translate this fact in an inequality in terms of the eigenvalues of $\Phi_i$ and $\tilde{\Phi}_i$ .

(IV.1) <u>Maximal sequences</u> $s_i$ are defined as follows. Take $s_i$ on $(W^s(q_i) \setminus W^u(p_i)) \cap D_i$ and $t_i$ such that

$$\Phi_i^{-t_i}(s_i) \to s \in W^s(p_0) \setminus \{p_0\} .$$

Assume that this sequence is maximal in the following sense. For any sequence $\bar{s}_i \in D_i$ there is no accumulation point $\bar{s} \in W^s(p_0)$ of $\Phi_i^{-t_i}(\bar{s}_i)$ which is contained in the component of $W^s(p_0) \setminus \{s\}$ that does not contain $p_0$; i.e. $\bar{s}$ should not be further away from $p_0$ than s. Clearly such a choice is possible. Similarly choose a maximal sequence $u_i$ on $W^u(p_i) \setminus W^s(q_i)) \cap D_i$ and numbers $v_i$ such that

$$\Phi_i^{+v_i}(u_i) \to u \in W^u(q_0) \setminus \{q_0\} .$$

Remark that since $h_\mu$ depends continuously on $\mu$, that $\tilde{s}_i = h_{\mu_i}(s_i)$ is also a maximal sequence for $\{\tilde{\Phi}_\mu\}$.

<u>Proposition (IV.2).</u> For a sequence of points $x_i \in D_i$ with $(\Phi_i)^{-k_i}(x_i) \to s$ and $(\Phi_i)^{+\ell_i}(x_i) \to u$ there are constants $\alpha, \beta$ such that :

$$\begin{aligned} \alpha.(a)^{+k_i} + \beta.(b)^{-\ell_i} &\lesssim \alpha.(a)^{t_i} . \\ \alpha.(a)^{+k_i} + \beta.(b)^{-\ell_i} &\lesssim \beta.(b)^{-v_i} . \end{aligned} \tag{IV.2}$$

This number $\alpha/\beta$ does not depend on the choice of the sequences, but only on s and u.

<u>Proof :</u> In a local $\varphi_\mu$ one can write $\varphi_\mu(W^u(p_\mu)) = \{y_1 = 0\}$ and $\varphi_\mu(W^s(q_\mu)) = \{(x,y) | y = x^2 - \varepsilon_\mu\}$. The invariant manifolds $W^u(p_\mu)$ and $W^s(q_\mu)$ depend $C^2$ on $\mu$. In fact $(0, p_0)$ is a partially hyperbolic fixed point for the $C^2$ diffeomorphism $(\mu, p) \to (\mu, \Phi_\mu(p))$ and hence this diffeomorphism has a $C^2$ centre unstable manifold.

Take for each $\mu$ a special metric $d_\mu$ which in the local coordinates induced by $\varphi_\mu$ coincides with the Euclidean metric. Let $d_{\mu,u}(x) = d_\mu(W^u(q_\mu), x)$ and denote the local maximum of the function $d_{\mu,u} | W^s(q_\mu)$ by $d_\mu$. From Assumption (II.1) it follows that we can choose a sequence of parameter values $\mu_i \to 0$ such that $d_{\mu_i} \neq 0$. As before denote $d_{\mu_i}$ by $d_i$.

<u>Lemma (IV.3)</u> For some $\alpha, \beta$ (depending on the metric d) one has :

$$\alpha.(a)^{t_i} \cong d_i \cong \beta.(b)^{-v_i} . \tag{IV.3}$$

<u>Proof of Lemma :</u> Since $\Phi_\mu$ is $C^2$ and dim(M) = 2 there are linearising $C^1$-coordinates at the periodic points $p_\mu$ and $q_\mu$, [Ha]. Via a backward iteration of $\varphi_\mu$ this induces a metric $\rho_\mu$ near $r_\mu$. Since the linearising coordinates near $p_\mu$ are $C^1$ one has as $z_i \to W^u(p_0)$ :

$$\frac{d_i(z_i, W^u(p_0))}{\rho_i(z_i, W^u(p_0))} \to \gamma \tag{IV.4}$$

for some constant $\gamma > 0$. Clearly one has that $\rho(s_i, W^u(p_i)). |a(\mu_i)|^{-t_i}$ converges to some constant, since by assumption $(\Phi_i)^{-t_i}(s_i) \to s$. But the separatrices $W^s(q_\mu)$ and $W^u(p_\mu)$

depend $C^2$ on $\mu$ and therefore $\rho(s_i, W^u(p_i)) \lesssim \mu_i$. Hence $t_i \lesssim |\log(\mu_i)|$ and since $\mu \to a(\mu)$ is $C^1$, as $i \to \infty$ :

$$|a(\mu_i)|^{-t_i} \cong (a)^{-t_i} .$$

Therefore

$$\rho(s_i, W^u(p_i)).(a)^{-t_i} \to \nu \tag{IV.5}$$

where $\nu$ only depends on s and on the choice of the metric $\rho_\mu$. Now take two sequences of points $z_{i,d}$ and $z_{i,\rho}$ contained in $(W^s(q_i) \setminus W^u(p_i)) \cap D_i$. Choose $z_{i,d}$ such that it is a maximum for $d(-, W^u(p_i))$ and $z_{i,\rho}$ such that $\rho(-, W^u(p_i))$ is maximal, i.e. :

$$\rho(z_{i,d}, W^u(p_i)) \leq \rho(z_{i,\rho}, W^u(p_i))$$
$$d(z_{i,\rho}, W^u(p_i)) \leq d(z_{i,d}, W^u(p_i))$$

Together with (IV.4) this implies :

$$\rho(z_{i,\rho}, W^u(p_i)) \cong \rho(z_{i,d}, W^u(p_i))$$

Therefore the sequence $z_{i,d}$ and $z_{i,\rho}$ are both maximal sequences as in (IV.1). By (IV.5) and (IV.4) ·

$$d_i := d(s_i, W^u(p_i)) \cong \gamma.\nu.(a)^{+t_i} \tag{IV.6}$$

Conclusion of the proof of Proposition (IV.2).

In the Euclidean metric d one has for any $x_i \in D_i$ $d(x_i, W^s(q_i)) + d(x_i, W^u(p_i)) \leq d_i$. Therefore as in equation (IV, 6) :

$$(\alpha + \nu_{1,i}).(a)^{+k_i} + (\beta + \nu_{2,i}).(b)^{-\ell_i} \leq (\alpha + \nu_{3,i}).(a)^{t_i}$$

where $\nu_{k,i} \to 0$ as $i \to \infty$, for $k = 1,2,3$. From this the Proposition follows.

Now we are going to choose nice sequences as above.

Proposition (IV.4). One can choose sequences $x_i$ and $\mu_i$ such that

(1) For certain subsequences

$$(\Phi_i)^{-(t_i+1)}(x_i) \to s \quad \text{and} \quad (\Phi_i)^{+\ell_i}(x_i) \to u ,$$

(2) $$\alpha.(a)^{t_i+1} + \beta.(b)^{-\ell_i} \cong \alpha.(a)^{t_i} . \qquad \text{(IV.7)}$$

Proof : Pick $\mu_i$ such that for the sequence $s_i$ as above $(\Phi_i)^{-i}(s_i) \to s$. Then consider the curve $c_i$ consisting of the set of points $x_i$ such that $d(x_i, W^u(p_i)) + d(x_i, W^s(q_i)) = d_i$ . Pick $x_i \in c_i$ such that $(\Phi_i)^{-i-1}(x_i) \to s$. Now we will show that some subsequence of the sequence $\{(\Phi_i)^\ell(x_i)\}_{i,\, \ell \in \mathbb{N}}$ converges to u. Clearly $d(x_i, W^u(p_i)) = \alpha.(a)^{i+1} + \nu_{1,i}.(a)^{i+1}$ . Therefore $d(x_i, W^s(q_i)) = \alpha.(1-a)(a)^i + \nu_{4,i}.(a)^i$. Since log(a)/log(b) is irrational there exists a subsequence $i_j$ and numbers $\ell_j$ such that

$$\alpha.(1-a).(a)^{i_j} + \nu_{4,i}.(a)^{i_j} \cong \beta.(b)^{-\ell_j} .$$

From this follows that statement (1) of the Proposition holds.

Finally we will show that from equation (IV.2) and Proposition (IV.4) follows that the number $\delta$ from equation (IV.1) is equal to 1 :

Proposition (IV.5). $\delta = 1$.

Proof : From equation (IV.7) :

$$(b)^{-\ell_i} . (a)^{-t_i} \cong \frac{\alpha}{\beta} . (1-a) .$$

The fact that $h_\mu$ depends continuously on $\mu$ implies e.g. that $(\tilde{\Phi}_i)^{-t_i-1}(h_i(x_i))$ converges to a point $\tilde{s} \in W^s(\tilde{p}_0)$. Therefore from Proposition (IV.2) :

$$(\tilde{b})^{-\ell_i} . (\tilde{a})^{-t_i} \geq \frac{\tilde{\alpha}}{\tilde{\beta}} . (1-\tilde{a}) .$$

And since $a = \tilde{a}^\delta$ and $b = \tilde{b}^\delta$ the two preceding equations yield :

$$\left(\frac{\tilde{\alpha}}{\tilde{\beta}} . (1-\tilde{a})\right)^\delta \leq \frac{\alpha}{\beta} . (1-\tilde{a}^\delta). \qquad \text{(IV.8)}$$

There is a similar expression in terms of $\tilde{b}$ :

$$\left(\frac{\tilde{\beta}}{\tilde{\alpha}} \cdot (\tilde{b}-1)\right)^{\delta} \leq \frac{\beta}{\alpha} \cdot (\tilde{b}^{\delta}-1) . \tag{IV.9}$$

Equations (IV.8) and (IV.9) imply that $\delta \geq 1$. By reversing the role of $\Phi$ and $\tilde{\Phi}$ one proves that $\delta \leq 1$. Q.E.D.

To complete the proof of Theorem A in this case remark that if $a = -\tilde{a}$ then $\Phi$ and $\tilde{\Phi}$ cannot be conjugate.

If for $\mu_i \to 0$ the separatrices $W^u(p_i)$ and $W^s(q_i)$ have no intersections near r then the function $d_u|W^s(q_i)$ (i.e. $x \to d(x, W^u(p_i))$ restricted to $W^s(q_i)$) has a local minimum. Therefore the proof of Theorem A in this case is essentially the same as the previous case.

## V. <u>The Proof of Theorem B.</u>

### V(a) <u>Properties preserved by the conjugacy.</u>

If dim(M) > 2 then it is possible that the function $d_s : x \to d(x, W^s(q_\mu))$ restricted to $W^u(p_\mu)$ does not have a unique maximum or minimum, but a saddle-point. In two ways the proof of Theorem A then breaks down. The comparison between the two metrics as in Lemma (IV.3) fails and furthermore there are no sets $D_\mu$ which are topologically invariant. But the essence of the proof of Theorem A was to show that the fact that h maps $D_\mu$ into $D_{\xi(\mu)}$ gave rise to four inequalities in terms of eigenvalues, see (IV.8) and (IV.9).

In higher dimensions we have to distinguish two cases and obtain either two equalities or four inequalities. In this way we get much stronger results than in Chapter III of [N.P.T.]. In a later paper we hope to use this in order to show that the conjugacy h is very rigid. (In fact see [M.P.S.].)

We have assumed that $W^s(q_0)$ and $W^u(p_0)$ intersect quasi-transversally in r. Let n = dim(M). For the time being assume that the dimensions s and u of resp. $W^s(q_\mu)$ and $W^u(p_\mu)$ are n - 1. In a neighbourhood U or r we have a normal form for $W^s(q_\mu)$ and $W^u(p_\mu)$ : There exists a $C^2$ local chart $\varphi_\mu$ such that

$$\varphi_\mu(W^u(p_\mu)) = \{x_1 = Q(x_2,\ldots,x_n) + \varepsilon_\mu\}, \tag{V.1}$$
$$\varphi_\mu(W^s(q_\mu)) = \{x_1 = 0\}.$$

We can also assume that $\varepsilon_{\mu_i} > 0$ for some sequence $\mu_i \to 0$. Since the index I of the quadratic function Q is determined by the intersection pattern of $W^s(q_0)$ and $W^u(p_0)$, the index I is a topological invariant.

Take the Euclidean metric d induced by the local coordinates near r and let $d_s(x) = d(x, W^s(q_\mu))$ and $d_u(x) = d(x, W^u(p_\mu))$. Let $d_\mu$ be the critical value of $d_s|W^u(p_\mu)$ (and therefore also of $d_u|W^s(q_\mu)$).

<u>Theorem (V.1).</u>

Let $s = u = n - 1$ and Q be definite and $\varepsilon_\mu > 0$. Then :

(i) $d_s|W^u(p_\mu)$ and $d_u|W^s(q_\mu)$ both have a unique minimum if $Q \geq 0$ (resp. maximum if $Q \leq 0$).

(ii) There is for each $\varkappa$, $0 \leq \varkappa \leq d_\mu$ a point $x_{\varkappa,\mu}$ such that $d_s(x_{\varkappa,\mu}) = \varkappa$, and furthermore

$$d_s(x) + d_u(x) = d_\mu,$$
$$\tilde{d}_s(h(x)) + \tilde{d}_u(h(x)) \geq \tilde{d}_{\xi(\mu)} \quad \text{if } Q \geq 0,$$
$$\text{(resp.} \quad \tilde{d}_s(h(x)) + \tilde{d}_u(h(x)) \leq \tilde{d}_{\xi(\mu)} \quad \text{if } Q \leq 0).$$

<u>Proof</u> : In the former case the proof is trivial. In the latter case the fact that the codimension of $W^s(q_\mu)$ and $W^u(p_\mu)$ is one implies that $W^s(q_\mu)$ and $W^u(p_\mu)$ bound a compact set $D_\mu$ as in Theorem A.

<u>Theorem (V.2).</u>

Let $s = u = n - 1$ and Q have a saddle point. Then :

(i) there exists a point $s_\mu \in W^s(q_\mu)$ such that $d_u(s_\mu) = d_\mu$ and $h_\mu(s_\mu) \in \tilde{W}^s(\tilde{q}_{\xi(\mu)})$, $\tilde{d}_u(h_\mu(s_\mu)) = \tilde{d}_{\xi(\mu)}$.

(ii) For each $\varkappa$ near 0 there exists a point $x_{\varkappa,\mu}$ with $d_u(x_{\varkappa,\mu}) = \varkappa$ and as $\mu \to 0$, $\varkappa \to 0$ :

$$|d_u(x_{\varkappa,\mu}) \pm d_s(x_{\varkappa,\mu})| \cong d_\mu,$$

$$|\tilde{d}_u(h_\mu(x_{\varkappa,\mu})) \pm \tilde{d}_s(h_\mu(x_{\varkappa,\mu})| \cong \tilde{d}_{\xi(\mu)} .$$

(Here ± means that, depending on $\varkappa$ and $\mu$, either both the equations are valid for + or both are valid for - .)

Proof : In the local coordinates near r we had : $W^s(q_\mu) = \{x_1 = 0\}$ and $W^u(p_\mu) = \{x_1 = Q(x_2,\ldots,x_n) + d_\mu\}$. Let

$$E = \{x | Q(x_2,\ldots,x_n) = 0\}$$

and $E_\varkappa = E \cap \{x_1 = \varkappa\}$. For $x \in E$ one has $d_s(x) = |x_1|$ and as $\varkappa,\mu \to 0$ : $d_u(x) \cong |Q + d_\mu - x_1| = |d_\mu - x_1|$, i.e. $|d_u(x) \pm d_s(x)| \cong ||d_\mu - x_1| \pm |x_1|| = d_\mu$. We are going to prove that the image $h_\mu(E_\varkappa)$ of the cone $E_\varkappa$ must have an intersection with the family of cones $\tilde{E}$. So suppose by contradiction that $h_\mu(E_\varkappa)$ is contained in the set $N_+(\tilde{E}_\varkappa) = \{\tilde{x} | \tilde{Q}(\tilde{x}) > 0\}$. Since Q has a saddle point one can choose the coordinates $(x_1,\ldots,x_n)$ such that $Q(x_2,\ldots,x_n) = x_2^2 + \ldots + x_{I+1}^2 - x_{I+2}^2 - \ldots - x_n^2$. Let $V = \{x_1 = 0,\ x_j = 0,\ j \geq I+2\}$ and $\pi: U \to V$ be the projection $\pi(x_1,\ldots,x_n) = (0,x_2,\ldots,x_{I+1},0,\ldots,0)$. Since one has for $\mu = 0$, $E_0 = W^u(p_0) \cap W^s(q_0)$, i.e. $h_0(E_0) = \tilde{E}_0$, the set $\pi(h_\mu(E_\varkappa))$ contains a neighbourhood of 0 in the plane $\tilde{V}$ for $\varkappa,\mu$ small. But $\pi(N_+(\tilde{E})) = \tilde{V}\backslash\{0\}$. Therefore we get a contradiction. Then finally the statement of the Theorem between the brackets follows since $W^u(p_\mu)$ and $W^s(q_\mu)$ are codimension one manifolds and divide U in several components.

(V)(b) Conclusion of the proof of Theorem B.

Step 1 : $s = u = n - 1$.

If we are in the case of Theorem (V.1) then the proof of Theorem B is exactly the same as the proof of Theorem A. In the case of Theorem (V.2) take $s_i \in W^s(q_{\mu_i})$ and parameter values $\mu_i \to 0$ such that $\Phi_{\mu_i}(s_i)$ converges to some point $s \in W^s(p_0)\backslash\{p_0\}$. Choose points $x_i$ and $y_i$ such that $\Phi_i^{-i-1}(x_i) \to s$ and $\Phi_i^{-i-2}(y_i) \to s$. Since log(a)/log(b) is irrational for certain sequences $\ell_i, m_i$ subsequences of $\Phi_i^{\ell_i}(x_i)$ and $\Phi_i^{m_i}(y_i)$ converge to the same point $u \in W^u(q_0)\backslash\{q_0\}$. If we choose these sequences $x_i$ and $y_i$ as in (ii) of Theorem (V.2) then there are constants $\alpha,\beta$ such that :

$$\alpha.(a)^{i+1} + \beta.(b)^{-\ell_i} \cong \alpha.(a)^i ,$$

$$\alpha.(a)^{i+2} + \beta.(b)^{-m_i} \simeq \alpha.(a)^i .$$

Since $h_\mu$ depends continuously on $\mu$ the sequence $\Phi_i^{\ell_i}(h(x_i))$ and similar sequences actually do converge. And by statement (ii) of Theorem (V.2) we have similarly :

$$\tilde{\alpha}.(\tilde{a})^{i+1} + \tilde{\beta}.(\tilde{b})^{-\ell_i} \cong \tilde{\alpha}.(\tilde{a})^i ,$$

$$\tilde{\alpha}.(\tilde{a})^{i+2} + \tilde{\beta}.(\tilde{b})^{-m_i} \cong \tilde{\alpha}.(\tilde{a})^i .$$

Since $a = \tilde{a}^\delta$ and $b = \tilde{b}^\delta$ these equations imply :

$$\frac{\beta}{\alpha} \cdot \left(\frac{\tilde{\alpha}}{\tilde{\beta}}\right)^\delta = \frac{1 - \tilde{a}^{k\delta}}{(1-\tilde{a}^k)^\delta}$$

for $k = 1, 2$. This is only possible for $\delta = 1$.

Step 2 : s, u general.

Take the invariant manifolds $W^{cu}(p_\mu)$ and $W^{cs}(q_\mu)$ from assumption (III.4). By the transversality assumption and since $a(\mu)$, $b(\mu)$ are real, the manifolds $W^u(p_\mu) \cap W^{cu}(p_\mu) \cap W^{cs}(q_\mu)$ and $W^s(q_\mu) \cap W^{cu}(p_\mu) \cap W^{cs}(q_\mu)$ are codimension one manifolds in $W^{cu}(p_\mu) \cap W^{cs}(q_\mu)$. The idea now is to restrict the problem to $W^{cu}(p_\mu) \cap W^{cs}(q_\mu)$ and use the preceding step then. This is done exactly as in Chapter III of [N.P.T.].

## References.


[Ha] P. Hartman, On local homeomorphisms of Euclidean spaces, Bol. Soc. Mat. Mexicana (2) 5, (1960).

[H.P.S.] M. Hirsch, C. Pugh & M. Shub, Lecture Notes in Math., Springer-Verlag, 583, (1977).

[M] W. de Melo, Moduli of stability of two-dimensional diffeomorphisms, Topology 19, (1980), 9-21.

[M.P.] W. de Melo & J. Palis, Moduli of stability for diffeomorphisms, Conference at Northwestern University, Lecture Notes in Math., Springer-Verlag, 819, (1980).

[M.P.S.] W. de Melo, J. Palis & S.J. van Strien, Characterising diffeomorphisms with modulus of stability one, this volume.

[N.P.T.] S. Newhouse, J. Palis & F. Takens, Stable families of diffeomorphisms, IMPA preprint, to appear.

[P] J. Palis, A differentiable invariant of topological conjugacies and moduli of stability, Astérisque 51, (1978), 335-346.

[P.S.] J. Palis & S. Smale, Structural stability theorems, Proc. Symp. Pure Math. A.M.S. 14, (1970), 223-232.

[T] F. Takens, Moduli of instability; non-transversal intersections of invariant manifolds of vector fields, preprint.

S.J. van Strien : Mathematics Institute, University of Utrecht, Utrecht, The Netherlands.

Detecting strange attractors in turbulence.

Floris Takens.

## 1. Introduction.

Since [19] was written, much more accurate experiments on the onset of turbulence have been made, especially by Fenstermacher, Swinney, Gollub and Benson [6,8,9,10]. These new experimental data should be interpreted according to [19] in terms of strange attractors, or they should falsify the whole picture given in that paper. For such interpretations one uses in general the so-called power spectrum. It is however not at all clear how to reconstruct the "strange attractors" from a power spectrum (with continu parts); even worse : how can one see whether a given power spectrum (with continuous parts) might have been "generated" by a strange attractor? In this paper I present procedures to decide whether one may attribute certain experimental data, as in the onset of turbulence, to the presence of strange attractors. These procedures consist of algorithms, to be applied to the experimental data itself and not to the power spectrum; in fact, I doubt whether the power spectrum contains the relevant information.

In order to describe the problems and results, treated in this paper, in more detail, I shall first review the ideas of [19], also comparing them with those exposed by Landau and Lifschitz [13], in relation with the flow between two rotating cylinders. It was this same experiment which was carried out to great precision by Swinney et.al. [6, 8, 10].

It should be noted that the discussion in [19] is not restricted to this situation but should also be applicable to other situations where an orderly dynamic changes to a chaotic one; see [8] for a discussion of some examples. Also, our present discussion should be applicable to these cases.

### The Taylor-Couette Experiment.

We consider the region D between two cylinders as indicated in figure 1. In this region we have a fluid. We study its motion when the outer cylinder, the top and bottom are at rest, while the inner cylinder has an angular velocity $\Omega$. p is some fixed point in the interior of D. For a number of values of $\Omega$, one component

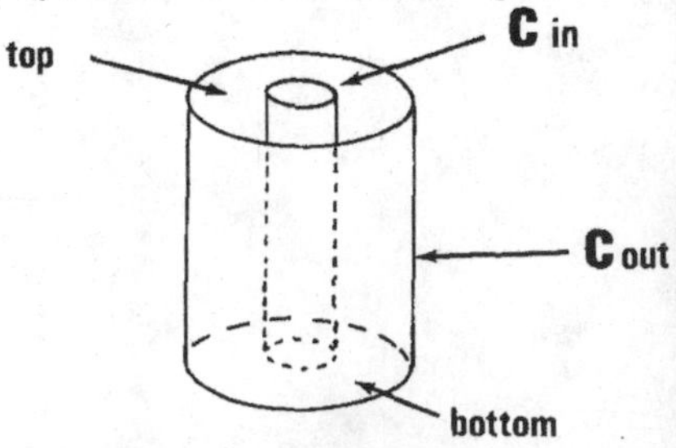

of the velocity of the fluid at p is measured as a function of time. In [19] the idea was the following : for each value of $\Omega$ the set of all "possible states" is a Hilbert space $H_\Omega$ consisting of (divergence free) vector fields on D satisfying the appropriate boundary conditions (these vector fields represent velocity distributions of the fluid). For each $\Omega$ there is an evolution semi-flow

$$\{\varphi_t^\Omega : H_\Omega \to H_\Omega\}_{t\in \mathbb{R}_+}, \mathbb{R}_+ = \{t \in \mathbb{R} | t \geq 0\} ,$$

such that if $X \in H_\Omega$ represents the state at time t = 0 then $\varphi_{t_0}^\Omega(X)$ represents the state at time $t_0$. We assume that for all values of $\Omega$ under consideration, there is an "attractor" $\Lambda_\Omega \subset H_\Omega$ to which (almost) all evolution curves $\varphi_t^\Omega(X)$ tend as $t \to \infty$. (At this point we don't want to specify the term "attractor".) $\Lambda_\Omega$ and $\varphi_t^\Omega | \Lambda_\Omega$ then describe the asymptotic behaviour of all evolution curves $\varphi_t^\Omega(X)$. Roughly the main assumptions in [19] could be rephrased as : $\varphi_t^\Omega | \Lambda_\Omega$ behaves just as an attractor in a finite dimensional differentiable dynamical system. In more detail, the assumption was that for all values of $\Omega$ under consideration there is a smooth finite dimensional manifold $M_\Omega \subset H_\Omega$, smoothly depending on $\Omega$, such that :

(i) $M_\Omega$ is invariant in the sense that for $X \in M_\Omega$, $\varphi_t^\Omega(X) \in M_\Omega$ ;

(ii) $M_\Omega$ is attractive in the sense that evolution curves $\varphi_t^\Omega(X)$, starting outside $M_\Omega$ tend to $M_\Omega$ for $t \to \infty$ ;

(iii) the flow, induced in $M_\Omega$ by $\varphi_t^\Omega$, is smooth, depends smoothly on $\Omega$ and has an attractor $\Lambda_\Omega$.

Some justification for this assumption was given by Marsden [15,16]. Apart from this we used genericity assumptions : if $Z_\Omega$ denotes the vector field on $M_\Omega$ which is the infinitessimal generator of $\varphi_t^\Omega | M_\Omega$, we assume $(M_\Omega, Z_\Omega)$ to be a generic one-parameter family of vector fields. (If however the physical system under consideration has symmetry, like the case of the Couette flow, then a same type of symmetry must hold for $M_\Omega, \varphi_t^\Omega$, and hence for $Z_\Omega$. In this case genericity should be understood within the class of vector fields having this symmetry; see [18].)

In the Landau-Lifschitz picture, one assumes that the limiting motion (or attractor) is quasi-periodic, i.e. of the form

$$\varphi_t^{\Omega}(X) = f^{\Omega}(X, a_1 e^{2\pi i \omega_1 t}, a_2 e^{2\pi i \omega_2 t}, \ldots )$$

where $\omega_i$, and $a_i$ depends on $\Omega$ and where for each $\Omega$ only a finite number of $a_i$ is non-zero. One can imagine that, as more and more $a_i$ become non-zero, the motion gets more and more turbulent.

Also in this last description we have a smooth finite dimensional manifold as attractor, namely an n-torus, but such attractors do not occur for generic parameter values of generic one-parameter families of vector fields. It should be noted however that for generic one-parameter families of vector fields there may be a set of parameter values with positive measure for which quasi periodic motion occurs; see [11].

This n-torus attractor has topological entropy zero and its dimension is an integer. On the other hand "strange attractors" have in general positive entropy and often non-integral dimension. Hence it would be important to determine entropy and dimension of attractors from "experimental data".

In view of the experiment just described, we have to add one more point to out formal description, namely we have to add the function (observable) from the state space to the reals giving the experimental output (when composed with $\varphi_t^{\Omega}(X)$ ). In the present example of the Taylor-Couette experiment, this function $y_{\Omega}: H_{\Omega} \to \mathbb{R}$ assigns to each $X \in H_{\Omega}$ the measured component of $X(p)$. As far as the asymptotic behaviour is concerned, we only have to deal with $y_{\Omega}|M_{\Omega}$ (or with $y_{\Omega}|\Lambda_{\Omega}$). Since $M_{\Omega}$ depends smoothly on $\Omega$ all $M_{\Omega}$ are diffeomorphic and so we may drop the $\Omega$.

Summarising, we have a manifold M with a smooth one-parameter family of vector fields $Z_{\Omega}$ and a smooth one-parameter family of functions $y_{\Omega}$. For a number of values of $\Omega$ the function $y_{\Omega}(\varphi_t^{\Omega}(x))$ is known by measurement (for some x in or near M which may depend on $\Omega$; $\varphi_t^{\Omega}$ denotes here the flow on M generated by $Z_{\Omega}$. The point is to obtain information about the attractor(s) of $Z_{\Omega}$ from these measurements, i.e. from the functions $t \mapsto y_{\Omega}(\varphi_t^{\Omega}(x))$. For this we shall allow ourselves to make genericity assumptions on $(M, Z_{\Omega}, y_{\Omega}, x)$.

We shall prove that under suitable genericity assumptions on $(M, Z_{\Omega}, y_{\Omega}, x)$ the positive limit set $L^+(x)$ of x is determined by the function $y_{\Omega}(\varphi_t^{\Omega}(x))$. In our "main theorem" in section 4 we describe algorithms which, when applied to a sequence

$\{a_i = y_\Omega(\varphi^\Omega_{\alpha,i}(x))\}_{i=1}^{\bar N}$, $\bar N$ sufficiently big, will give an approximation for the dimension of $L^+(x)$, respectively for the topological entropy of $\varphi^\Omega_\alpha|L^+(x)$. This leads in principle to a possibility of testing and comparing the hypothesis made by Landau-Lifschitz [13] and Ruelle-Takens [19]; see the observation at the end of section 4. The author wishes to acknowledge the hospitality of the department of mathematics of Warwick University and the many discussions with participants of the turbulence and dynamical systems symposium there during the preparation of this paper.

## 2. Dynamical systems with one observable.

Let M be a compact manifold. A dynamical system on M is a diffeomorphism $\varphi: M \to M$ (discrete time) or a vector field X on M (continuous time). In both cases the time evolution corresponding with an initial position $x_0 \in M$ is denoted by $\varphi_t(x_0)$ : in the case of discrete time $t \in \mathbb{N}$ and $\varphi_i = (\varphi)^i$; in the case of continuous time $t \in \mathbb{R}$ and $t \mapsto \varphi_t(x_0)$ is the X integral curve through $x_0$.

An observable is a smooth function $y: M \to \mathbb{R}$. The first problem is this : if, for some dynamical system with time evolution $\varphi_t$, we know the functions $t \mapsto y(\varphi_t(x))$, $x \in M$, then how can we obtain information about the original dynamical system (and manifold) from this. The next three theorems deal with this problem. (After the research for this paper was completed, the author was informed that this problem, or at least parts of it, was also treated by other authors, see [1,17]. Since out results are in some sense somewhat more general we still give here a treatment of the problem independent of the results in the above papers.)

<u>Theorem 1.</u> Let M be a compact manifold of dimension m. For pairs $(\varphi, y)$, $\varphi: M \to M$ a smooth diffeomorphism and $y: M \to \mathbb{R}$ a smooth function, it is a generic property that the map $\Phi_{(\varphi,y)}: M \to \mathbb{R}^{2m+1}$, defined by

$$\Phi_{(\varphi,y)}(x) = (y(x), y(\varphi(x)), \dots, y(\varphi^{2m}(x)))$$

is an embedding; by "smooth" we mean at least $C^2$.

<u>Proof.</u> We may, and do, assume that if x is a point with period k of $\varphi$, $k \leq 2m+1$, all eigenvalues of $(d\varphi^k)_x$ are different and different from 1. Also we assume that no two different fixed points of $\varphi$ are in the same level of y. For $\Phi_{(\varphi,y)}$ to be an immersion near a fixed point x, the co-vectors $(dy)_x, d(y\varphi)_x, \dots, d(y\varphi^{2m})_x$ must span $T^*_x(M)$. This is the case

for generic y if $d\varphi$ satisfies the above condition at each fixed point.

In the same way one proves that $\Phi_{(\varphi,y)}$ is generically an immersion and even an embedding when restricted to the periodic points with period $\leq 2m+1$. So we may assume that for generic $(\bar{\varphi},\bar{y})$ we have : $\Phi_{(\bar{\varphi},\bar{y})}$, restricted to a compact neighbourhood V of the set of points with period $\leq 2m+1$ is an embedding; for some neighbourhood $\mathfrak{U}$ of $(\bar{\varphi},\bar{y})$, $\Phi_{(\varphi,y)}|V$ is an embedding whenever $(\varphi,y) \in \mathfrak{U}$. We want to show that for some $(\varphi,y) \in \mathfrak{U}$, arbitrarily near $(\bar{\varphi},\bar{y})$, $\Phi_{(\varphi,y)}$ is an embedding.

For any point $x \in M$, which is not a point of period $\leq 2m+1$ for $\bar{\varphi}$, the co-vectors $(d\bar{y})_x, d(\bar{y}\bar{\varphi})_x, d(\bar{y}\bar{\varphi}^2)_x, \ldots, d(\bar{y}\bar{\varphi}^{2m})_x \in T^*(M)$ can be perturbed independently by perturbing $\bar{y}$. Hence arbitrarily near $\bar{y}$ there is $\bar{\bar{y}}$ such that $(\bar{\varphi},\bar{\bar{y}}) \in \mathfrak{U}$ and such that $\Phi_{(\bar{\varphi},\bar{\bar{y}})}$ is an immersion. Then there is a positive $\varepsilon$ such that whenever $0 < \rho(x,x') \leq \varepsilon$, $\Phi_{(\bar{\varphi},\bar{\bar{y}})}(x) \neq \Phi_{(\bar{\varphi},\bar{\bar{y}})}(x')$; $\rho$ is some fixed metric on M. There is even a neighbourhood $\mathfrak{U}' \subset \mathfrak{U}$ of $(\bar{\varphi},\bar{\bar{y}})$ such that for any $(\varphi,y) \in \mathfrak{U}'$, $\Phi_{(\varphi,y)}$ is an immersion and $\Phi_{(\varphi,y)}(x) \neq \Phi_{(\varphi,y)}(x')$ whenever $x \neq x'$ and $\rho(x,x') \leq \varepsilon$. From now on we also assume that each component of V has diameter smaller than $\varepsilon$.

Finally we have to show that in $\mathfrak{U}'$ we have a pair $(\varphi,y)$ with $\Phi_{(\varphi,y)}$ injective. For this we need a finite collection $\{U_i\}_{i=1}^{N}$ of open subsets of M, covering the closure of $M \setminus \{\bigcap_{j=0}^{2m} \varphi^j(V)\}$, and such that :

(i) for each $i = 1,\ldots,N$ and $k = 0,1\ldots,2m$, diameter $(\bar{\varphi}^{-k}(U_i)) < \varepsilon$;

(ii) for each $i,j = 1,\ldots,N$ and $k,l = 0,1,\ldots,2m$, $\bar{\varphi}^{-k}(U_i) \cap U_j \neq \emptyset$ and $\bar{\varphi}^{\ell}(U_i) \cap U_j \neq \emptyset$ imply that $k = \ell$;

(iii) for $\bar{\varphi}^{j}(x) \in M \setminus (\bigcup_i U_i)$, $j = 0,\ldots,2m, x' \notin V$ and $\rho(x,x') > \varepsilon$, no two points of the sequence $x, \bar{\varphi}(x), \ldots, \bar{\varphi}^{2m}(x), x', \bar{\varphi}(x'), \ldots, \bar{\varphi}^{2m}(x')$ belong to the same $U_i$.

Note that (ii) implies, but is not implied by

(ii)' no two points of the sequence $x, \bar{\varphi}(x), \ldots, \bar{\varphi}^{2m}(x)$ belong to the same $U_i$.

We take a corresponding partition $\{\lambda_i\}$ of unity, i.e., $\lambda_i$ is a non-negative function with support $\bar{U}_i$ and $\sum_{i=1}^{N} \lambda_i(x) = 1$ for all $x \in \overline{M \setminus V}$. Consider the map

$\Psi: M \times M \times \mathbb{R}^N \to \mathbb{R}^{2m+1} \times \mathbb{R}^{2m+1}$ which is defined in the following way $\Psi(x,x',\varepsilon_1,\ldots,\varepsilon_N) = (\Phi_{(\bar{\varphi},\bar{\bar{y}}_\varepsilon)}(x), \Phi_{(\bar{\varphi},\bar{\bar{y}}_\varepsilon)}(x'))$, where $\varepsilon$ stands for $(\varepsilon_1,\ldots,\varepsilon_N)$ and $\bar{\bar{y}}_\varepsilon = \bar{\bar{y}} + \sum_{i=1}^{N} \varepsilon_i\lambda_i$. We define $W \subset M \times M$ as $W = \{(x,x') \in M \times M \mid \rho(x,x') \geq \varepsilon$ and not both $x$ and $x'$ are in $\mathrm{int}(V)\}$. $\Psi$, restricted to a small neighbourhood of $W \times \{0\}$ in $(M \times M) \times \mathbb{R}^N$, is transverse with respect to the diagonal of $\mathbb{R}^{2m+1} \times \mathbb{R}^{2m+1}$. This transversality follows immediately from all the conditions imposed on the covering $\{U_i\}_{i=1}^N$. From this transversality we conclude that there are arbitrarily small $\bar{\varepsilon} \in \mathbb{R}^N$ such that $\Psi(W \times \{\bar{\varepsilon}\}) \cap \Delta = \emptyset$. If also for such an $\bar{\varepsilon}$, $(\bar{\varphi}, \bar{\bar{y}}_{\bar{\varepsilon}}) \in \mathcal{U}'$ then $\Phi_{(\bar{\varphi},\bar{\bar{y}}_\varepsilon)}$ is injective and hence an embedding.

This proves that for a dense set of pairs $(\varphi, y)$, $\Phi_{(\varphi,y)}$ is an embedding. Since the set of all embeddings is open in the set of all mappings, there is an open and dense set of pairs $(\varphi, y)$, for which $\Phi_{(\varphi,y)}$ is an embedding. This proves the theorem.

Remark. This theorem also works for M non-compact if we restrict our observables to be proper functions.

Theorem 2. Let M be a compact manifold of dimension m. For pairs $(X, y)$, X a smooth (i.e., $C^2$) vector field and y a smooth function on M, it is a generic property that $\Phi_{X,y}: M \to \mathbb{R}^{2m+1}$, defined by $\Phi_{X,y}(x) = (y(x), y(\varphi_1(x)), \ldots, y(\varphi_{2m}(x))$ is an embedding, where $\varphi_t$ is the flow of X.

Proof. The proof of this theorem is almost the same as the proof of theorem 1. In this case we impose the following generic properties on X :

(i) if $X(x) = 0$ then all eigenvalues of $(d\varphi_1)_x : T_x(M) \to T_x(M)$ are different and different from 1;

(ii) no periodic integral curve of X has integer period $\leq 2m + 1$.

In this case $\varphi_1$ satisfies the same conditions as $\bar{\varphi}$ in the previous proof. The rest of the proof carries over immediately.

The next theorem is only included for the sake of completeness; it will not be used in the sequel of this paper.

Theorem 3. Let M be a compact manifold of dimension m. For pairs (X, y), X a smooth vector field and y a smooth function on M, it is a generic property that the map $\tilde{\Phi}_{X,y}: M \to \mathbb{R}^{2m+1}$, defined by

$$\tilde{\Phi}_{X,y}(x) = \left(y(x), \frac{d}{dt}(y(\varphi_t(x)))\Big|_{t=0}, \ldots, \frac{d^{2m}}{dt^{2m}}(y(\varphi_t(x)))\Big|_{t=0}\right)$$

is an embedding. Here $\varphi_t$ again denotes the flow of X; this time, smooth means at least $C^{2m+1}$.

Proof. Also this proof is quite analogous to that of theorem 1. First we may, and do, assume that a generic vector field X has the property that whenever X(x) = 0, all eigenvalues of $(dX)_x$ are different and different from zero. Sing(X) denotes the set of points where X is zero; this set is finite.

As in the proof of theorem 1, for such a vector field X the set of functions $y: M \to \mathbb{R}$ such that $\tilde{\Phi}_{X,y}$ is an immersion and, when restricted to a small neighbourhood of Sing(X), an embedding, is residual.

Finally, to obtain an embedding for $(X, \bar{y})$, $\bar{y}$ near y, we don't need an open covering in the present case. One can construct directly a map $y_v$, v in some finite dimensional vector space V, which is the analogue of $y_\varepsilon$, with the following properties :

(i) $y_0 = y$;

(ii) for $x \in$ Sing(X), the 1-jet of $y_v$ is independent of v;

(iii) for $x, x' \notin$ Sing(X), $x \neq x'$ the map

$$j_x^{2m} \times j_{x'}^{2m} : V \to J_x^{2m}(M) \times J_{x'}^{2m}(M)$$

has a surjective derivative for all (x, x') in v = 0; $J_x^{2m}(M)$ is the vector space of 2m-jets of functions on M in x; $j_x^{2m}(v)$ is the 2m-jet of $y_v$ in x.

Using $y_v$ one defines a map

$$\tilde{\Phi}: M \times M \times V \to \mathbb{R}^{2m+1} \times \mathbb{R}^{2m+1} \quad \text{as before .}$$

The rest of the proof of theorem 1 now carries over to the present situation.

From the last three theorems it is clear how a dynamical system with time evolution $\varphi_t$ and observable y is determined generically by the set of all functions $t \to y(\varphi_t(x))$. In practice the following situation may occur : we have a dynamical system with continuous time, but the value of the observable y is only determined for a discrete set $\{0, \alpha, 2\alpha, \ldots\}$ of values of t; $\alpha > 0$. This happens e.g. in the measurements of the onset of turbulence [6, 8, 9, 10]. Also instead of all sequences of the form $\{y(\varphi_{i\alpha}(x))\}_{i=0}^{\infty}$, $x \in M$, we only know such a sequence for one, or a few values of x (depending on the number of experiments) and these sequences are not known entirely but only for $i = 1, \ldots, \bar{N}$ for some finite but big $\bar{N}$ (in [6], $\bar{N} = 8192 = 2^{13}$). In this light we should know whether, under generic assumptions, the topology of, and dynamics in the positive limit set

$$L^+(x) = \{x' \in M \mid \exists t_i \to \infty \text{ with } \varphi_{t_i}(x) \to x'\}$$

of x is determined by the sequence $\{y(\varphi_{i,\alpha}(x))\}_{i=0}^{\infty}$. This question is treated in the next theorem and its corollary; in later sections we come back to the point that these sequences are only known up to some finite $\bar{N}$.

<u>Theorem 4.</u> Let M be a compact manifold, X a vector field on M with flow $\varphi_t$ and p a point in M. Then there is a residual subset $C_{X,p}$ of positive real numbers such that for $\alpha \in C_{X,p}$, the positive limit sets of p for the flow $\varphi_t$ of X and for the diffeomorphism $\varphi_\alpha$ are the same. In other words, for $\alpha \in C_{X,p}$ we have that each point $q \in M$ which is the limit of a sequence $\varphi_{t_i}(p)$, $t_i \in \mathbb{R}$, $t_i \to +\infty$, is the limit of a sequence $\varphi_{n_i.\alpha}(p)$, $n_i \in \mathbb{N}$, $n_i \to \infty$.

<u>Proof.</u> Take $q \in L^+(p)$. For $\varepsilon$ a (small) positive real number define $C_{\varepsilon,q} = \{\alpha>0 \mid \exists\, n \in \mathbb{N}, \text{ such that } \rho(\varphi_{n.\alpha}(p),q) < \varepsilon\}$, $\rho$ is some fixed metric on M. Clearly $C_{\varepsilon,q}$ is open; it is also dense. To prove this last statement we observe that for any $\bar{\alpha} > 0$ and $\bar{\varepsilon} > 0$, there is a point of $C_{\varepsilon,q}$ in $(\bar{\alpha}, \bar{\alpha} + \bar{\varepsilon})$ if and only if there is a $t \in (n.\bar{\alpha}, n.(\bar{\alpha}+\bar{\varepsilon}))$ with $\rho(\varphi_t(p), q) < \varepsilon$ for some integer n. The existence of such t follows from the fact that for big n the intervals $(n.\bar{\alpha}, n.(\bar{\alpha} + \bar{\varepsilon}))$ overlap (in the sense that for big n, $n.(\bar{\alpha} + \bar{\varepsilon}) > (n+1).\bar{\alpha}$ and the fact that there are arbitrary big values of t with $\rho(\varphi_t(p), q) < \varepsilon$.

Since $C_{\varepsilon,q}$ is open and dense we can take for $C_{X,p} \subset \mathbb{R}_{++}$ the following residual set $C_{X,p} = \bigcap_{i,j=1}^{\infty} C_{\frac{1}{i},q_j}$ where $\{q_j\}$ is a countable dense sequence in $L^+(p)$.

Corollary 5. Let M be a compact manifold of dimension m. We consider quadruples, consisting of a vector field X, a function y, a point p, and a positive real number $\alpha$. For generic such $(X, y, p, \alpha)$ (more precisely : for generic $(X, y)$ and $\alpha$ satisfying generic conditions depending on X and p), the positive limit set $L^+(p)$ is "diffeomorphic" with the set of limit points of the following sequence in $\mathbb{R}^{2m+1}$ :

$$\{(y(\varphi_{k,\alpha}(p)), y(\varphi_{(k+1).\alpha}(p)), \ldots, y(\varphi_{(k+2m).\alpha}(p)))\}_{k=0}^{\infty} .$$

The meaning of "diffeomorphic" should be clear here : it means that there is a smooth embedding of M into $\mathbb{R}^{2m+1}$ mapping $L^+(p)$ bijectively to this set of limit points.

For further reference we remark that the metric properties of $\overline{\{\varphi_{i.\alpha}(p)\}_{i=0}^{\infty}} \subset M$, with $\{\varphi_{i.\alpha}(p)\}$ as a sequence of distinguished points are the same as $\overline{\{b_i\}_{i=1}^{\infty}} \subset \mathbb{R}^{2m+1}$ with $\{b_i\}$ as a sequence of distinguished points :

$$b_i = (y(\varphi_{i.\alpha}(p)), \ldots, y(\varphi_{(1+2m).\alpha}(p))) \in \mathbb{R}^{2m+1} .$$

These metric properties are the same in the sense that distances in M and the corresponding distances in $\mathbb{R}^{2m+1}$ have a quotient which is uniformly bounded and bounded away from zero.

## 3. Limit capacity and dimension.

There are several ways to define the notion of dimension for compact metric spaces. The definition which we use here gives the so-called limit capacity. Some information on this notion can be found in [14]. Since this limit capacity is not well known we treat here some of its basic properties.

Let $(S, \rho)$ be a compact metric space. For $\varepsilon > 0$ we make the following definitions

- $s(S, \varepsilon)$ is the maximal cardinality of a subset of S such that no two points have distance less than $\varepsilon$; such a set is called a maximal $\varepsilon$-separated set;
- $r(S, \varepsilon)$ is the minimal cardinality of a subset of S such that S is the union of all the $\varepsilon$-neighbourhoods of its points; such a set is also called a minimal $\varepsilon$-spanning set.

Note that

$$r(S, \varepsilon) \leq s(S, \varepsilon) \leq r(S, \tfrac{\varepsilon}{2}) \quad \ldots\ldots\ldots\ldots \quad (1)$$

The first inequality follows from the fact that a maximal $\varepsilon$-separated set is $\varepsilon$-spanning. The second inequality follows from the fact that in an $\frac{\varepsilon}{2}$-neighbourhood of any point (of a minimal $\frac{\varepsilon}{2}$-spanning set) there can be at most one point of an $\varepsilon$-separated set.

Next we define the limiting capacity D(S) of S as

$$D(S) = \liminf_{\varepsilon\to 0} \frac{\ln (r(S,\varepsilon))}{-\ln \varepsilon} = \liminf_{\varepsilon\to 0} \frac{\ln (s(S,\varepsilon))}{-\ln \varepsilon} ;$$

the fact that the last two expressions are equal follows from (1). The notion of capacity, or rather $\varepsilon$-capacity, was originally used for $s(S,\varepsilon)$. This limit capacity is strongly related to the Hausdorff dimension, see [5 or 12], which is clear from the following equivalent definition. Let $\mathfrak{U}$ be a finite covering $\{U_i\}_{i\in I}$ of S. Then for a > 0 $D_{a,\mathfrak{U}} = \sum_{i\in I} (\text{diam}\,(U_i))^a$. Next we define $D_{a,\varepsilon}$ as the infinum of $D_{a,\mathfrak{U}}$ where $\mathfrak{U}$ runs over all finite covers of S each of whose elements has diameter $\varepsilon$. Notice that $D_{a,\varepsilon} \in [r(S,\varepsilon).\varepsilon^a, r(S,\frac{\varepsilon}{2}).\varepsilon^a]$. It is not hard to see that there is a unique number, which is in fact the limit capacity D(S), such that for a > D(S), resp. a < D(S), $\liminf_{\varepsilon\to 0} D_{a,\varepsilon}$ is zero, resp. infinite. This last definition of limit capacity goes over in the definition of Hausdorff dimension if we replace "each of whose elements has diameter $\varepsilon$" by "each of whose elements has diameter $\le\varepsilon$.

For later reference we indicate a third definition of limit capacity. Let $\{b_i\}_{i=0}^{\infty}$ be some countable dense sequence in S. For $\varepsilon > 0$ we define the subset $J_\varepsilon \subset \mathbb{N}$ by :

$0 \in J_\varepsilon$; for $i > 0$ :

$i \in J_\varepsilon$ if and only if for all j with $0 \le j < i$ and $j \in J_\varepsilon$, we have $\rho(b_i, b_j) \ge \varepsilon$.

$C_\varepsilon$ denotes the cardinality of $J_\varepsilon$. From these definitions it easily follows that whenever $0 < \varepsilon < \varepsilon'$,

$$r(S,\varepsilon') \le C_\varepsilon \le s(S,\varepsilon).$$

Hence we may also define D(S) by $D(S) = \liminf_{\varepsilon\to 0} \frac{\ln C_\varepsilon}{-\ln \varepsilon}$. From the literature, see [12], we know that the Hausdorff dimension is greater than or equal to the topological dimension and from the above considerations it is clear that the limit capacity is greater than or equal to the Hausdorff dimension. Both the Hausdorff dimension and the limit capacity depend on the metric (and not only on the topology). If however $\rho$ and $\rho'$ are metrics on S such that for some constant C and any $x, y \in S$, $C.\rho(x,y) \ge \rho'(x,y) \ge C^{-1}.\rho(x,y)$, then the limit capacity and the Hausdorff dimension are the same for the metrics $\rho$ and $\rho'$. In this case the metrics $\rho$ and $\rho'$ are called metrically equivalent. Finally if S is a compact manifold with a metric $\rho$

which is metrically equivalent with a metric induced by a Riemannian structure, then the limit capacity equals the topological dimension.

Examples where the Hausdorff dimension is different from the limit capacity were given by Mañé [14]. It seems to be an open question whether a difference between Hausdorff dimension and limit capacity can occur for positive limit sets of smooth vector fields on compact manifolds; if the answer is no then for all our purposes the Hausdorff dimension and the limit capacity are the same.

Contrary to the topological dimension, the Hausdorff dimension and the limit capacity need not be integers. If we take for example for $\tilde{S}$ a Cantor set in $\mathbb{R}$, define as $\tilde{S} = \bigcap_{i=0}^{\infty} S_i$ where : $S_0 = [0,1]$; $S_{i+1} \subset S_i$; $S_i$ has $2^i$ intervals of length $\alpha^i$, $\alpha < \frac{1}{2}$; and $S_{i+1}$ is obtained fron $S_i$ by removing in the middle of each segment of $S_i$ a segment of length $\alpha^i.(1-2\alpha)$. We takeas countable dense subset S the union of the left endpoints of the intervals of $S_i$ for all i. If we compute $C_\varepsilon$ for $\varepsilon = \alpha^i$, we find $C_{\alpha^i} = 2^i$. From this it is not hard to deduce that

$$D(S) = -\frac{\ln 2}{\ln \alpha}.$$

In determining the limit capacity of a closed subset of a compact manifold it is important to note that there is only one metric equivalence class on the manifold which contains a metric induced by a Riemannian structure. Limit capacity is always assumed to be defined with respect to a metric in this class.

## 4. Determination of dimension and entropy.

We consider the following situation : M is a compact manifold with a smooth vector field X, a smooth function $y: M \to \mathbb{R}$ and a point $p \in M$. We assume that p is part of its own positive limit set $L^+(p)$; also we assume that for some fixed $\alpha > 0$, the sequence $\{\varphi_{i\alpha}(p)\}_{i=0}^{\infty}$ is dense in $L^+(p)$ and that $(\varphi_\alpha, y)$ is generic in the sense of theorem 1; $\varphi_t$ denotes the flow of X. Note that the only non-generic assumption we made on $(M, X, y, p, \alpha)$ is $p \in L^+(p)$. This assumption can in some sense be justified : if the orbit $\varphi_t(q)$ goes to an "attractor for $t \to \infty$", then if we replace q by $\varphi_T(q) = \tilde{q}$, $T \gg 1$, it is almost true that $\tilde{q} \in L^+(\tilde{q})$. So the assumption $p \in L^+(p)$ can be seen as a way to include in the description (see the introduction) the fact that we can only start measuring after the experiment is already going for quite a time (with fixed $\Omega$).

In this situation we have the sequence $\{a_i = y(\varphi_{i.\alpha}(p))\}_{i=0}^{\infty}$ which represents the experimental output (so for the moment we assume the experiment has been carried out for an infinite amount of time). From this sequence we obtain subsets $J_{n,\varepsilon} \subset \mathbb{N}$ by the following inductive definition (see also the end of section 3) :

$$0 \in J_{n,\varepsilon}; \text{ for } i > 0 :$$

$$i \in J_{n,\varepsilon} \text{ if and only if for all } 0 \leq j < i, \text{ with } j \in J_{n,\varepsilon},$$
$$\max(|a_i - a_j|, |a_{i+1} - a_{j+1}|, \ldots, |a_{i+n} - a_{j+n}|) \geq \varepsilon .$$

$C_{n,\varepsilon}$ denotes the cardinality of $J_{n,\varepsilon}$ .

Main theorem. The limit capacity of $L^+(p)$ equals

$$D(L^+(p)) = \lim_{n\to\infty} (\liminf_{\varepsilon\to 0} \frac{\ln C_{n,\varepsilon}}{-\ln \varepsilon}),$$

where $\lim_{n\to\infty}$ reaches the limit value for every $n \geq 2(\dim(M))$.

The topological entropy of $\varphi_\alpha | L^+(p)$ equals

$$H(L^+(p)) = \lim_{\varepsilon\to 0} (\limsup_{n\to 0^+} (\frac{\ln C_{n,\varepsilon}}{n})) ,$$

where $\lim_{\varepsilon\to 0}$ often (e.g. if $L^+(p)$ is an expansive basic set [3]) reaches the limit value for every $0 < \varepsilon < \varepsilon_0$ for some $\varepsilon_0$.

Proof. We take some $N \geq 2.\dim(M)$. The map $\Phi: M \to \mathbb{R}^{N+1}$, defined by

$$q \mapsto (y(q), y(\varphi_\alpha(q)), \ldots, y(\varphi_{N.\alpha}(q))$$

is an embedding. On $\Phi(M)$ we use the metric

$$\rho((x_0, \ldots, x_N), (x'_0, \ldots, x'_N)) = \max_i |x_i - x'_i| .$$

This metric is equivalent in the metric sense to any metric on $\Phi(M)$ derived from a Riemannian metric.. Se we may use $\rho$ to compute $D(L^+(p)) = D(\Phi(L^+(p)))$.

The first statement in the main theorem now follows by applying section 3 to to the sequence $\{\varphi_{i.\alpha}(p)\}_{i=1}^{\infty}$ in $L^+(p)$.

Next we come to the determination of the topological entropy of $\varphi_\alpha|L^+(p)$. For this we have to find the cardinality of a minimal $\varepsilon$-spanning set of orbits of length n, see Bowen [2]. A minimal $\varepsilon$-spanning set of orbits of length n of $\varphi_\alpha$ is a finite set $\{q_i\}_{i\in I}$ in $L^+(p)$ such that :

(i) for every $q \in L^+(p)$ there is some $i_0 \in I$ such that $\rho(\varphi_{i.\alpha}(q), \varphi_{i.\alpha}(q_{i_0})) < \varepsilon$ for all $0 \leq i \leq n$ ;

(ii) among all subsets of $L^+(p)$ satisfying (i), $\{q_i\}_{i\in I}$ has minimal cardinality.

Let $r(n,\varepsilon)$ be this cardinality. There is also a maximal cardinality of $\varepsilon$-separated orbits of length n, denoted by $s(n,\varepsilon)$ (see [1]). The entropy can now be defined as

$$H(L^+(p)) = \lim_{\varepsilon\to 0} (\limsup_{n\to\infty}(\frac{\ln r(n,\varepsilon)}{n})) = \lim_{\varepsilon\to 0} (\limsup_{n\to\infty}(\frac{\ln(s(n,\varepsilon))}{n}).$$

If we use the metric $\rho$, defined above, we can replace $s(n,\varepsilon)$ or $r(n,\varepsilon)$ by $C_{n+N,\varepsilon}$; see section 3. From this we obtain :

$$H(L^+(p)) = \lim_{\varepsilon\to 0} (\limsup_{n\to\infty} (\frac{\ln(C_{n+N,\varepsilon})}{n})) = \lim_{\varepsilon\to 0} (\limsup_{n\to\infty} (\frac{\ln(C_{n,\varepsilon})}{n})) .$$

<u>Observation.</u> Application of this main theorem to the output of the Taylor-Couette experiment, described in the introduction, gives some complications due to the fact that $\{a_i\}_{i=1}^{\bar{N}}$ is finite in this case. For such a finite sequence one should proceed as follows: for $n,\varepsilon,m$ with $n + m \leq \bar{N}$ we define subsets $J_{n,\varepsilon,m} \subset \mathbb{N}$ as follows :

(i) $0 \in J_{n,\varepsilon,m}$; for $i > 0$ :

(ii) $i \in J_{n,\varepsilon,m}$ if and only if both :

  (a) $i \leq m$;

  (b) for all $j < i$, $j \in J_{n,\varepsilon,m}$, $\max_{0\leq k\leq n} |a_{i+k} - a_{j+k}| \geq \varepsilon$.

$C_{n,\varepsilon,m}$ denotes the cardinality of $J_{n,\varepsilon,m}$. For $\bar{N} = \infty$, one would have $\lim_{m\to\infty} C_{n,\varepsilon,m} = C_{n,\varepsilon}$. $C_{n,\varepsilon,m}$ is non-decreasing in m. Hence it seems reasonable to take $C_{n,\varepsilon,\bar{N}-n}$ as an approximation of $C_{n,\varepsilon}$ provided the difference between $C_{n,\varepsilon,\bar{N}-n}$ and say, $C_{n,\varepsilon[\frac{1}{2}(\bar{N}-n)]}$

is sufficiently small, say of the order of 1 or 2%. In this way we have the possibility of calculating $C_{n,\varepsilon}$ in a certain region of the $(n,\varepsilon)$- plane; also one should consider these values for $C_{n,\varepsilon}$ only reliable if $\varepsilon$ is well above the expected errors in the measurement. From these numerical values for $C_{n,\varepsilon}$ one should decide, on the basis of the main theorem what the values of $D(L^+(p))$ and $H(L^+(p))$ are or whether the limits defining these values "do not exist numerically".

If, in the calculation of $D(L^+(p))$, the $\lim_{n\to\infty}$ would have the tendency of going to infinity, this would imply that representing the evolution on a finite dimensional manifold is a mistake. If on the other hand this limit would go to a non-integer, this would be evidence in favour of a strange attractor. Namely, as we have seen in section 3, for a Cantor set C we may have D(C) a non integer, and strange attractors have in general a Cantor set like structure, e.g. see [3].

If the experimental data do not clearly indicate the limits in the calculation of $D(L^+(p))$ and $H(L^+(P))$ to exist and to be finite, then both the Landau-Lifschitz and the Ruelle-Takens picture are to be rejected as explanation of the experimental data.

Final remarks.

1. It does not seem to be known whether, for differentiable dynamical systems the "inf" and "sup" in the definition of limit capacity and entropy can be omitted. If they can omitted, one has a better test on the validity of the assumptions "finite dimensional and deterministic" : also the first limit has "to exist numerically".

2. Yorke pointed out to the author that he and others had made calculations of limit capacities in relation with a conjecture on Lyapunov numbers and dimension for attractors, see [7]. His calculating scheme is different from ours and probably faster. The calculations indicate that the computing time rapidly increases with dimension, which probably also holds for our computing scheme.

3. It should be noticed that the defining formulas for dimension and entropy become more alike when we write them in the following form .

$$D(L^+(p)) = \lim_{n\to\infty} \left(\lim_{\varepsilon\to 0} \inf \left(\frac{\ln C_{n,\varepsilon}}{n \cdot \ln \varepsilon}\right)\right)$$

$$H(L^+(p)) = \lim_{\varepsilon\to 0} \left(\lim_{n\to\infty} \sup \left(\frac{\ln C_{n,\varepsilon}}{n \cdot \ln \varepsilon}\right)\right) .$$

If we denote $\frac{\ln C_{n,\varepsilon}}{n - \ln \varepsilon}$ by $Z(n, -\ln \varepsilon)$ and regard both n and $-\ln \varepsilon$ as continuous variables one can see from a few examples (Anosov automorphisms on the torus and horseshoes) that often $\lim_{\substack{\alpha,\beta\to\infty \\ \alpha/\beta\to\gamma}} Z(\alpha,\beta)$ exists for all positive $\gamma$, forming a one-parameter family of "topologically invariants" connecting entropy with limit capacity. It would be interesting to investigate the existence of these limits for more general attractors. This might be connected with the above mentioned conjecture of Yorke.

## References.


1. D. Aeyels, Generic observability of differentiable systems, preprint, April 1980, Dept. of System Dynamics, State Univ. Gent.

2. R. Bowen, Entropy of group endomorphisms and homogeneous spaces, Trans. A.M.S., 153 (1971), 401-414.

3. R. Bowen, On Axiom A diffeomorphisms, Regional Conference Series in Mathematics, 35, A.M.S. Providence, 1977.

4. M. Denker, C. Grillenberger, & K. Sigmund, Ergodic theory on compact spaces, Lecture Notes in Mathematics, 527, Springer-Verlag, Berlin, 1976.

5. H. Federer, Geometric measure theory, Springer-Verlag, Berlin, 1969.

6. P.R. Fenstermacher, J.L. Swinney & J.P. Gollub, Dynamical instability and the transition to chaotic Taylor vortex flow, Journal Fluid Mech. 94 (1979) (1) 103-128.

7. P. Frederickson, J.L. Kaplan & J.A. Yorke, The dimension of the strange attractor for a class of difference systems, preprint, June 1980, University of Maryland.

8. J.P. Gollub, The onset of turbulence : convection, surface waves, and oscillations, in Systems far from Equilibrium, Proc. Sitges Int. School and Symposium on Statistical Mech., Ed. L. Garrido, J. Garcia, Lecture Notes in Physics, Springer-Verlag, Berlin, to appear.

9. J.P. Gollub, & S.V. Benson, Time-dependent instability and the transition to turbulent convection, preprint, Physics Dept, Haverford College, Haverford, Pa. 19041, USA.

10. J.P. Gollub & H.L. Swinney, Onset of turbulence in a rotating fluid, Phys. Rev. Lett. 35 (1975), 927-930.

11. M. Herman, Mesure de Lebesgue et nombre de rotation, in Geometry and Topology, ed. J. Palis and M. do Carmo, Lecture Notes in Mathematics 597, Springer-Verlag, Berlin, 1977.

12. W. Hurewicz & H. Wallman, Dimension theory, Princeton University Press, 1948, Princeton, N.J.

13. L. Landau & E. Lifschitz, Méchanic des Fluides, ed. MIR, Moscow, 1971.

14. R. Mañé, On the dimension of the compact invariant sets of certain non-linear maps, preprint IMPA, Rio de Janeiro, 1980.

15. J.E. Marsden, The Hopf bifurcation for non-linear semi groups, Bull. A.M.S., 79 (1973), 537-541.

16. J.E. Marsden & M. McCracken, The Hopf bifurcation and its applications, Appl. math. sci. 19, Springer-Verlag, Berlin, 1976.

17. N.H. Packerd, J.P. Crutchfield, J.D. Farmer & R.S. Shaw, Geometry from the time series, preprint, May 1979, University of California, Santa Cruz, (Dynamical Systems Collective).

18. D. Rand, The pre-turbulent transitions and flows of a viscous fluid between concentric rotating cylinders, preprint Warwick University, June 1980.

19. D. Ruelle & F. Takens, On the nature of turbulence, Comm. math. Phys. 20 (1971), 167-192; 23 (1971), 343-344.

F. Takens : Mathematisch Instituut, P.O. Box 800, Groningen, Holland.

# Local and simultaneous structural stability of certain diffeomorphisms.

Marco A. Teixeira.

## Introduction.

The present paper is devoted to the study of local and simultaneous structural stability of a pair of involutions, both of them defined on the plane. We shall just deal with those involutions which are germs of $C^\infty$ diffeomorphisms (at 0) $\varphi: \mathbb{R}^2, 0 \to \mathbb{R}^2, 0$ satisfying $(\varphi \circ \varphi) = Id$ and $\det(\varphi'(0)) = -1$.

The simultaneous structural stability of a pair of involutions is reached by the following definition :

Definition - Two pairs of involutions $(\varphi, \psi)$ and $(\tilde{\varphi}, \tilde{\psi})$ are equivalent if there is a germ of a homeomorphism $h: \mathbb{R}^2, 0 \to \mathbb{R}^2, 0$ satisfying $h\varphi = \tilde{\varphi}h$ and $h\psi = \tilde{\psi}h$.

The motivation for studying such stability comes from the study of discontinuous vector fields (see Application I below).

The main result is :

Theorem - A pair of involutions $(\varphi, \psi)$ is locally and simultaneously structurally stable at 0 (under $C^1$ perturbations of $\varphi$ and $\psi$) if and only if 0 is a hyperbolic fixed point of the composition $\varphi \circ \psi$. Moreover, the structural stability in the space of pairs of involution is not generic.

All results here are established in the 2-dimensional case although it is expected that similar results are true for higher dimensions.

The theorem stated above has the following applications :

Application 1 - Discontinuous Vector Fields in $\mathbb{R}^3$.

Let Z be a germ of a vector field (at 0) in $\mathbb{R}^3$ given by

$$Z(x,y,z) = \begin{cases} X(x,y,z) & \text{if } z > 0 \\ Y(x,y,x) & \text{if } z < 0 \end{cases}$$

where X and Y are germs of $C^\infty$ vector fields at 0 in $\mathbb{R}^3$. This means that Z can have discontinuities on z = 0. We denote Z by (X, Y).

We are going to consider the following generic situation :

Let $t \to \gamma_X(t)$, $t \to \gamma_Y(t)$ be parametrisations of the trajectories of X and Y passing through 0, respectively. Assume that $\gamma_X(0) = \gamma_Y(0) = 0$, $\pi(\gamma_X'(0)) = \pi(\gamma_Y'(0)) = 0$, $\pi(\gamma_X''(0)) < 0$ and $\pi(\gamma_Y''(0)) > 0$, $\pi$ being the canonical projection of $\mathbb{R}^3$ to z-axis (see picture below).

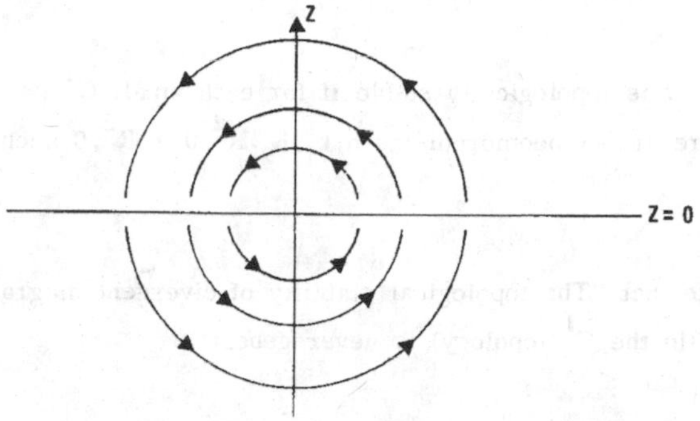


Associated with X (resp. Y) there is a smooth curve $L_X$ (resp. $L_Y$) constituted by the elements of z = 0 of non transversal contact between X (resp. Y) and z = 0. Suppose $L_X$ is transverse to $L_Y$ at 0.

Now X (resp. Y) induces on z = 0 a $C^\infty$ diffeomorphism $\varphi_X$ (resp. $\varphi_Y$) given by

$$\varphi_X(x, y, 0) = (x, y, 0) \quad \text{if } (x, y, 0) \in L_X\text{; otherwise } \varphi_X(x, y, 0)$$

is the point (different from (x, y, 0)) where the trajectory of X passing through (x, y, 0) meets z = 0.

Similarly we define $\varphi_Y$.

It is easy to find coordinates x, y on z = 0 such that $L_X$ = {x-axis} and $L_Y$ = {y-axis}. Moreover $\varphi_X$ and $\varphi_Y$ are involutions on $\mathbb{R}^2, 0$; moreover the simultaneous stability of both is closely related to the stability of Z = (X, Y).

We can conclude that "The structural stability of Z = (X, Y) (under $C^1$ perturbations of X and Y) is never generic".

We remark that in the definition of equivalence of such discontinuous vector fields, we are imposing that the homeomorphism must preserve the set of discontinuity.

Application II - Divergent Diagrams of Differentiable Mappings.

Consider a divergent diagram of germs of $C^\infty$ mappings, as follows :

(D) $\mathbb{R}^2,0 \xleftarrow{f} \mathbb{R}^2,0 \xrightarrow{g} \mathbb{R}^2,0$ where f and g have singularities at 0 (so that generical both are folds).

We say that D is topologically stable if for each small $C^1$ perturbations $\tilde{f},\tilde{g}$ of f, g respectively there are homeomorphisms $h, k_1, k_2 : \mathbb{R}^2,0 \to \mathbb{R}^2,0$ such that $k_1 f = h\tilde{f}$ and $k_2 g = h\tilde{g}$.

We can state that "The topological stability of divergent diagrams of type $\mathbb{R}^2 \leftarrow \mathbb{R}^2 \to \mathbb{R}^2$ (in the $C^1$ topology) is never generic.

To see this it is enough to consider 0 as being a fold point of both mappings f and g such that $\Sigma_f$ (singular set of f) is transverse to $\Sigma_g$ at 0. The mappings f and g induce involutions $\varphi_f$ and $\varphi_g$, respectively, which satisfy $f \circ \varphi_f = f$ and $g \circ \varphi_g = g$. It is not difficult to check that the instability of the pair $(\varphi_f, \varphi_g)$ implies the instability of the present diagram.

In addition, take for example, the involution $\varphi$, given by $\varphi(x, y) = ((4x + 3y)/5, (3x - 4y)/5)$ and a fold g associated with $\varphi$ (we can find g by combining a smooth change of variables and Lemma 2.1). One can prove that the diagram $D : \mathbb{R}^2 \xleftarrow[f]{} \mathbb{R}^2 \xrightarrow[g]{} \mathbb{R}^2$ where $f(x, y) = (x, y^2)$ is not stable under $C^1$-perturbations of f and g. One can also conclude that this instability is generic by Proposition 3.1 of §3, Remark 4.2 of §4 and observing that 0 is not a hyperbolic fixed point of the composition $\varphi_0 \circ \varphi$ where $\varphi_0(x, y) = (x, -y)$.

In Section 1 we give definitions and establish the notation. Section 2 contains a study of a single involution. One shows that any involution is $C^\infty$ conjugate to $(x, y) \to (x, -y)$.

In Section 3 we exhibit an invariant of hyperbolicity of a pair of involutions. We see that this concept is not generic.

In Section 4 we investigate the stability of a pair of linear involutions.

In Section 5 we prove the main result of this paper.

The author wishes to thank J.P. Dufour for many helpful conversations.

The referee has pointed out that there is a simpler proof of Proposition 2.2 of §2, as a special case of the Slice Theorem in Riemannian Manifolds.

## §1. Preliminaries.

We are going to study a class of germs of diffeomorphisms defined as follows :

1.1 Definition - An involution is a $C^\infty$ diffeomorphism $\varphi : \mathbb{R}^2 \to \mathbb{R}^2$ satisfying $\varphi(0) = 0$, $\varphi(\varphi(x)) = x$ and $\det [\varphi'(0)] = -1$.

Denote $G^r$ the space of germs of involutions at 0 with the $C^r$ topology. Consider $W^r = G^r \times G^r$ with the natural product topology.

To simplify the notation, we will make no distinction between a germ of an involution (at 0) and anyone of its representatives.

1.2 Definition - Let $\varphi = (\varphi_0, \varphi_1)$, $\psi = (\psi_0, \psi_1) \in W^r$. Then $\varphi$ and $\psi$ are topologically equivalent (at 0) if there exists a germ of a homeomorphism $h: \mathbb{R}^2, 0 \to \mathbb{R}^2, 0$ which satisfies $h\varphi_0 = \psi_0 h$ and $h\varphi_1 = \psi_1 h$. So the (local) Structural Stability in $W^r$ is defined in a natural way.

## §2. The Normal Form of an Involution.

Let $\varphi: \mathbb{R}^2, 0 \to \mathbb{R}^2, 0$ be a germ of a $C^\infty$ diffeomorphism which satisfies $\varphi \circ \varphi = \mathrm{Id}$ with components $\varphi = (\alpha, \beta)$.

2.1 Lemma - Assume $\varphi$ satisfies the conditions :

i) $\beta(x, 0) = 0$ and ii) $y\beta(x, y) < 0$ if $y < 0$.

Then there exists a germ of a $C^\infty$ fold $f: \mathbb{R}^2, 0 \to \mathbb{R}^2, 0$ such that $f \circ \varphi = f$.

Proof - We have

$$\varphi'(x,0) = \begin{pmatrix} \frac{\partial\alpha}{\partial x}(x,0) & \frac{\partial\alpha}{\partial y}(x,0) \\ 0 & \frac{\partial\beta}{\partial y}(x,0) \end{pmatrix}$$

By a straightforward calculation we get

$$\left(\frac{\partial\alpha}{\partial x}(x,0)\right) = 1, \quad \left(\frac{\partial\alpha}{\partial y}(x,0)\right)\cdot\left(\frac{\partial\alpha}{\partial x}(x,0) + \frac{\partial\beta}{\partial y}(x,0)\right) = 0 \text{ and } \left(\frac{\partial\beta}{\partial y}(x,0)\right) = -1.$$

The required function is given by

$$f(x,y) = (x + \alpha(x,y), \quad x\alpha(x,y) + y\beta(x,y)) .$$

Since

$$f' = \begin{pmatrix} 1 + \frac{\partial\alpha}{\partial x} & \frac{\partial\alpha}{\partial y} \\ \alpha + x\frac{\partial\alpha}{\partial x} + y\frac{\partial\beta}{\partial x} & \beta + y\frac{\partial\beta}{\partial y} + x\frac{\partial\alpha}{\partial y} \end{pmatrix}$$

we have

$$f'(0,0) = \begin{pmatrix} 2 & b \\ 0 & 0 \end{pmatrix} \quad \text{with } b = \frac{\partial\alpha}{\partial y}(0,0) .$$

The function $\Delta(x,y) = \det [f'(x,y)]$ satisfies

$$\frac{\partial\Delta}{\partial x}(0,0) = 0 \qquad \text{and} \qquad \frac{\partial\Delta}{\partial y}(0,0) < 0 .$$

So the curve $K = \mathrm{Kern}(f'(0,0)) = \{(x,y): 2x + by = 0\}$ is transverse to $\Sigma_0(f) = \{(x,y): \Delta(x,y) = 0\}$ at 0.

This finishes the proof. □

2.2 Proposition - Under the hypotheses of Lemma 2.1 there exists a germ of a $C^\infty$ diffeomorphism $h:\mathbb{R}^2,0 \to \mathbb{R}^2,0$ such that $\varphi h = h\varphi_0$ where $\varphi_0:\mathbb{R}^2,0 \to \mathbb{R}^2,0$ is given by $\varphi_0(x,y) = (x,-y)$.

Proof - Let f be the fold associated to $\varphi$. (via Lemma 2.1.)

One knows that the following diagram is commutative

$$\begin{array}{ccc} \mathbb{R}^2,0 & \xrightarrow{f} & \mathbb{R}^2,0 \\ h\downarrow & & \downarrow k \\ \mathbb{R}^2,0 & \xrightarrow[f_0]{} & \mathbb{R}^2,0 \end{array}$$

where $f_0(x,y) = (x, y^2)$ and for selected $C^\infty$ diffeomorphisms h and k.

Now the proof of the proposition follows immediately. □

§3. Invariant of Hyperbolicity.

Given $\varphi = (\varphi_0, \varphi_1)$ in $W^r$ we may choose coordinates x, y around 0 in $\mathbb{R}^2$ such that $\varphi_0(x,y) = (x, -y)$. Then for some a, b, c $\in \mathbb{R}$ we have

$$\varphi_1'(0) = \begin{pmatrix} a & b \\ c & -a \end{pmatrix} \quad \text{with } a^2 + bc = 1 .$$

By a direct calculation we can prove

3.1 Proposition - 0 is a hyperbolic fixed point of $\Phi = \varphi_0 \circ \varphi_1$ if and only if $a^2 > 1$.

3.2 Remark - The eigenvalues of $\Phi'(0)$ are $\lambda = a \pm (a^2-1)^{1/2}$ .

§4. Linear Involutions.

Consider the following linear involutions $\varphi_0(x,y) = (x, -y)$ and $\varphi_1(x,y) = (ax + by, cx - ay)$ with $a^2 + bc = 1$.

Let J be the set of germs of homeomorphisms $h: \mathbb{R}^2, 0 \to \mathbb{R}^2, 0$ such that $h\varphi_0 = \varphi_0 h$. So a diffeomorphism h $\in$ J if and only if $h(x,y) = (h_0(x,y^2), yh_1(x,y^2))$.

Suppose 0 is a hyperbolic fixed point of $\Phi = (\varphi_0 \circ \varphi_1)$. This implies, in particular, that $bc < 0$. Consider the following steps :

Step 1 - By considering an element $h_0 \in J$ given by $h_0(x,y) = (x, \frac{B}{b}y)$ where $B = (-bc)^{1/2}$ we may take $\Phi$ as being equal to $(ax + By,\ Bx + ay)$ with $a^2 - B^2 = 1$.

Consider the following objects

i) $\lambda = (a+B)$ (so $\lambda^{-1} = a-B$),

ii) the rotation $r(x,y) = (x+y, x-y)$ ,

iii) the isomorphism $f:\mathbb{R}^2,0 \to \mathbb{R}^2,0$ given by $f(x,y) = (\lambda x, \lambda^{-1}y)$.

We have the following relation

$$r^{-1}fr = \Phi .$$

Corresponding to another such linear involution $\tilde{\varphi}_1(x,y) = (\tilde{a}x + \tilde{b}y,\ \tilde{c}x - \tilde{a}y)$, with $\tilde{b}\tilde{c} < 0$, there are the following similar objects : $\tilde{\Phi}(x,y) = (\tilde{a}x + \tilde{B}y,\ \tilde{B}x + \tilde{a}y)$ with $\tilde{a}^2 - \tilde{B}^2 = 1$, the real number $\tilde{\lambda}$ and the isomorphism $\tilde{f}$ satisfying $r^{-1}\tilde{f}r = \tilde{\Phi}$.

Step 2 - Let $p:\mathbb{R}_+,0 \to \mathbb{R}_+$ be given by $p(0) = 0$ and $p(x) = x^k$ where $k = (\log \lambda)/(\log \tilde{\lambda})$.

Define $P:\mathbb{R},0 \to \mathbb{R}$ by $P(0) = 0$, $P(x) = p(x)$ for $x > 0$ and $P(x) = -p(-x)$ for $x < 0$.

Step 3 - Define $K:\mathbb{R}^2,0 \to \mathbb{R}^2,0$ by $K(x,y) = (P(x),P(y))$. We claim that $K \circ \tilde{f} = f \circ K$. In fact,

$$K(\tilde{f}(x,y)) = K(\tilde{\lambda}x, \tilde{\lambda}^{-1}y) = (P(\tilde{\lambda}x),\ P(\tilde{\lambda}^{-1}y)) =$$
$$= (\lambda P(x),\ \lambda^{-1}P(y)) = f(K(x,y)) .$$

Step 4 - Define $h:\mathbb{R}^2,0 \to \mathbb{R}^2,0$ by $h = r^{-1}Kr$. We have the following result :

4.1 Proposition - i) $h \in J$, ii) $h\Phi = \tilde{\Phi}h$ and iii) $h\varphi_1 = \varphi_1 h$.

Proof - The following relation gives part ii) :

$$r^{-1}K\tilde{f}r = r^{-1}fKr = (r^{-1}fr)(r^{-1}Kr)$$

The proofs of i) and ii) follow immediately. □

4.2 Remark - If 0 is not a hyperbolic fixed point of $\Phi = (\varphi_0 \circ \varphi_1)$ then $\Phi$ is equivalent to $\Phi_1(x,y) = (ax + By, -Bx + ay)$ with $a^2 + B^2 = 1$ (this equivalence being in J). So $a = \cos\alpha$, $B = \sin\alpha$ for some $\alpha \in \mathbb{R}$. It is clear that if $\alpha \neq \tilde{\alpha} + 2n\pi$ then $\Phi_1$ is not equivalent to $\tilde{\Phi}_1(x,y) = (\tilde{a}x + \tilde{B}y, -\tilde{B}x + \tilde{a}y)$ with $\tilde{a} = \cos\tilde{\alpha}$, $\tilde{B} = \sin\tilde{\alpha}$.

## §5. Structural Stability in $W^1$.

The following lemma has an easy proof :

5.1 Lemma - Let $\varphi \in G^r$. Given $\varepsilon > 0$ there exists a neighbourhood U of 0 and an extension $\theta:\mathbb{R}^2 \to \mathbb{R}^2$ of $\varphi|_U$ of the form $\varphi'(0) + \alpha$ where $\alpha \in C^0_b(\mathbb{R}^n)$ (space of bounded continuous mappings) is Lipschitz with bounded constant by $\varepsilon$. Furthermore $\theta \circ \theta = \mathrm{Id}$.

Let $(\varphi_0, \varphi_1) \in W^1$.

5.1 Proposition - If 0 is a hyperbolic fixed point of $\Phi = \varphi_0 \circ \varphi_1$ then $(\varphi_0, \varphi_1)$ is equivalent to $(A_0, A_1)$ where $A_0$ (resp. $A_1$) is the linear part (at 0) of $\varphi_0$ (resp. $\varphi_1$).

Proof - Let $W^s$ and $W^u$ be the stable and unstable manifold associated to $\Phi'(0)$ respectively.

Since $\varphi^2 = \psi^2 = \mathrm{Id}$, we have $(\varphi_0\varphi_1)^{-1} = \varphi_1\varphi_0, \varphi_1\varphi_0 = \varphi_0(\varphi_0\varphi_1)\varphi_0^{-1}$ and $\varphi_1\varphi_0 = \varphi_1(\varphi_0\varphi_1)\varphi_1^{-1}$. So $\varphi_0$ and $\varphi_1$, both interchange $W^s$ and $W^u$.

Let $\theta_0, \theta_1$ be the extension of $\varphi_0, \varphi_1$ respectively (via Lemma 5.1). We have that $\theta_0 \circ \theta_1$ is $C^1$ near linear. By Hartman's Theorem [2] there exists a unique homeomorphism of the form $h = \mathrm{Id} + g$, g being a bounded mapping, which is a conjugacy between $(\theta_0\theta_1)$ and $(A_0A_1)$. This means that

$$h\,\theta_0\theta_1 h^{-1} = A_0A_1$$

But then

$$A_0(h\,\theta_0\theta_1 h^{-1})A_0^{-1} = A_1A_0 \quad , \quad \text{and}$$

. .

$$A_1 A_0 = A_0 ((h\theta_0)(\theta_1 \theta_0)(\theta_0^{-1} h^{-1}) A_0^{-1} .$$

So

$$A_1 A_0 = (A_0 h \theta_0)(\theta_1 \theta_0)(\theta_0^{-1} h^{-1} A_0^{-1}).$$

and $A_0 h \theta_0$ is a conjugacy between

$$(\theta_0 \theta_1)^{-1} = \theta_1 \theta_0 \quad \text{and} \quad (A_0 A_1)^{-1} = A_1 A_0$$

of the form Id + $\tilde{g}$, with $\tilde{g}$ being a bounded mapping. This implies that

$$h = A_0 h \theta_0 \quad \text{and finally} \quad A_0 h = h\theta_0 .$$

Similarly one shows that $A_1 h = h\theta_1$. This finishes the proof. □

The proof of the following theorem is an immediate consequence of 4.1, 4.2 and 5.2.

5.3 Theorem - $(\varphi, \psi)$ is structurally stable in $W^1$ if and only if 0 is a hyperbolic fixed point of $\varphi \circ \psi$.

## References.


1. J.P. Dufour, Diagrammes D'Applications Differentiables, Tese Université du Montpellier, France, (1979).
2. J. Palis & W. de Melo, Introdução aos Sistemas Dinâmicos - Projeto Euclides, IMPA, Rio de Janeiro, (1978).


M.A. Teixeira : Universidade Estadual de Campinas, Instituto de Metematica, Estatistica e Ciência da Computação, IMECC- UNICAMP, Campinas, Brazil.